Growth and Optical Properties of Wide-Gap II–VI Low-Dimensional Semiconductors

NATO ASI Series

Advanced Science Institutes Series

A series presenting the results of activities sponsored by the NATO Science Committee, which aims at the dissemination of advanced scientific and technological knowledge, with a view to strengthening links between scientific communities.

The series is published by an international board of publishers in conjunction with the NATO Scientific Affairs Division

A	Life Sciences	Plenum Publishing Corporation
B	Physics	New York and London

C	Mathematical and Physical Sciences	Kluwer Academic Publishers Dordrecht, Boston, and London
D	Behavioral and Social Sciences	
E	Applied Sciences	

F	Computer and Systems Sciences	Springer-Verlag
G	Ecological Sciences	Berlin, Heidelberg, New York, London,
H	Cell Biology	Paris, and Tokyo

Series B: Physics

Growth and Optical Properties of Wide-Gap II–VI Low-Dimensional Semiconductors

Edited by

T. C. McGill

California Institute of Technology
Pasadena, California

C. M. Sotomayor Torres

University of Glasgow
Glasgow, United Kingdom

and

W. Gebhardt

University of Regensburg
Regensburg, Federal Republic of Germany

Plenum Press
New York and London
Published in cooperation with NATO Scientific Affairs Division

Proceedings of the NATO Advanced Research Workshop on
Growth and Optical Properties of Wide-Gap
II–VI Low-Dimensional Semiconductors,
held August 2–6, 1988,
in Regensburg, Federal Republic of Germany

Library of Congress Cataloging in Publication Data

NATO Advanced Research Workshop on Growth and Optical Properties of Wide-Gap II–VI Low-Dimensional Semiconductors (1988: Regensburg, Germany)
  Growth and optical properties of wide-gap II–VI low-dimensional semiconduc-tors / edited by T. C. McGill, C. M. Sotomayor Torres, and W. Gebhardt.
      p.   cm. — (NATO ASI series. Series B, Physics: v 200)
  "Proceedings of the NATO Advanced Research Workshop on Growth and Op-tical Properties of Wide-Gap II–VI Low-Dimensional Semiconductors, held August 2–6, 1988, in Regensburg, Federal Republic of Germany"—T.p. verso.
  Bibliography: p.
  Includes index.
  ISBN 0-306-43221-8
  1. Compound semiconductors—Congresses. 2. Semiconductors—Optical pro-perties—Congresses. 3. Crystals—Growth—Congresses. I. McGill, T. C. II. Sotomayor Torres, C. M. III. Gebhardt, W. IV. North Atlantic Treaty Organization. Scientific Affairs Division. V. Title. VI. Series: NATO advanced science institutes series. Series B, Physics: v. 200.
TK7871.99.C65N37   1988   89-8490
621.381'52—dc20   CIP

© 1989 Plenum Press, New York
A Division of Plenum Publishing Corporation
233 Spring Street, New York, N.Y. 10013

Printed in the United States of America

SPECIAL PROGRAM ON CONDENSED SYSTEMS OF LOW DIMENSIONALITY

This book contains the proceedings of a NATO Advanced Research Workshop held within the program of activities of the NATO Special Program on Condensed Systems of Low Dimensionality, running from 1983 to 1988 as part of the activities of the NATO Science Committee.

Other books previously published as a result of the activities of the Special Program are:

PREFACE

This volume contains the Proceedings of the NATO Advanced Research Workshop on "Growth and Optical Properties of Wide Gap II–VI Low Dimensional Semiconductors", held from 2 – 6 August 1988 in Regensburg, Federal Republic of Germany, under the auspices of the NATO International Scientific Exchange Programme.

Semiconducting compounds formed by combining an element from column II of the periodic table with an element from column VI (so called II–VI Semiconductors) have long promised many optoelectronic devices operating in the visible region of the spectrum. However, these materials have encountered numerous problems including: large number of defects and difficulties in obtaining p– and n–type doping. Advances in new methods of material preparation may hold the key to unlocking the unfulfilled promises. During the workshop a full session was taken up covering the prospects for wide–gap II–VI Semiconductor devices, particularly light emitting ones. The growth of bulk materials was reviewed with the view of considering II–VI substrates for the novel epitaxial techniques such as MOCVD, MBE, ALE, MOMBE and ALE–MBE. The controlled introduction of impurities during non–equilibrium growth to provide control of the doping type and conductivity was emphasized.

Non linear optical processes in wide gap II–VI semiconductors have drawn much attention for their many potential applications. These are explored and reviewed including semiconductor–doped glasses. Growth of low dimensional II–VI structures by various techniques occupied a substantial part of the workshop with special consideration to interface strain and to III–V substrates. The trend points to the careful design of a buffer layer. Various aspects of the 2D optical properties were discussed including exciton self–trapping, Raman scattering, magneto–optical processes and acoustoelectric characterisation, which, together with optically–detected magnetic resonance provide a powerful set for the study of optical processes in II–VI's.

The Organizing Committee would like to express its sincere thanks to A. Graf, S. Schmitzer, B. Langen, R. Creuzburg and J. Ploner for their assistance with the running of the workshop. The invaluable link provided by Marcia Hudson across the Atlantic is also greatly appreciated. The Committee would like to thank the following companies and institutions for additional financial assistance: SERC–UK, IBM–Munich, Telefunken Electronics Heilbronn, AIXTRON Aachen, Siemens AG Munich, the US Defense Advanced Research Projects Agency, the US–AFOR and the US–ONR.

Finally, we would like to express our appreciation of the help provided by Janis Whitelaw, Morag Watt and Kenneth Melvin in the preparation of this volume.

T. C. McGill
C. M. Sotomayor Torres
W. Gebhardt

Autumn 1988

CONTENTS

PART 4 : NON-LINEAR OPTICAL PROPERTIES OF WIDE GAP II–VI SEMICONDUCTORS

PART 5 : 2D GROWTH OF II–VI SEMICONDUCTORS

OPTOELECTRONIC DEVICES FROM

WIDE BAND GAP II-VI SEMICONDUCTORS

R.N. Bhargava

Philips Laboratories
Division of North American Philips Corporation
345 Scarborough Road
Briarcliff Manor, NY 10510

ABSTRACT

The recent advances made in the reproducible growth of the high quality wide band gap II-VI materials, particularly the thin films and the quantum structures grown by MBE and MOCVD are discussed. The combination of the low temperature growth and the choice of purer source materials has resulted in decreased background impurity concentration to the extent that meaningful optical and electrical experiments can be performed. These improvements in the materials are expected to yield devices, an area of research presently being pursued actively. Some of these devices are (i) a blue Laser/LED based on p-n junction; (ii) electron beam pumped visible lasers; and (iii) novel optical devices based on non-linear properties of II-VI's. Both the concepts and the key recent research results related to the above devices obtained by us and others are reviewed.

1. INTRODUCTION

The wide band gap II-VI semiconductors such as ZnS, ZnSe, CdS, ZnTe and CdSe possess large, direct energy band gaps and can, in principle, yield efficient visible optoelectronic devices including visible lasers. Although these materials are currently used for efficient TV phosphors and for thin-film electroluminescent devices, their extensive use as optoelectronic devices has been hampered by poor control of the electrical properties. This limitation has been associated with material properties (e.g. intrinsic and extrinsic point defects, line defects, precipitates, solubility of shallow acceptors, donors, and compensating impurities, and their relationship to generation of deep states, etc.). The resurgence in II-VI materials research in the last several years[1-6] has been triggered by the expectation that current semiconductor growth technology, in particular MBE and MOVPE, developed for III-V compounds when applied to II-VI materials will result in better and novel devices. The significant improvements in the growth of thin films have recently been succeeded by the growth of multi-quantum wells and superlattices. This breakthrough in the growth technology has produced II-VI materials on which meaningful optical and electrical measurements can be performed.

1

This review presents the key developments in the growth of high-purity ZnSe, as a case study, and discuss as to how these improvements in ZnSe material can impact the development of different optoelectronic devices. At present, the main driving force behind II-VI semiconductor device research is the desire to develop visible lasers, particularly a blue laser, for applications in displays, optical memories (e.g. video disc), printers, and communications. Future applications might also include optical computing and optical memories.

2. MATERIALS PREPARATION

During the past ten years, efforts have primarily been oriented towards the study of the issue of p-n junction formation by the incorporation of shallow acceptors and donors, by different growth techniques[1-6] as well as by careful control of stoichiometry[7]. In all these studies, although several donors and acceptors were identified and their ionization energies were measured[1], their role in controlling the conductivity was never unequivocally specified. The reason for this was that the key properties such as majority and minority carrier concentration, conductivity, and mobility and their temperature-dependence are difficult to ascertain from the present measurements in materials which frequently are highly resistive and heavily compensated. The poor "electrical" quality of the material is recognized to be due to our inability to incorporate appropriate dopants reproducibly in sufficiently high concentration. This results from (i) a lack of control of stoichiometry and its relation to the density of intrinsic and extrinsic defects, and (ii) the presence of unwanted impurities. This, in particular, becomes critical because the electrically active background impurities (e.g. Li, Na, Cu, Cl, etc.) are incorporated in II-VI compounds[8-11]much more easily in large numbers than in Group III-V or IV semiconductors.

In recent years, attempts have been made to prepare materials under stoichiometric conditions[7,12,13]. The control of stoichiometric conditions to obtain p-type conductivity in ZnSe has recently been studied by Nishizawa and co-workers[7,12]. They incorporated Group Ia elements with control of stoichiometry of ZnSe during bulk crystal growth using a temperature-difference method under controlled vapor pressure. When doped with Li, p-type ZnSe with conductivity up to 0.5 $(\Omega \text{ cm})^{-1}$ has been obtained. The material shows a hole mobility of about 50 $\text{cm}^2/\text{V.s}$ at 300 K. The realization of a blue LED with a brightness of 2mcd operating at 2.7V and 2mA was reported by Nishizawa[12]. Since this initial exciting report, no new results have been reported by the above group.

Significant increase of activity in MOVPE and MBE of ZnSe to achieve p-n junctions has occurred in the last five years. The key objective is to prepare materials at low temperature using MOVPE and MBE, where defects and incorporation of shallow donors and acceptors can be controlled.

MOVPE growth of ZnSe is currently being pursued to grow high-quality layers and to incorporate appropriate dopants to obtain conducting device-grade material[11,14]. One promising modification of the MOVPE is photo-induced MOVPE[15]. With irradiation, high-quality thin films can be grown at lower temperatures as compared to the growth temperatures without irradiation. This results in materials with lower defect density. Similar modifications are observed in photo-assisted MBE growth of thin films[16].

To date, the most "pure" ZnSe is reported by MBE[9,10]. Here "pure" is being referred to in relation to the ratio of the intensity of free excitonic transition to bound excitonic transitions as well as their overall intensity ratio to the deep-band emission. Under high purity conditions, ZnSe grown by VPE[17] and MOCVD[11], as well as bulk single crystals[13] have also shown dominant free excitonic emission. However, the MBE grown ZnSe demonstrates high reproducibility, particularly when ultra pure Zn and Se sources are used[10]. II-VI compounds, in general, are suitable for MBE because they can be congruently evaporated from a binary or ternary source[18], and the low growth temperature (300-350°C) minimizes generation of native defects and incorporation of background impurities[3,10]. Recent advances in MBE growth of ZnSe[2,3,10,17-20] and superlattices and multi-quantum well structures of ZnSe with ZnSSe[17,18], ZnMnSe[21], and ZnTe[22] continue to stimulate a wide range of physical ideas for future devices. Structural studies by TEM and optical spectroscopy are the key characterization tools used for understanding the physics of heterostructures.

High quality ZnSe on GaAs has been grown by MBE[3,10,18-20]. The lattice mismatch of 0.27% between ZnSe and GaAs results in a compressive stress in the ZnSe for layer thicknesses up to 0.15μm[23,24]. These layers contain stacking fault defects[24]. Above layer thickness of 0.15μm misfit dislocations are generated from surface sources and the stacking fault defects, to accommodate the lattice mismatch. For layer thickness > 1μm, the films exhibit biaxial tension associated with the difference in thermal expansion coefficient of ZnSe and GaAs as revealed by photoluminescence[23], TEM and X-ray techniques[24,25]. Strain must be controlled and minimized in order to reduce the density of dislocations and background impurities, as shown by several groups recently[10,19,26,27].

An alternative process to MBE is to grow epilayers by atomic layer epitaxy (ALE). In ALE, constituent elements are alternately deposited onto the substrate to achieve step-wise film growth[2,18,28]. Two key advantages of ALE over MBE are that (i) dopant incorporation can be enhanced by δ–doping in a more controlled fashion and (ii) growth temperature can be significantly reduced (150-200°C)[28].

3. DEVICES

P-N Junction

The primary goal in wide-band-gap II-VI's is to fabricate p-n junction LEDs and lasers. Partial success and shortcomings in achieving this device will be discussed in light of the results to date. I believe that the situation has reached a stage where "p-n junction in wide-band-gap II-VI is indeed a possibility." However, to achieve **reliable** and **repeatable** p-n junctions, technological breakthroughs are still needed but, above all, the current limitations need to be understood in a systematic manner. The evidence, in recent years, seem to indicate that the limited achievable p-type conductivity in ZnSe is associated with extrinsic defects (background impurities) rather than intrinsic defects (e.g. vacancies and self interstitials)[1]. However, an increase in deep-level concentration is frequently observed with increased incorporation of acceptors. This deep level is associated with vacancy-impurity complexes. The above phenomena is, in particular, quite pronounced in p-type ZnSe. For example, in N-doped ZnSe, Ohki, et al.[11] have observed that the concentration of deep levels increases when total N concentration exceeds 10^{17}cm^{-3}. Yao and co-workers[29] had observed a shallow acceptor state due to phosphorus incorporation only when its concentration was low. Increased incorporation of the P in ZnSe generated the well known deep state associated with P-doping[10,29]. When Na[30] and As[31] were incorporated, similar increases in the deep level concentration were observed. This observation, when combined with the fact that resulting hole concentration can decrease by orders of magnitude for a small increased fraction of deep levels in a close compensation regime[32], can explain our inability to obtain p-type ZnSe even with enhanced acceptor incorporation.

It is concluded from the above results that increased p-type conductivity may not be achieved by simply increasing the acceptor concentration. One must understand the effect of increased dopant incorporation on the resultant active acceptor concentration N_A, in par-

3

ticular, the solubility limit of acceptors and its interaction with the other background impurities and native defects to form complexes and deep levels.

Table I lists recent results where good p-type ZnSe has been reported. From this table, we conclude that although several researchers have claimed p-type material and p-n junctions, yet at present all such results reflect severe limitations and are not reproduced.

Table I

SUMMARY OF P-TYPE ZnSe/P-N JUNCTION (1975-88)

	Growth Method & Substrate	Acceptor	p-Resistivity Carrier Concentration, Mobility	p-n Junction	Light Output & Ext. Efficiency	Comments
(a) Robinson & Kun 1975	VPE ZnSe:I	Ga,In,Tℓ and Li	1.0 Ω-cm 2×10^{17} cm^{-3} ~10 cm^2/V•s	Yes	3×10^3 FL at 100 mA 1%	Thin Film (1000 Å) Not reproduced
(b) Nishizawa et al. 1986	TDM-CVP ZnSe	Li	0.5 Ω-cm ~10^{16} cm^3 ~50 cm^2/V•s	Yes	2 mcd at 2 mA	Not reproduced
(c) Yasuda et al. 1988	MOVPE GaAs	Li, N	0.2 Ω-cm 9×10^{17} cm^{-3} 40 cm^2/V•s	Yes	?	Not reproduced
(d) Cheng et al. 1988	MBE GaAs	Li	6×10^4 Ω-cm 3×13^{13} cm^{-3} ---	No	?	Limited to p<3×10^{13} cm^{-3}
(e) Ohki et al. 1988	MOVPE GaAs	N	10^2-10^3 Ω-cm 10^{14} cm^{-3} 50 cm^2/Vsec	No	?	Difficult to achieve p > 10^{14} cm^{-3}
(f) Stücheli & Bucher 1988	VPE ZnSe:I	Unknown	0.1 Ω-cm ~ 10^{18} cm^{-3} 100 cm^2/Vsec	No	?	?

(a) Ref. 33, (b) Ref. 12, (c) Ref. 14, (d) Ref. 30, (e) Ref. 11, (f) Ref. 30A.

To explain the above results, associated with the incorporation of shallow acceptors in Table I, key comments are mentioned which are based on the explanations provided by the authors. In the case of Robinson and Kun[33], a non-equilibrium quenching process led to a p-layer of 1000A thickness on the surface of a n-type ZnSe substrate, and this was identified with the presence of low concentrations of Li in the substrate. Nishizawa, et al.[12] also reported p-type ZnSe doped with Li; however, the general reproducibility issue still remained a problem. It is implied in the above experiments that the Li-acceptor is mobile, and under certain heat treatment conditions, it can lead to conducting p-type ZnSe.

In recent results reported by Yasuda et al.[14], co-doping of Li and N of ZnSe grown by MOVPE resulted in highly conducting p-type ZnSe with carrier concentration of up to 9×10^{17}cm^{-3}. This result, if it can be reproduced and understood, would open the door for all kinds of optoelectronic devices. Ohki et al.[11] have attempted to obtain p-ZnSe with nitrogen using high rates of NH$_3$ flow in MOVPE growth. They obtained 100Ω-cm p-type ZnSe for N incorporation of ~10^{17}cm^{-3}. Increased N incorporation resulted in increased resistivity as well as enhanced deep level emission. Similar increased resistivity was reported by Cheng et al. [30] using Li doping in MBE grown ZnSe. In both Ohki et al. [11] and Cheng et al. [30], it seems that the incorporated acceptor concentration N_A is in the range of ~10^{17} - 10^{18} cm^{-3} while the carrier concentration at room temperature is only of the order of ~10^{14}cm^{-3}. From semiconductor statistics for acceptor level at E_A=110 mev and for N_A= 10^{18} cm^{-3}, a hole carrier concentration of 8×10^{17}cm^{-3} is estimated. This suggests

4

that with increased concentration of dopants there must be an additional compensation mechanism which is preventing the p-type characteristics. This could be attributed to enhanced incorporation of shallow donors, such as the interstitials of Li, Na and possibly Cu [34]. It seems that the combination of generation of deep levels and interstitial donors is hampering the fabrication of p-type ZnSe.

Based on the above data, what are the significant improvements made so far in the electrical properties of ZnSe? For the n-type ZnSe, MBE grown layers on GaAs show a net room temperature carrier concentration $n_{RT}{\sim}10^{14}cm^{-3}$ [10] and a peak mobility $\mu_n$max~8000 cm^2/v.s. at about 50K [10,35]. These, when compared to value obtained by Aven [36] in Zn-annealed bulk ZnSe are $n_{RT}{\sim}6x10^{13}cm^{-3}$ and $\mu_n$max~12,000 cm^2/v.s. Although bulk mobility is better, the key advantage of MBE over bulk is its reproducibility. It is expected that further refinement of purity and control of defects (e.g. ALE, lattice matched MBE, etc.) will increase mobility of MBE grown material to the best obtained in the bulk. In the case of p-type ZnSe, the issue is much more complex since high quality p-type ZnSe with controlled and reasonable conductivity is not yet available. A sample with $p_{RT}{\sim}10^{15}$ - $10^{16}cm^{-3}$ and $\mu_p^{RT} \geq 50$ cm^2/v.s. would be considered a good sample.

What are the possibilities to overcome the above limitation to obtain p-type ZnSe? In my opinion, (i) the enhanced incorporation of shallow acceptors from Group V (e.g. N, P and As) must be understood; (ii) purer sources for growth should be used; (iii) non-equilibrium incorporation of acceptors such as δ-doping, photo-assisted incorporation, rapid thermal annealing, etc. at low temperature must be attempted; (iv) incorporation of acceptors in layers grown on different substrates such as bulk ZnSe, other lattice matched substrates and differently oriented GaAs substrates should be studied; (v) effect of strain on the incorporation of acceptors and donors should be understood; and (vi) the theory of acceptors and donors and their complexes must be studied. The current research effort by several groups is being carried out to address the above issues, and it is expected to culminate into a reliable p-type conductivity in ZnSe.

Electron Beam Pumped Lasers

Visible lasers from ZnSe, CdS, CdSSe, ZnCdSe and CdSe have been demonstrated under excitation of high energy electron beams[37-40]. The electron beam pumped (EBP) lasers offer several potential applications in display areas and laser printer technology. Although these lasers operate at room temperature, considerable improvements of the materials is needed for reliable operation. At room temperature, the threshold of EBP lasers of longitudinal geometry from bulk material typically is around 25A/cm^2 for an accelerating potential of 35 keV. MBE grown layers of thickness ~4 μm have resulted in thresholds of 5A/cm^2 [17] in transverse geometry. Optical and carrier confinements based on, for example, ZnSe-ZnSSe heterostructures, are expected to decrease this latter value to about 1-2A/cm^2 leading to a reliable room temperature EBP blue laser.

To modify the band gap and achieve carrier and optical confinement for lower laser thresholds, as well as other non-linear optical devices, strained layer superlattices (SLS) of ZnSe-ZnSSe have been grown[17,18,41]. In these SLS, quantum confinements and strain effects have been studied in detail[42,43]. In SLS, a blue shift was observed with increased quantum confinements as the layer thickness decreases. Photoluminescence experiments and theoretical calculations reveal that ZnSe-ZnSSe interface exhibit a very small conduction band offset[43].

Non-Linear Optical Devices

II-VI's possess large polarizability and non-linear coefficients. For example, the second harmonic coefficient for ZnSe and ZnS is comparable to that of well known SHG materials such as lithium iodate and barium sodium niobate[44]. II-VI's are not being used for SHG devices because of the lack of conventional phase matchable conditions due to the absence of birefringence. Because of this, the current effort on non-linear optical devices from II-VI's is concentrated on artificially grown structures such as waveguides, superlattices and multi-quantum wells, where the non-linear properties can be modified. On the other hand, optical band-edge associated non-linearities could be modified by the absorption process. In this process, generation of high density of carriers strongly renormalize the optical properties through many body effects and cause effects such as plasma screening of coulombic interaction; reduction of band gap and band filling[45]. Experiments such as pump-probe spectroscopy four-wave mixing and non-linear interferometry are being done to study and fabricate optical-bistable devices [45-47]. For optical bistability experiments, the measured transmission as a function of intensity can be used to perform gating operations, if one alters the operating conditions at the input side. Recently, optical non-linearities of extremely small isolated semiconductor particles in glass has been studied to measure quantum size effects[48,49]. In recent years, with the availability of superlattice and multi-quantum well structures, strong third order non-linearities were reported in GaAs-GaAsAs QW and SL structures[50] The quantum confined stark effect was used to demonstrate successfully a self electro-optic effect (SEED) devices[51]. These results in III-V semiconductors can be extended to II-VI materials.

Recently, second order non-linearities in some II-VI structures has been studied theoretically in asymmetric coupled quantum well structures[52]. These calculations potentially offer an alternate way to fabricate devices based on frequency mixing, second harmonic generation and linear electro-optic (Pockels) effect. Yokogawa et.al.[53] have grown low loss optical ridge waveguides using ZnSe-ZnS strained-layer superlattices. These waveguides showed a large difference in the propagation loss between TE and TM polarizations. This may be associated with the birefringence for TE and TM polarizations due to the slight anisotropy of the refractive index in the superlattice structure.

The distinct advantage of II-VI's over III-V's is that non-linear devices could be used in the visible range. An extremely desirable device would be a blue laser generated by doubling the laser frequency of GaAs-GaAlAs by quantum well or superlattice structure of II-VI materials.

4. CONCLUSION

The progress made in the last few years has demonstrated that high quality layers of II-VI semiconductors are ready to be exploited for different devices. P-N junctions and other novel devices based on superlattices and multi-quantum well structures are possible in the near future as demonstrated by several preliminary results, and provides further impetus to research in wide gap II-VI semiconductors.

5. ACKNOWLEDGMENT

The author has greatly benefitted from discussions with Sel Colak.

6. REFERENCES

[1] R.N. Bhargava, J. Crystal Growth 59, 15 (1982)

[2] T. Yao, in: The Technology and Physics of Molecular Beam Epitaxy, Eds. M.G. Dowsett and E.H.C. Parker (Plenum, New York, 1985) pp. 313-345, and references therein

[3] T. Yao, J. Crystal Growth 72, 31 (1985)

[4] W. Stutius, J. Crystal Growth 59, 1 (1982)

[5] P.J. Dean, Phys. Stat-Solidi (a) 81, 625 (1984)

[6] R.N. Bhargava, J. Crystal Growth 86, 873 (1988)

[7] J. Nishizawa, K. Itoh, Y. Okuno and F. Sakurai, J. Appl. Phys. 57, 2210 (1985)

[8] C.J. Werkhoven, B.J. Fitzpatrick, S.P. Herko, R.N. Bhargava and P.J. Dean, Appl. Phys. Lett. 38. 540 (1981)

[9] K. Yoneda, Y. Hishida, T. Toda, H. Ishii and T. Niina, Appl. Phys. Lett. 45, 1300 (1984)

[10] J.M. DePuydt, T.L. Smith, J.E. Potts, H. Cheng and S.K. Mohapatra, J. Crystal Growth 86, 318 (1988); J.M. DePuydt, H. Kheng, J.E. Potts, T.L. Smith and S.K. Mohapatra, J. Appl. Phys. 62, 4756 (1988)

[11] A. Ohki, N. Shibata and S. Zembutsu, Jpn. Appl. Phys. 27, L909 (1988)

[12] J. Nishizawa, R. Suzuki and Y. Okuno, J. Appl. Phys. 59, 2256 (1986)

[13] M. Isshiki, J. Crystal Growth 86, 615 (1988)

[14] T. Yasuda, I. Mitsuishi and H. Kukimoto, Appl. Phys. Lett. 52, 57 (1988)

[15] S.J.C. Irvine, J.B. Mullin, H. Hill, G.T. Brown and S.J. Barnett 86, 188 (1988)

[16] N.C. Giles, R.N. Bicknell, R.L. Harper, S. Hwang, K.A. Harris and J.F. Schetzina, J. Crystal Growth 86, 348 (1988)

[17] T. Matsumoto, N. Kobayashi and T. Ishida, Jpn. J. Appl. Phys. 26, L209 (1987)

[18] D.A. Cammack, R.J. Dalby, H.J. Cornelissen and J. Khurgin, J. Appl. Phys. 62, 3071 (1987); H.J. Cornelissen, D.A. Cammack and R.J. Dalby, J. Vac. Sci. Technology B6, 769 (1988)

[19] H. Cheng, S.K. Mohapatra, J.E. Potts and T.L. Smith, J. Crystal Growth 81, 512 (1987); R.M. Park, H.A. Mar and N.M. Salansky, Appl. Phys. Lett. 46, 386 (1985)

[20] M. Tamargo, J.L. de Miguel, F.S. Turco, B.J. Skromme, D.M. Hwang, R.E. Nahory and H.H. Farrell, this proceeding; B.J. Skromme, M.C. Tamargo, J.L. de Miguel and R.E. Nahory, Mat. Res. Soc. Sym. Proc. 102, 577 (1988)

[21] L.A. Kolodziejski, R.L. Gunshor, N. Otsuka, S. Datta, W.M. Becker and A.V. Nurmikko, IEEE J. Quant. Electron. QE-22, 1666 (1986)

[22] M. Konagai, M. Kobayashi, R. Kimura and K. Takahashi, J. Crystal Growth 86, 290 (1988); M. Konagai, M. Kobayashi, N. Teraguchi, S. Dosho, Y. Takemura, R. Kimura and K. Takahashi, this proceeding

[23] K. Mohammed, D.A. Cammack, R. Dalby, P. Newbury, B.L. Greenberg, J. Petruzzello and R.N. Bhargava, Appl. Phys. Lett. 50, 37 (1987); K. Shahzad, Phys. Rev. B38 (1988) to be published

[24] J. Petruzzello, B.L. Greenberg, D.A. Cammack and R. Dalby, J. Appl. Phys. 63, 2299 (1988)

[25] T. Yokogawa, H. Soto and M. Ogura, Appl. Phys. Lett. 52, 1678 (1988)

[26] N. Shibata, A. Ohki, H. Nakanishi and S. Zembutsu, J. Crystal Growth 86, 268 (1988)

[27] K. Ohmi, I. Suemune, T. Kanda, Y. Kau and M. Yamanishi, J. Crystal Growth 86, 467 (1988)

[28] T. Yao and T. Takeda, Appl. Phys. Lett. 48, 160 (1986); and T. Yao (private communication 1988)

[29] T. Yao and Y. Okada, Jpn. J. Appl. Phys. 25, 821 (1986)

[30] H. Cheng, J.M. DePuydt, J.E. Potts, T.L. Smith, Appl. Phys. Lett. 52, 147 (1988); J.E. Potts, H.A. Mar and C.T. Walker (unpublished work and final report ONR Contract N00014-85-C-0552 Dec. 1987)

[30A] N. Stucheli and E. Bucher, this proceeding

[31] M. Okajima, M. Kawachi, T. Sato, K. Hirahara, A. Kamata and T. Beppu in Extended Abs. 18th Conf. on Solid State Devices & Materials, Tokyo 1986 pp.645

[32] G.F. Neumark, J. Appl. Phys. 57, 3383 (1980); ibid Phys. Rev. B26, 2250 (1982)

[33] Robinson and Z. Kun, Appl. Phys. Lett. 27, 74 (1976)

[34] G.F. Neumark and S.P Herko, J. Crystal Growth 59, 604 (1982)

[35] M. Vaziri, R. Reifenberger, L.A. Kolodziejski, R.L. Gunshor and R. Holzer, Bulletin of Am. Phys. Soc. 33, 694 (1988); These authors have measured μ peak ~8000 cm^2/v.s. in Ga doped ZnSe with $n_{RT}= 1\times10^{15}cm^{-3}$. M. Vaziri (private communication)

[36] M. Aven, J. Appl. Phys. 42, 1204 (1971)

[37] V.I. Koslovskii, A.S. Nasibov, A.N. Pechenov, Y.M. Popov, O.N. Talenskii and P.V. Shapkin, Sov. J. Quant. Electron. 7, 194 (1977); N.G. Basov, O.V. Bogdankevich, A.S. Nasibov, V.I. Koslovskii, V.P. Papusha and A.N. Pechenov, Soviet J. Quant. Electron. 4, 1408 (1975)

[38] S. Colak, B.J. Fitzpatrick and R.N. Bhargava, J. Crystal Growth 72, 504 (1985); R.N. Bhargava, S. Colak, B.J. Fitzpatrick, D. Cammack and J. Khurgin, in: Proc. Intern. Display Research Conf. (IEEE, New York 1985) pp. 200

[39] J.E. Potts, T.L. Smith and H. Cheng, Appl. Phys. Lett. 50, 7 (1987)

[40] S. Colak, J. Khurgin, W. Seemungal and A. Hebling, J. Appl. Phys. 62, 2639 (1987)

[41] S. Fujita, Y. Matsuda and A. Sasaki, Appl. Phys. Lett. 47, 955 (1985)

[42] K. Mohammed, D.J. Olego, P. Newbury, D.A. Cammack, R.J. Dalby and H.J. Cornelissen, Appl. Phys. Lett. 50, 1820 (1987)

[43] K. Shahzad, D. Olego and C.G. van de Walle, Phys. Rev. B 38, 1417 (1988)

[44] c.f. S. Singh, "Non-linear Optical Materials", page 489, CRC Handbook of Lasers, Editor: R.J. Pressley, The Chemical Rubber Co., Cleveland, Ohio (1971)

[45] S.W. Koch, N. Peyghambarian and H.M. Gibbs, J. Appl. Phys. 63, R1 (1988) and references cited therein

[46] C. Klingshirn, U. Becker, C. Dornfeld, H. Kalt, M. Kunz, M. Lambsdorff, V.G. Lyssenko, F.A. Magumder, R. Renner, S. Sherel, H.E. Swoboda, C, Weber and M. Wegenerr, Proc. 18th Int'l. Conf. on Phys. of Semiconductors World Scientific Publishing Co., pp. 1667 (1987)

[47] J.Y. Bigot, A. Daunois, R. Leonelli, M. Sence, J.G.H. Mathew, S.D. Smith and A.C. Walker, Appl. Phys. Lett. 49, 844 (1986)

[48] G.R. Olbright, N. Peyghambarian, S.W. Koch and L. Banyai, Optics Lett. 12, 413 (1987)

[49] B.G. Potter and J.H. Simmons, Phys. Rev. 37, 10838 (1988)

[50] D.A.B. Miller, J.S. Weiner and D.S. Chemla, IEEE J. Quant. Electron. QE-22, 1816 (1986)

[51] D.A.B. Miller, D.S. Chemla, T.C. Damen, A.C. Gossard, W. Wiegmann, T.H. Wood and C.A. Burrus, Appl. Phys. Lett. 45, 13 (1984)

[52] J. Khurgin, Appl. Phys. Lett. 51, 2100 (1987); ibid Phys. Rev., August 1988 (to be published)

[53] T. Yokogawa, M. Ogura and T. Kajiwara, Appl. Phys. Lett. 52, 120 (1988)

RECENT DEVELOPMENT TRENDS IN THIN FILM

ELECTROLUMINESCENT DISPLAYS

Runar Törnqvist

Lohja Corporation, Finlux Display Electronics
Olarinluoma 9
SF-02200 Espoo Finland

INTRODUCTION

During the last few years thin film electroluminescence (TFEL) has emerged as one important technical application of optical properties in II-VI compounds. The thin film electroluminescent structure, schematically shown in fig. 1, is almost ideal for a flat panel display. Light is generated in a less than 2 µm thick thin film stack, which is deposited on a glass substrate. The light emitting film is deposited between two insulating films, and the light emitting area is defined by two electrodes that complete the structure. Light is emitted when the electric field exceeds about 1 MV/cm in the phosphor film. The thickness of the display is thus completely determined by the driving electronics and the thickness of the glass substrate. Crisp images and good contrast are further advantages. Since the decay time of the emission is about one millisecond or less, full video capability is achieved. Among monochromatic displays, the ZnS:Mn based yellow TFEL display panel is widely recognized to have excellent visual appearance. According to marketing forecasts TFEL is expected to gain a reasonable share in the fast growing flat panel display market in the 90'ies.

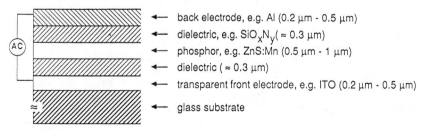

Fig. 1. Schematic illustration of the cross-section of the thin film electroluminescent structure.

The TFEL structure is in principle easy to make. Poly-crystalline or amorphous films are deposited on a glass sub-strate. Compared to wafer processing lithografy is less de-manding because typical dimensions are a few hundred microme-ters. There are now three established TFEL manufactures in the world. One in Japan, one in the USA, and one in Finland. Bearing in mind the advantages mentioned above this number might seem small. A fact is, however, that the technology is rather delicate although it appears simple in theory. In addition the need of high voltages, 150 V to 250 V peak, makes the driving electronics expensive in comparison to liquid crystal displays.

Facing the challenge there is a lot of development work going on. The presently marketed ZnS:Mn based structure is improved in order to meet the demand for larger sizes, lower power consumption, grey scales etc. Another direction of development is aiming at a full-colour TFEL display. In the following two chapters we highlight some of the main issues in these development works.

DEVELOPMENT OF YELLOW EMITTING ZnS:Mn TFEL DISPLAYS

The largest TFEL display manufactured at present is a 640 x 400 matrix. Average brightness is about 50 cd/m^2 and power consumption is around 20 W. Although this half page display meets the requirements for many applications, there is a demand for larger sizes, grey scales, higher bright-ness and lower power consumption.

The address time per line is a key factor when multiplex-ing a matrix. As the matrix is scanned at 60 Hz (or more) this limits the address time to approximately 1/Nf where N is the number of lines and f is the frame frequency. Figure 2 illustrates the time dependence of the driving voltage pulse, U, brightness for the two polarities of the driving voltage, B_+ and B_-, and the dissipative current in the ZnS:Mn layer. It is seen that luminance rises to its maximum value within 40 us from the onset of the voltage pulse. This is a consequence of the fast rise and decay of the current density in the ZnS:Mn layer that is nearly proportional to the excitation rate of the light emitting Mn^{2+} ions. Only little luminance is lost if the voltage is cut off in 25 us. To obtain fast rising, short current pulses it is essential that the onset of the current in the ZnS:Mn layer is well defined. This depends on the tunneling threshold for the electron injection at the two ZnS:Mn-insulator interfaces, and a good control of the interface properties is becoming an important topic. This applies as well for the materials, deposition techniques and for possible impurities. It seems possible to multiplex matrixes with more than 500 lines, and development of the interfaces may allow almost 1000 lines in future. Another constraint arises, however, from the current supply of the row drivers. The capacitance of the TFEL structure is typically between 50 pF/mm^2 and 100 pF/mm^2. Present drivers provide about 100 mA, which consequently limits the area, and thus the length of the rows to some 30 cm.

At present every second electrode is contacted from the same side of the display. If every electrode is contacted from the same side, a matrix can naturally be divided allowing multiplexing of even larger sizes. Nevertheless, one line at a time multiplexing always implies that the whole matrix is charged when each line is addressed. Therefore most of the power consumption is "unnecessarily" lost in electrodes and drivers. Much hope has been given to hysteresis in the brightness vs. voltage characteristics of ZnS:Mn TFEL structures [2,3]. In principle it should be possible to drive the display at a constant (sinusoidal) sustain voltage, and voltage transients are only needed when information is changed. Stability problems have so far prevented the practical use of this effect. It might be that a deeper understanding in the interface properties mentioned above induces progress in making use of the hysteresis effect.

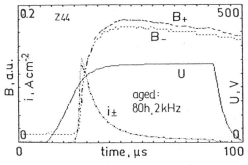

Fig. 2. Time dependence of the driving voltage, U, the brightness emitted for the two polarities of the driving voltage, B_+ and B_-, and the corresponding current densities in the ZnS:Mn layer, i_+ and i_-. The structure has been aged for 80 hours at 2 kHz, and the insulating layers were made of zirconium oxide [1]. (Permission for Reprint, courtesy Society for Information Display).

Another approach to generate hysteresis in the brightness vs. voltage characteristic is to incorporate a photoconductor sensitive to the yellow emission [4]. This is an interesting solution provided that the impact of ambient light can be suppressed and that the processing of the photoconductor does not cause other inconveniences. A third way is to make use of thin film transistors. Since two transistors will be needed for each pixel this is a complicated solution.

13

There is another reason why the electron injection at the ZnS:Mn-dielectric interfaces is important. It was mentioned above that short current pulses in the ZnS:Mn layer are desirable. But in addition a well defined threshold field for the electron injection is needed to achieve a good discrimination ratio. That is, in order to minimize power consumption and to achieve high brightness, there should be

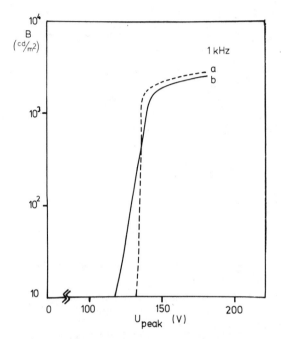

Fig. 3. Brightness vs. voltage characteristics measured at 1 kHz for a TFEL structure grown with ALE [8].
a. as-grown structure,
b. stabilized by operation at 175 V (peak) at 2 kHz for 4 h. (Permission for Reprint, courtesy Society for Information Display).

a steep rise in the brightness vs. voltage characteristic [5,6]. The brightness vs. voltage characteristics of an as-grown and a stabilized (aged) TFEL structure is shown in fig. 3. This sample was grown using Atomic Layer Epitaxy [7], but similar shifts have been observed for assymetric driving in structures deposited with other techniques.

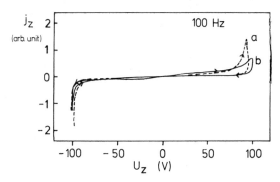

Fig. 4. The current density in the ZnS:Mn layer as
function of the voltage drop across the layer
measured at 100 Hz sinusoidal voltage applied to
the back Al electrode [8]. a and b as in fig. 3.
(Permission for Reprint, courtesy Society for
Information Display).

The reason for the softening is illustrated in fig. 4,
which shows the current density in the ZnS:Mn layer as a
function of the voltage drop across the ZnS:Mn layer. The
threshold field for the electron injection seems to be substantially lower near the first grown dielectric-ZnS:Mn interface than at the last grown dielectric-ZnS:Mn interface,
where a very well defined threshold is observed. This behaviour is enhanced during the stabilization procedure.

Progress has recently been made in affecting the electron injection properties. Müller et al. [1] improved the
ageing behaviour by using ZrO_2 dielectrics. Nishikawa et al.
[9] reported on a steep and stable brightness vs. voltage
characteristic by depositing thin layers of CaS on both sides
of the ZnS:Mn layer. Whether the softening of the brightness
vs. voltage characteristic is only due to degradation of the
interface because of continuous hot electron bombardment,
or if there is also e.g. sodium diffusion involved [10] is not
settled yet.

Except for ALE grown structures, the softening of the
brightness vs. voltage characteristic is apparently predominant when using assymetric driving voltage in the sense that
the delay between the two polarities of the driving voltage
is not the same. This is the case when using a refresh driving scheme, where modulation only occurs at the other
polarity of the driving voltage. As a consequence off pixels
tend to become visible on the first few and last few rows of
the matrix, where the difference in time delay between the
pulses is largest. Problems with retained images because of

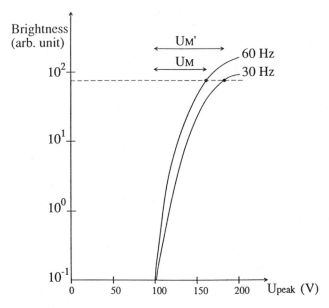

Fig. 5. Brightness vs. voltage at 60 Hz and 30 Hz.
A higher modulation is required in the latter case
to obtain the same brightness.

the "differential ageing" [11] phenomenon have been avoided
using a symmetric driving scheme. A smaller modulation voltage can be used since modulation occurs on both polarities
of the driving voltage. Because only one light pulse is
emitted per frame time (1/60 Hz), compared to two pulses in
the refresh drive, brightness is reduced, see fig. 5. The
total modulation, or the "swing", must therefore exceed the
modulation used in refresh driving. Symmetric driving can
reduce power consumption without sacrificing brightness.

Although symmetric driving brings some advantages,
there is a price one has to pay. In refresh driving it does
not matter if the light pulses emitted for the two polarities of the driving voltage have different intensity. But if
the light pulses do not have the same magnitude, flicker
will be observed under symmetric driving. The problem is
highly reduced by addressing every second line at the same
polarity and every other second line at the other polarity.
But to fully avoid flicker, symmetrical light emission is
needed. Essentially this requirement goes back again to the
electron injection at the interfaces. It is crucial that the
same luminous efficiency is obtained for both polarities of
the driving voltage, and consequently it must be assured
that conduction electrons in the ZnS:Mn layer for both
polarities only are liberated in an electric field sufficient for accelerating the electrons up to optical energies.
Other effects that destroy the symmetry are inhomogenous
excitation and non-radiative relaxation of Mn^{2+} ions in the
thickness direction of the ZnS:Mn layer.

Grey scales implies a very tight thickness control of the thin films in the TFEL structure. Since operation voltages are on the steep portion of the brightness vs. voltage characteristic, thickness variations should not be more than a few percent. In addition it is important that the tunneling thresholds at the ZnS:Mn-dielectric interfaces are under control. No doubt that this puts considerable requirements on processing of the TFEL structure.

3. OTHER COLOURS

Until now sufficient luminous efficiency and brightness for commercial displays has only been achieved with the yellow emitting ZnS:Mn TFEL structure. There has, however, been significant progress in the development of TFEL structures emitting other colours during the last few years. The progress of green emitting ZnS:Tb is illustrated in fig. 6, which shows how maximum brightness has increased during the last years. The performance of $ZnS:TbF_x$ ($x \approx 1$) is actually not far behind ZnS:Mn.

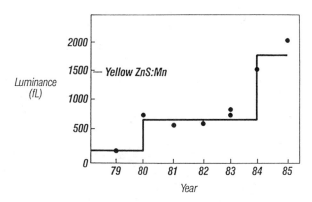

Fig. 6. The development of green ZnS:Tb luminance measured at 5 kHz during the last few years (1 fL = 3.43 cd/m^2) [12]. (Permission for Reprint, courtesy Society for Information Display).

In addition to efficient colour TFEL structures, it is also necessary to develop a multicolour display structure. That is, investigations must be made how to place the different phosphor layers. Since the insulator-phosphor-insulator structure is transparent to visible light it is possible to stack the phosphors. But it may be more advantageous to put at least two phosphor layers in the same plane.

The colour thin film electroluminescent technology has recently been reviewed by C. King [12], and for a more detailed treatment of the subject we refer to this article.

Phosphors that have been successful to obtain green or blue in cathode ray tubes (CRT), such as ZnS:Cu,Al and ZnS:Ag, are very poor in TFEL. This is most likely due to high field ionization of donors and acceptor states, which consequently reduce the luminous efficiency. Better results have been achieved using inner shell transitions of dopants. The most well-known example is ZnS:Mn where manganese substitutes zinc forming an iso-electronic impurity that only marginally distorts the lattice. Other colours have been obtained using rare earth dopants (RE) in which the light emission originates from $4f^n$ state transitions.

When compared to ZnS:Mn, there are two fundamental problems in the ZnS:RE system; geometric and valence mismatch. Development has therefore been focused on charge compensation and making use of other host materials such as CaS and SrS that better allow incorporation of rare earth atoms.

Best results have been achieved for the green emitting ZnS:TbF$_x$ where x is close to one depending to some extent on the terbium concentration. Although sufficient brightness and luminous efficiency for practical applications has been reported, see table I, there are still stability problems that must be solved.

TABLE I [12]

LUMINANCE OF TFEL PHOSPHORS AT 60Hz

Color	Material	Luminance	CIE X	CIE Y
Red	ZnS:Sm, F	3.5fL	0.60	0.38
	ZnS:Sm, P	3.5fL	0.63	0.36
	ZnS:Sm, Cl	3.5fL	0.64	0.35
	CaS:Eu	3.0fL	0.68	0.31
	CaS:Eu, F, Cu, Br	6.3fL	0.66	0.34
Green	ZnS:Tb, F	40.0fL	0.31	0.59
Blue	ZnS:Tm	0.04fL	0.15	0.20
	SrS:Ce, F	10.0fL	0.20	0.36

(Permission for Reprint, courtesy Society for Information Display.)

The progress in the development of a red emitting TFEL structure has not been as good. ZnS:Sm has traditionally been the best candidate for red, and research has aimed at optimizing charge compensation because of the +3 valence of samarium. Flourine, chlorine and phophorus have been used. Another key issue has been the improvement of the chromaticity because of the orange shade of $ZnS:SmF_3$. Best results have been achieved using chlorine as a co-activator and brightness of 1000 cd/m^2 at 5 kHz has been reported [13].

CaS:Eu is another candidate for red. Because of better lattice match and the +2 valence of europium in CaS this system is expected to be more stable. Another advantage is the superior chromaticity. A drawback is that CaS is more sensitive to moisture which makes processing more difficult and which may give rise to stability problems. Significant improvement with CaS:Eu,F,Cu,Br has recently been presented by Ohnishi and co-wokers [14], who reported a brightness of 350 cd/m^2 at 1 kHz and better stability by placing thin layers of SiO_2 on both sides of the light emitting layer.

ZnS:Tm has for a long time been the best blue emitting TFEL phosphor. Brightness and luminous efficiency are, however, far from what is required, and there has only been an improvement by a factor of two during the last seven years. Since ZnS:Tm has received rather little attention lately, one might expect improvement. During the last few years much hope has been given to SrS:Ce which at present has about twenty-five times more brightness than ZnS:Tm. The colour is, however, blue-green, and filtering it to blue would imply a decrease in brightness by a factor of ten. At least the same processing problems are involved with SrS as with CaS.

Before a full-colour TFEL display is introduced on the market, it is likely that monochromatic black and white displays appear for PC applications. Tanaka et al. [15] reported recently on a white TFEL structure with a brightness of 30 – 50 cd/m^2 at 60 Hz. The brightness is thus approximately one half of the brightness required for practical applications. The phosphor layer was SrS:Ce,K,Eu. This is as such a remarkable result. Using colour filters full-colour capability was demonstrated. The brightness of white was in this case reduced to 9 cd/m^2 (60 Hz). For full-colour applications it would be desirable that brightness were increased almost by a factor of ten. Yet this brightness is to the best of our knowledge superior to the brightness so far reported on full-colour TFEL structures involving three different colour phosphors.

It is too early to state whether full-colour TFEL will be realized using one white phosphor and colour filters, or three different colour phosphors. As already mentioned, the latter case requires the development of the three phosphor display structure where the placing of the phosphors is a key issue. For a discussion of layered, patterned and hybrid structures we refer to King's paper [12].

19

CONCLUSION

Present development trends in thin film electroluminescence (TFEL) are further improvement of the yellow emitting ZnS:Mn based structure and the research aiming at a full-colour TFEL display. The stability and the shape of the brightness vs. voltage characteristic has become an important subject in the development of the ZnS:Mn TFEL structure in order to satisfy the demand for reduced power consumption, grey scales and lower cost. A better control of the properties of the ZnS:Mn-dielectric interfaces is rising as a fundamental question. Colour development relies on the process improvement of different phosphors. A white light phosphor allowing full-colour capability can be very important. Most colour phosphors are not as ideal as ZnS:Mn where manganese substitutes zinc as an iso-electronic impurity causing no significant lattice distortion. Lattice matching and charge compensation in colour phosphors are likely to give rise to stability problems that have to be addressed before full-colour displays will appear on the market.

REFERENCES

1. G. O. Müller, R. Mach, R. Reetz, and G. U. Reinsperger, Degradation Mechanisms of ac Thin-Film Electroluminescent Displays, Digest 1988 SID Int. Symp., p. 23 (1988)

2. Y. Yamauchi, M. Takeda, Y. Kakihara, M. Yoshida, J. Kawaguchi, H. Kishishita, Y. Nakata, T. Inoguchi, and S. Devices, Proc. 6th Conf. on Solid State Devices, Tokyo, 1974, Suppl. J. Jap. Soc. Appl. Phys. vol. 44, p. 348 (1974)

3. P. M. Alt, D. B. Dove, and W. E. Howard, Experimental Results on the Stability of Thin-Film Electroluminescent Devices, J. Appl. Phys. vol. 53, p. 5186 (1982)

4. P. Thioulouse, and C. Gonzalez, Thin-Film Photoconductor Electroluminescent Memory Device with a High Brightness and a Wide and Stable Hysteresis, Appl. Phys. Lett. vol. 50, p. 1203 (1987)

5. P. M. Alt, Thin-Film Electroluminescent Displays: Device Characteristics and Performance, Proc. SID 1984 vol. 25/2, p. 123 (1984)

6. R. Törnqvist, Properties and Performance of TFEL Structures, J. Crystal Growth vol. 72, p. 538 (1985)

7. T. Suntola, J. Antson, A. Pakkala, and S. Lindfors, Atomic Layer Epitaxy for Producing EL-Thin Films, Digest 1980 SID Int. Symp., p. 108 (1980)

8. J. Karila, J. Hyvärinen, and R. Törnqvist, Electron Injection from Al_2O_3-ZnS:Mn Interfaces in ALE Grown TFEL Structures, Proc. Eurodisplay Conf., London, 1987, p. 236 (1987)

9. M. Nishikawa, T. Matsuoka, T. Tohda, Y. Fujita, J. Kuwata, and A. Abe, A Highly Stable TFEL Panel with CaS Buffer Layer, Digest 1988 SID Int. Symp., p. 19 (1988)

10. J. Watanabe, M. Wakitani, S. Sato, and S. Miura, Analysis of Deterioration in Brightness-Voltage Characteristics of TFEL Devices, Digest 1987 SID Int. Symp., p. 288 (1987)

11. R. T. Flegal, and C. N. King, Differential Aging Effects in AC-TFEL Displays, Proc. Int. EL Workshop, Kah-Nee-Ta, Oregon 1986

12. C. N. King, Color TFEL Technology, SID 1988 Seminar Lecture Notes vol. I, p. 2A.1 (1988)

13. K. Hirabayashi, H. Kozawaguchi, and B. Tsujiyma, Color Electroluminescent Devices Prepared by Metal Organic Chemical Vapor Deposition, Jap. J. Appl. Phys. vol. 26, p. 1472 (1987)

14. H. Ohnishi, R. Iware, and Y. Yamasaki, Red-Color ACTFEL Devices Using Sputtered CaS:Eu Thin Films, Digest 1988 SID Int. Symp., p. 289 (1988)

15. S. Tanaka, H. Yoshoyama, J. Nishiura, S. Ohshio, H. Kawakami, and H. Kobayashi, Bright White-Light Electro-luminescent Devices with New Phosphor Thin-Films Based on SrS, Digest 1988 SID Int. Symp., p. 293 (1988)

CONSIDERATIONS FOR BLUE-GREEN OPTOELECTRONICS BASED ON EPITAXIAL ZnSe/ZnS

K.P. O'Donnell and B. Henderson

Department of Physics and Applied Physics
University of Strathclyde
John Anderson Building
Glasgow G4 ONG
Scotland, UK

1. INTRODUCTION

The blue-emitting junction laser diode is the Holy Grail of the wide bandgap II-VI community. Such a device would permit a range of improvements in the current applications of laser diodes to the recording and reproduction of data. In this paper, we discuss the prospects for realising a "blue shift" in technology, away from III-V and towards II-VI materials. Recent rapid advances[1] in the chemical and structural control of II-VI compounds, superlattices and ordered alloys encourage us to assess these prospects. Characterisation of new material rests very firmly on a literature which goes back over twenty years.[2] Of particular relevance to light emitting devices is the very thorough optical spectroscopic studies pioneered in the '60s and '70s by the late Paul Dean and his co-workers. It is a great pity that Paul Dean did not live to see this work bear fruit.

In terms of the realisation of II-VI optoelectronic devices, it is encouraging to detect parallels in the growth and maturation of the III-V laser diodes, from its cold, dim and short-lived beginnings to the ubiquitous room-temperature device that we know today. In what follows, we will enlarge upon these themes. We also consider alternative strategies for the creation and technical exploitation of compact blue laser light sources and discuss a few of them briefly.

2. CHARACTERISATION OF EPITAXIAL MATERIAL

The University of Strathclyde has begun a collaboration with the University of Hull and the Royal Signals and Radar Establishment at Malvern on a programme of growing wide bandgap II-VI epilayers and devices by metal organic chemical vapour deposition (MOCVD). The Hull programme (see John Davies' paper in this volume) has advanced to the growth of CdSe/CdS hexagonal superlattices. The Strathclyde programme has focused on the analogous (but cubic) zinc

compounds. Physical properties of our optimised-growth ZnSe films
(~7 μm) are detailed in Table 1. Material of comparable quality
has been reported by many other workers. But what do we mean here
by "quality"?

Low temperature (say less than about 20K) photoluminescence (PL)
spectroscopy provides a cheap and easy way of comparing different
samples. In the case of ZnSe, a lot more work has been done in
the frequency or wavelength domain than in the time domain, but
these two aspects of PL spectroscopy are expected to give
complementary information.

In energy-domain PL, the emission spectrum shows narrow lines
(~ 1 meV) which have been assigned to recombination processes
involving free excitons, donor-bound excitons, acceptor-bound
excitons, donor-acceptor pairs and broad bands (10-100 meV) due to
deep-centre recombinations or carrier capture. The photon energies
at which these processes occur are well-known[3] and can help to
identify, for example, (residual) impurities. The relative
intensity of the free-exciton emission to that of the rest is often
used as a measure of sample purity. The half width of the exciton
lines gives some measure of the root mean square strain present in
the sample, according to a rule of thumb which states that unit
strain will shift the exciton by one electron-volt.

Time-domain PL of ZnSe is at present a somewhat underdeveloped
field. The times involved are relatively short and the intermediate
processes involved between the "bang" (incidence of a suprabandgap
photon) and the "glow" (emergence of a photon at the exciton energy)
may be quite complex, particularly if the incident power densities
are moderately high.

What really matters as far as blue light devices are concerned?
First of all, it is important to know the relative probability of
emission into the blue spectral region compared to the total light
emission from a given sample when using a particular method of
pumping, at room temperature. Methods of pumping include
photoexcitation above the bandgap, cathodoexcitation and

Table 1. Properties of optimised-growth film

ZnSe (7 μm) on GaAs	
Residual donors N_D-N_A	10^{15} cm^{-3}
X-ray double crystal width	100 arc seconds
Electron mobility	500 cm^2 v^{-1} s^{-1} (300K)
	6000 cm^2 v^{-1} s^{-1} (77K)
Photoluminescence: shallow/deep ratio	10^2-10^3
	no deep iron band

electron/hole injection using homo- or hetero-junctions, including MIS structures. Secondly, we should like to know what is the quantum yield of the material or the efficiency of the device. In other words we want to know the ratio of the usable output of blue light to the input. We must be sure that we are detecting all of the emitted radiation. A very common impurity in bulk ZnSe is the transition metal, iron. An interesting sequential capture process[4] by Fe^{2+} results in broad-band infra-red emission at 0.98 μm and 3.3 μm. Now, these wavelengths are beyond the range of conventional blue-sensitive photomultiplier tubes. Consequently, these emission bands are easily missed. In fact, there are no reports as far as we are aware of iron in epitaxial ZnS and ZnSe. However, strong iron PL was observed in our first grown layers and also in layers grown previously by the team at RSRE. (It has been eliminated by use of a novel zinc source.) Iron is not simply a case of an impurity which might be missed through IR blindness. It is a competitive impurity which shunts excitation from the blue to the IR and so acts as a "killer centre" for blue luminescence in ZnS and ZnSe. A correct optical investigation of any material should extend to the wavelength region where non-radiative multiphonon processes take over from purely optical ones. If one hundred electron-hole pairs are created at room temperature in a II-VI sample or device, how many will recombine to give blue light? In general this figure is difficult to estimate for a particular sample but it may at best be typically about one.[5] To characterise our material, we look for more easily measured quality factors. The following three suggest themselves:

   i) the PL shallow/deep ratio measures the ratio of total blue light
      emission (of whatever origin) to all the other emission (green
      red and infra-red).
  ii) the blue PL cold/hot ratio compares the total blue emission at
      say, liquid nitrogen temperature to that at room temperature.
 iii) the free exciton radiative lifetime at low temprature is self-
      explanatory.

These three QF's are interdependent. A good sample will have PL shallow/deep in excess of 100, blue cold/hot of order 5-10 and a free exciton radiative lifetime in the 10 nS range. These figures are for undoped material.

3. BLUE LIGHT EMITTING DEVICES

There have been several reports in the literature of blue LED's incorporating ZnSe homojunctions. (See the papers of Barghava, Stücheli and Kukimoto in this volume.) No injection laser, however, has been made to operate. On the other hand, blue lasers have been produced by several groups using electron beam pumping.

A naive comparison of cathodoexcitation and injection can be made as follows. A single 50 keV electron will produce about 10^4 electron-hole pairs in ZnSe. Taking a realistic figure of 25% for the internal quantum yield[6] (it is above 90% for GaAs), we find that one electron produces 2500 blue photons. A single injected electron, on the other hand, can yield only a single photon at best. In other words, a $1A/cm^2$ e-beam yields the same photon flux as 2500 A/cm^2 injection current. It is interesting to note that the lasing threshold for electron beam pumping[7] is about $4A/cm^2$. This corresponds to 10^4 A/cm^2 injection current. A similar figure is obtained by simply invoking bandgap scaling ($j_{th} \sim E_g^3$) to state of the art III-V injection lasers.

There is the question now of whether an injection device in ZnSe can support such a high level of current injection without cooling becoming necessary to preserve the structure. In particular we might start to worry about the stability of thermodynamically unstable acceptor centres in the presence of massive recombination. Even in a cooled diode, recombination-enhanced diffusion might act to neutralise the acceptors and destroy the device.

These fears may be groundless. The first III-V lasers proved to be extremely fragile. Crucial improvements in the stability of these devices resulting from the application of bandgap engineering to lower the threshold current. In what seems now to be a natural progression, improved growth techniques led to single heterojunction and double heterojunction devices and finally to quantum well devices. We now examine whether similar techniques, already well advanced for II-VI materials, can be expected to improve the prospects for II-VI devices.

4. BAND OFFSETS IN ZnSe/ZnSSe SUPERLATTICES

A quantum well modifies the emission properties of a material in three distinct, and yet closely related ways. Firstly, the conduction and valence band offsets may act to enhance carrier confinement and so encourage radiative recombination. Secondly, the dielectric discontinuity leads to light wave guiding in the well. Thirdly, quantum confinement modifies the allowed energies of the confined particles and blue-shifts the emission spectrum. These effects have all been observed and utilised in ZnSe/ZnSSe superlattices.

Unfortunately, however, experiment suggests that for these superlattices the partitioning of the band offset to the conduction and valence band is not optimal. The CB offset is in fact very small so that the holes alone are appreciably confined. In SLS e-beam pumped lasers for example, the threshold lowering compared to that of single layers may be nullified by the poorer optical quality of the SLS due to the presence of misfit dislocations. It appears that more sophisticated bandgap engineering is required to press home the advantages of SLS over single layers in this case. Promising routes of investigation include the use of atomic layer epitaxy (ALE) to produce ordered alloys, and isoelectronic doping.

5. COMPETITIVE TECHNOLOGIES

Are there realistic alternatives to II-VI materials in the quest for an efficient compact blue or green laser diode? Cubic and hexagonal SiC have bandgaps of the correct magnitude and type and both may be doped p- or n-type. There are large bandgap III-Vs such as BN which also form diodes and emit light. The materials control of these refactory compounds is difficult, but not overwhelming and development continues. There are also possibilities with bulk and thin CVD grown diamond films. Among **alternative strategies is the efficient frequency doubling of the** red emission from a high power GaAs laser: a prototype device has recently been reported by Matsushita.[8] In fact, all IR laser diodes are also blue lasers by virtue of the so-called "self-doubling light". This is sufficient for conventional optical absorption spectroscopy using temperature tuning of the emission wavelength.[9]

Yet another approach is that of two-photon pumping. Rare-earth ions in a solid host are pumped to a high-energy state via an intermediate storage level.[10] This involves a careful choice of materials but can be surprisingly efficient.

6. FINAL WORDS

The world awaits the development of a compact blue laser. It wants a box, as small as possible, with a switch at one end and a well-defined strong blue beam at the other. It does not really care what is inside the box. But we would be happy if it were a ZnSe homojunction. Or a II-VI superlattice. Or an ordered alloy. Or.....?

REFERENCES

1. For examples, see the papers by Kukimoto, Konagai, Kolodziejski and Barghava in this volume: also Volume 86 of the Journal of Crystal Growth.
2. For a review of early work, see: Physics and Chemistry of II-VI Compounds by M Aven and J.S. Prener (Eds.) North-Holland, Amsterdam, 1967.
3. R.N. Barghava, Journal of Luminescence 40 and 41 (1988) 24-27 and references therein.
4. K.P. O'Donnell, K.M. Lee and G.D. Watkins, J.Phys.C 16 (1983) L723-L728.
5. D.R. Wight, P.J. Wright and B. Cockayne, Electron Letters 18 (1982) 14.
6. Estimate of Toshiba group, communicated by T. Uemoto.
7. D.A. Cammack, R.J. Dalby, H.J. Cornelissen and J. Khurgin J Appl. Phys. 62 (1987) 3071-3074.
8. Reported in New Scientist 21 July 1988 p. 38.
9. K. Sakurai, N. Yamada and H. Baba, Technical Digest of IQEC '88, pp. 340-341.
10. R.M. Macfarlane, F. Tong and W. Lenth, Technical Digest of IQEC '88, pp. 570-571.

BLUE ELECTROLUMINESCENT DEVICES BASED ON LOW RESISTIVE p-TYPE ZnSe

N. Stücheli and E. Bucher

Universität Konstanz, Fakultät für Physik
Postfach 5560
7750 Konstanz, Fed. Rep. of Germany

Blue electroluminescent metal/p-ZnSe/n-GaAs heterojunctions were prepared by vapor phase epitaxy in an open system using iodine and hydrogen as transport agents. The ZnSe layers exhibited a p-type conductivity up to 50 $(\Omega cm)^{-1}$ and a net carrier concentration of 4×10^{18} cm^{-3} together with a Hall mobility of 100 cm^2/Vs at room temperature. These values are the highest ones reported so far.

INTRODUCTION

There has been a considerable interest in growing high quality ZnSe material due to its potential applications in optoelectronic devices. Especially, ZnSe is one of the most promising materials for blue light emitting diodes because of its wide band gap of 2.7 eV at room temperature. However, blue LED's with high efficiency have not been realized mainly because of the difficulty in obtaining p-type conduction. Several investigators have attempted to incorporate elements of group Ia and Va such as Li or Na as well as N, P or As, respectively, in ZnSe by molecular beam epitaxy (MBE) [1-3] or metalorganic chemical vapor deposition (MOCVD) [4,5]. Furthermore, the successful growth of low resistive p-type ZnSe single crystals has been reported [6].

Our efforts tend to minimize contamination and to avoid the so-called self-compensation effect which prevent the formation of p-type conduction in ZnSe. We are using a preparation method which has fallen into oblivion during the last few years; however, our promising results may lead to its possible come-back. We are describing here the successful growth of heteroepitaxial ZnSe layers. One objective of this work was to study the influence of the growth conditions on the electrical and optical properties of these layers. In a second part this paper deals with the preparation and characterisation of p–ZnSe/n–GaAs electroluminescent heterojunctions.

GROWTH AND PREPARATION TECHNIQUE

Our ZnSe layers were prepared by vapor phase epitaxy (VPE) in an open system using hydrogen and iodine as transport agents. The schematic representation of our setup is given in fig. 1. For the transport of the iodine to the ZnSe source hydrogen is flowing through the iodine source located outside of the epitaxy furnace. The iodine source is kept at a constant temperature T_I between $-20\ ^{\circ}C$ to $+20\ ^{\circ}C$ to control the iodine concentration. In the epitaxial furnace iodine and hydrogen are reacting with the ZnSe powder at $800\ ^{\circ}C$. The gaseous reaction products are transported in a hydrogen flow to the substrate zone. Within an inner quartz tube the substrates are exposed to a fixed linear temperature gradient ranging from $550\ ^{\circ}C$ to $650\ ^{\circ}C$ over a length of 10 cm. As substrate material we are using GaAs wafers due to its excellent lattice match of only 0.27 %. Depending on the iodine concentration ZnSe single

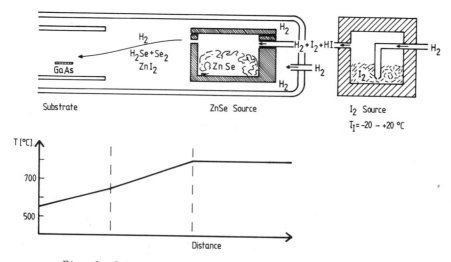

Fig. 1. Schematic representation of the growth system.

crystal layers up to 4 μm thickness could be grown at a growth rate of about 0.5 μm/h. The thickness of the films was estimated from the weight increase. Gold was evaporated for the ohmic contacts to the p–ZnSe.

RESULTS AND DISCUSSION

Electrical Transport Properties

For the electrical transport measurements ZnSe layers were grown on semi-insulating GaAs substrates with a resistivity higher than 10^8 Ωcm. The electrical properties were determined by the four–point resistivity and Hall technique. The resistivity of p–ZnSe layers grown at different substrate temperatures is shown in fig. 2 as a function of the reciprocal temperature. By increasing the growth temperature T_S from 583 °C to 648 °C the resistivity decreases by 6 orders of magnitude from 6×10^5 Ωcm to 0.1 Ωcm. This behaviour is observed for layers grown at the lowest iodine concentration corresponding to an iodine source temperature T_I of –20 °C. In fig. 3 the resistivity is plotted versus the growth temperature T_S for different iodine source temperatures. For the lowest iodine concentration and a substrate temperature of about 600 °C a drastic increase in the resistivity can be observed. For higher iodine source temperatures the resistivity remains below 100 Ωcm over the whole substrate temperature range. It is interesting that for an intermediate iodine concentration

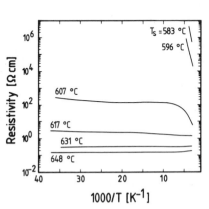

Fig. 2. Resistivity of p–ZnSe grown at different substrate temperatures T_S vs. T. ($T_I = -20$ °C)

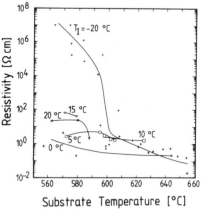

Fig. 3. T_S dependence of the resistivity at different T_I.

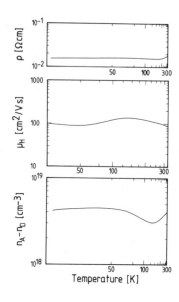

Fig. 4. Temperature dependence of the resistivity ρ, the Hall mobility μ_H and the net carrier concentration n_A-n_D of a higly conductive p-ZnSe layer.

(e.g. T_I = 0 ^oC) the conductivity remains high even at low substrate temperatures. The transport properties of a low resistive p-type ZnSe layer are illustrated in fig. 4. The sample was prepared at a growth temperature of 648 ^oC and an iodine source temperature of -20 ^oC. The minimum resistivity of 2 x 10^{-2} Ωcm achieved by our preparation technique lies one order of magnitude below the best published values for p-type ZnSe [4]. The resistivity is almost independent on temperature. Hall voltages yield p-type conduction of our films over the whole temperature range. The Hall mobility of 100 cm^2/Vs lies well above the highest values given by literature [4] for p-type material (40 cm^2/Vs). The net carrier concentration of 4 x 10^{18} cm^{-3} indicates that our p-ZnSe is degenerated. The dip in the carrier concentration and Hall mobility curve at 130 K is not elucidated until now. High conduction of the ZnSe layers could also be achieved by the homoepitaxial growth on high resistivity ZnSe substrates. Thus, we can exclude an influence of the GaAs substrate.

<u>Photoluminescence</u>

Low temperature photoluminescence (PL) spectra of our layers exhibit various emission lines (fig. 5). At the highest photon energy (2.8031 eV) there is the free exciton followed by two well-known bound excitons. The I_2 emission peak at 2.7968 eV originates from an exciton bound to a neutral donor such as Al or Ga and the I_1 (2.7941 eV) to a neutral acceptor such as Li or Na [7-9]. The I_b (2.7887 eV) and the I_c (2.7797 eV) lines may be caused by the radiative recombination of bound excitons at neutral acceptors judging from its energy position. The origin of the I_b and the I_c excitons has not been elucidated so far. The band emission at 2.695 eV results from the recombination of the donor acceptor pair band (DAP) Q_o with Li as acceptor and Al or Ga as donor, respectively [9]. Fig. 5 demonstrates the substrate temperature dependence of the emission lines of our layers. By increasing the growth temperature the DAP Q_o and the I_b line intensities are decreasing whereas the I_c peak is increasing. The I_1 and I_2 excitons are not changing their intensity in respect to the other lines. With the increase of the I_c line a new DAP appears at 2.717 eV. It seems that the acceptor involved in the I_c line may cause the high p-type conduction of our layers.

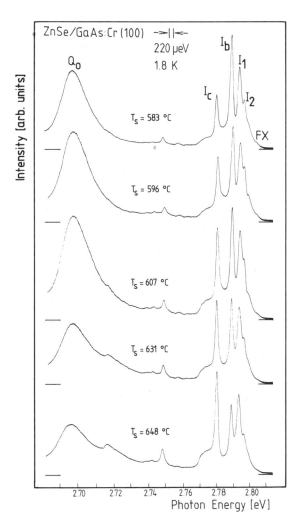

Fig. 5. Low temperature PL spectra for p-ZnSe layers grown at different T_S.

p-ZnSe layers were grown on n-GaAs (Se as dopant) to form a p/n hetero-junction. The energy band diagram (fig. 6) is calculated according to the Anderson model assuming a net carrier concentration of 2×10^{18} cm^{-3} for the p-ZnSe. Therefore, the Fermi-level lies about 40 meV above the ZnSe valence band. The n-GaAs is degenerated ($n_D = 4.5 \times 10^{18}$ cm^{-3}). Taking the published values [10] for the electron affinity χ and work function Φ into account a spike of 1.27 eV in the valence band is impeding the hole flow. The spectral response curve of a Au/p-ZnSe/n-GaAs diode (fig. 7) can be explained by this spike: Holes in the GaAs which are generated by the incoming light with an energy between the energy gap of the GaAs ($E_{G/GaAs}$) and the ZnSe ($E_{G/ZnSe}$) are not able to overcome the spike in the valence band. Only at light energies higher than $E_{G/ZnSe}$ the photocurrent is increasing.

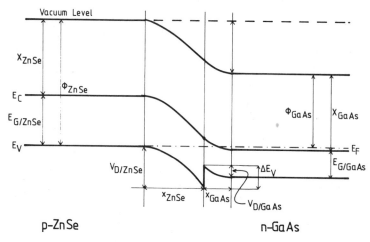

Fig. 6. Energy band diagram for a p-ZnSe/n-GaAs heterojunction.

Au/p-ZnSe/n-GaAs

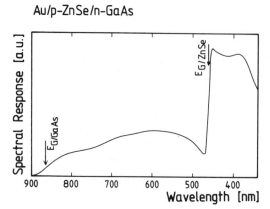

Fig. 7. Spectral response for a Au/p-ZnSe/n-GaAs hetero-junction at zero bias.

Metal/p-ZnSe/n-GaAs heterojunctions exhibit rectification with forward conduction occurring when the ZnSe is positively biased. Typical current voltage curves have been reported elsewere [11].

The forward I-V curves of a Au/p-ZnSe/n-GaAs heterojunction are plotted in fig. 8 for different temperatures. Between room temperature and 150 K the quality factor is increasing with decreasing temperature. Below this temperature the quality factor reaches a minimum at 136 K and then increases again. This quality factor minimum seems to be correlated to the dip of the Hall mobility and the carrier concentration curves depicted in fig. 4. The reverse currents at a constant reverse bias V_r of -6 V and -9V as a function of the reciprocal temperature exhibit a change in the slope at 160 K and 120 K, respectively (fig. 9). From these slopes one can calculate the partial built-in voltage $V_{D/ZnSe}$ within the ZnSe region applying the Anderson model: with decreasing temperature we found 1.50 V or 1.44 V at -6V or 2.15 V and 2.13 V at -9V, respectively.

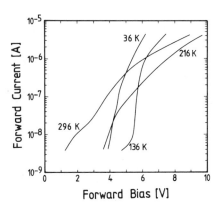

Fig. 8. Forward I-V curves of a Au/p-ZnSe/n-GaAs diode at different temperatures.

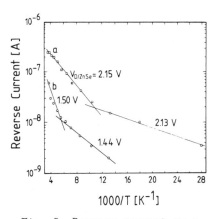

Fig. 9. Reverse current as a function of temperature T at a constant reverse bias: (a) V_r=-9V, (b) V_r=-6V

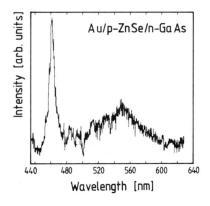

Fig. 10. EL spectrum of a Au/p-ZnSe/n-GaAs diode at room temperature.

In forward direction these diodes exhibit electroluminescence (EL) at 1 mA current. At room temperature the observed intensity is relatively weak and the light is emitted by single spots under the metal film. A typical spectrum at room temperature is shown in fig. 10 for a device at 2.4 V bias and 0.9 mA current. It is dominated by a blue peak at 2.7 eV due to band to band recombination.

SUMMARY AND CONCLUSIONS

In summary we have shown the preparation of p-type ZnSe layers with the highest conductivity of 50 $(\Omega cm)^{-1}$ reported so far. Hall effect as well as I-V measurements yielded p-type conduction. The resistivity could be varied over 6 orders of magnitude by selecting the appropriate growth temperature. The temperature dependence of transport properties and I-V characteristics indicated that at 130 K the electrical transport mechanism is changing. An application as electroluminescent heterojunctions was presented. As we have demonstrated earlier [11] the reduction of the green, yellow and red emission of the spectra is only attainable if the layer thickness is more than 3 μm. Therefore, for efficient blue LED's we intend to prepare homojunctions to overcome the problem of the deep levels in the strained interface region.

ACKNOWLEDGEMENT

We would like to thank Siemens AG in Munich for supporting this work in particular Dr.C. Weyrich for many constructive comments.

REFERENCES

1.   H. Cheng, J.M. DePuydt, J.E. Potts and T.L. Smith,
     Growth of p-type ZnSe:Li by molecular beam epitaxy,
     Appl. Phys. Lett. 52(2):147 (1988).
2.   K. Ohkawa, T. Mitsuyu and O. Yamazaki,
     Molecular beam epitaxial growth of nitrogen-doped ZnSe with ion doping technique,
     J. Cryst. Growth 86:329 (1988).

3. R.M. Park, H.A. Mar and N.M. Salansky,
    Photoluminescence properties of nitrogen-doped ZnSe grown by molecular beam epitaxy,
    J. Appl. Phys. 58(2):1047 (1985).

4. T. Yasuda, I. Mitsuishi and H. Kukimoto,
    Metalorganic vapor phase epitaxy of low-resistivity p-type ZnSe
    Appl. Phys. Lett. 52(1):57 (1988).

5. A. Yoshikawa, S. Muto, S. Yamaga and H. Kasai,
    The dependence on growth temperature of the photoluminescence properties of nitrogen-doped ZnSe grown by MOCVD,
    J. Cryst. Growth 86:279 (1988).

6. J. Nishizawa, R. Suzuki and Y. Okuno,
    p-type conduction in ZnSe grown by temperature difference method under controlled vapor pressure,
    J. Appl. Phys. 59(6):2256 (1986).

7. J.L. Merz, K. Nassau and J.W. Shiever,
    Pair spectra and the shallow acceptors in ZnSe,
    Phys. Rev. B8:1444 (1973).

8. R.N. Bhargava,
    The role of impurities in refined ZnSe and other II-VI semiconductors
    J. Cryst. Growth 59:15 (1982).

9. R.N. Bhargava, R.J. Seymour, B.J. Fitzpatrick and S.P. Herko,
    Donor-acceptor pair bands in ZnSe,
    Phys. Rev. B20:2407 (1979).

10. A.G. Milnes and D.L. Feucht,
    "Heterojunctions and metal-semiconductor junctions,"
    Academic Press, New York and London (1972).

11. N. Stücheli, G.G. Baumann and E. Bucher,
    p-ZnSe/n-GaAs heterojunctions for blue electroluminescent cells,
    Proceedings of the 18th International Conference on the Physics of Semiconductors 1:223 (1986).

PROSPECTS FOR II-VI HETEROJUNCTION LIGHT EMITTING DIODES

J.O. McCaldin

T. J. Watson, Sr., Laboratory of Applied Physics
California Institute of Technology
Pasadena, California 91125

The difficulty of making pn junctions by conventional processing in the wider band-gap II-VIs has been attacked over the years in several ways. One way, which is the subject of this paper, is to join a p-type II-VI to an n-type II-VI to form a heterojunction (HJ). The HJ approach has been tried for over 20 years and has been quite successful in making pn junctions in these materials.

The application of such pn junctions to light-emitting-diodes (LEDs), however, is subject to significant limitations, particularly due to the band offsets at HJ, as is illustrated in Fig. 1. A conventional pn homojunction, as in commercial III-V LEDs, is shown at the left of the figure. Current flow is limited by the Fermi potentials associated with carrier concentrations, ϕ_n for electron flow originating at the left and ϕ_p for hole flow originating at the right, and by band bending, which is a common barrier to both current flows. The relative contribution of the two currents is principally controlled by choosing carrier concentrations, i.e. ϕ_n and ϕ_p. Such homojunction diodes fabricated of III-V materials have been very successful and have resulted in commercial LEDs reaching into the green. In this paper, we ask whether wide-gap II-VIs, reaching across the whole visible spectrum, might be employed in a similar simple way, except using HJ.

An analagous band diagram for a pn HJ appears in Fig. 1b, which treats the case of a HJ of type I, i.e. with the smaller gap material "nested" within the larger gap. Current originating in the smaller band gap material is limited, not only by the factors just discussed, but also by the barrier that the band offset ΔE_v represents. Note, however, that the electron current originating in the wider gap material sees its corresponding offset in the opposite sense, i.e. as *not* an energy barrier. Furthermore, the part of the total band bending that occurs in the smaller gap material will often be smaller than the offset ΔE_c and hence not act as a barrier. Usually this part of the band bending is relatively small, since the smaller gap material is the more heavily doped. We will not consider it further in the following discussion. Thus, in this type I example, current injection into the smaller gap material is favored and indeed is limited only by the factors mentioned above for homojunctions.

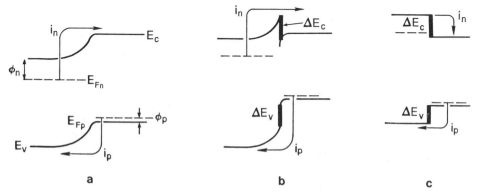

FIG. 1.   pn junctions under forward bias. E_c is conduction band edge, E_v valence band edge, E_{Fn} and E_{Fp} are quasi-Fermi levels in n and p material, respectively, i_n and i_p are majority carrier currents originating in n and p material. ϕ are Fermi potentials, and ΔE are offsets. (a) homojunction. (b) type I HJ approaching flat band and (c) type I HJ at flat band.

The situation just discussed can be represented by the "flat band" diagram of Fig. 1c, in which the bias is increased so much that no band bending remains. Since flat band diagrams are easy to construct and to read, we will use them in the remainder of the paper, though, of course, the simple diodes discussed here could not actually be driven so far into forward bias.

Fairly often, HJ adopt the staggered, or type II, relationship shown in Fig. 2, however. In this case, both offsets have the same sense, i.e. both will be current barriers, as in Fig. 2a, or neither will be barriers, as in Fig. 2b. The former, or Fig. 2a case, is, unfortunately, by far the more common. We will encounter one example of the more favorable Fig. 2b case later in this paper. In this latter case, the HJ is rather similar to a homojunction: only Fermi potentials and band bending serve to limit either of the currents.

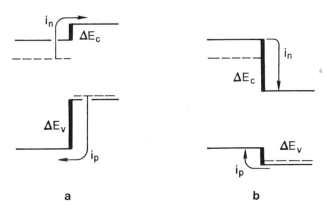

FIG. 2. Type II HJ biased to flat band. (a) the more common case where both offsets are positive, i.e. present barriers to the currents i_n and i_p. (b) the more unusual case where both offsets are negative, i.e. present no barrier.

The HJ of type II shown in Fig. 2a can be forward biased beyond flat band without drawing enormous currents. However, in such bias the quasi-Fermi levels approach and can pass the band edges, so that very heavy carrier accumulations develop and drive any recombination centers near the interface. We will not consider this possibility further.

GENERAL PREDICTIONS OF BAND OFFSETS

A rather large literature of offset values has accumulated in recent years, as the accompanying references attest. One way to deal with this is to use "selected" values as did Kroemer[1] . Objections to such selectivity have been raised by Margaritondo[2], however, and here we opt to present all the relevant data of which we are aware.

Some offset data is presented as ΔE_v values for specific HJ, a procedure that is particularly appropriate when strain due to the HJ is taken into account. This data will be presented later in connection with specific HJ. The present section, however, discusses more general values, the so-called "linear" predictions, in which an E_v value is assigned to each semiconductor, from which ΔE_v readily follows.

Of the many linear theories, some have yielded extensive lists of E_v values, most notably those of Harrison[3] and Harrison and Tersoff[4], which deal with some 36 substances. Results of the latter theory are shown in Fig. 3, which exhibits only the more common II-VI and III-V semiconductors. The plotting scheme used shows, for any HJ, the band offsets, degree of lattice mismatch, and dopability of each material. For example, the familiar ZnSe/ZnTe HJ is seen to be about 7% lattice mismatched and predicted to be type I with moderate ΔE_v and ΔE_c and with one member p-dopable and one n-dopable. On the other hand, the HJ ZnTe/AlSb appears to be more unusual. This about 0.5% lattice mis-matched HJ is predicted to be type II with no offset barriers to current flow, i.e. the less common case discussed above in connection with Fig. 2b. Such a HJ, of course, would have the advantage that the absolute magnitude of the offsets poses no limitation on the injection currents that can be drawn.

Most other theories tend to be less optimistic for the present application, for example, those of Tersoff[5] and of Harrison[3]. Plots similar to Fig. 3, may be used to compare trends in the theories and this has been done[6] for Harrison[3] vs Harrison and Tersoff[4]. Yet other theoretical predictions will be cited when specific HJ are discussed. The references cited are so extensive, however, that a system of acronyms, shown in Table I, will often be used for easier identification of references.

Turning now to experimental results, the question arises as to which are the more general or extensive. Five such data collections will be cited in the present section: the aforementioned "selected" data of Kroemer[1], the results of Katnani and Margaritondo[16], the electron affinity data of Freeouf and Woodall[13], the so-called "common anion" data of McCaldin, McGill and Mead[17], and the "common cation" data of O. von Ross[18]. The last three data collections need some comment, however. The electron affinity data has been strongly criticized in recent years[24,25], yet is still cited in practice and, at least in the present context, is not notably discordant with other data. The "common anion" data (MMM[17]) is seldom used as such, but is cited in connection with the large ΔE_v(AlAs/GaAs) and similar HJ. The original papers[17] stated clearly that AlAs and AlSb did *not* follow the "rule", which calls into question the appropriateness of this referencing. On the other hand, MMM[17] related E_v values to Au-Schottky barrier heights, an idea currently back in vogue with the work of Tersoff[5]. Furthermore, the actual E_v values of MMM[17], which are presented in numerical form in Table II, are not out of line from current values, as will become apparent. So we shall present this data, which has often been omitted in

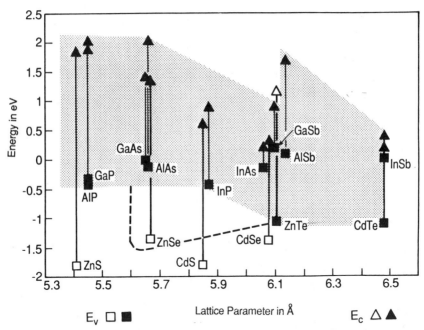

FIG. 3. Energy relationships, according to Harrison and Tersoff, for band edge of the common II-VI and III-V semiconductors. Abscissa is the lattice parameter of the cubic structures, or, for Wurtzite structures, equivalent parameter on basal plane. Vertical solid lines denote band gaps. Material that conventionally cannot be made usefully p-type or n-type is denoted by the appropriate open symbol; otherwise, filled symbols are used. Stipling highlights regions free of open symbols. Dashed line highlights regions in which useful p-type doping has recently been reported. Zero of energy is arbitrarily set at $E_v(\text{GaAs})$.

previous comparisons. For completeness, the "common cation" data of O. von Ross[18] is included at this place in the paper also.

Figure 4, which includes more data than the original KM[16] paper, illustrates the way in which intercomparison is made between the five sets of data. For each semiconductor that two authors make predictions about the deviation in eV of the E_v positions is plotted in the bar graph. Since the zero of energy is chosen for each comparison so that the sum of the deviations is zero, root-mean-square deviations may be calculated between the authors. Such rms deviations intercomparing the five sets of data are presented in Table III. Rather low rms values occur for all the comparisons involving

Table I. Acronyms used to refer to sources of data.

Acronym	AN	CC	CMB	DHF	EK	FK	FW	GBV	H	HT	KCK	K	KM
Reference	7	8	9	10	11	12	13	14	3	4	15	1	16

Acronym	MMM	OR	RMW	SS	T	WM	WSO	WZ
Reference	17	18	19	20	5	21	22	23

Table II. Energies of Valence Band Maxima E_v in eV relative to $E_F(\mathrm{Au}) = 0$, according to MMM.

E_v	−0.03	−0.45	−0.76	−0.77	−1.28	−1.64
SEMI-CONDUCTORS	InSb	InAs	InP	CdTe	CdSe	CdS
	GaSb	GaAs	GaP	ZnTe	ZnSe	ZnS
			BP			

Table III. Root-mean-square deviations in eV between experiment-based estimates of E_v. Number in parenthesis is number of materials in common between two investigators predictions.

	OR	MMM	FW	KM
MMM	0.50 (11)			
FW	0.46 (11)	0.34 (12)		
KM	0.42 (13)	0.11 (11)	0.26 (11)	
K	0.18 (6)	0.14 (4)	0.14 (5)	0.15 (6)

Kroemer[1]; however, the number of semiconductors on which such comparison can be made is relatively small. Overall, the rms values vary from rather good agreement, 0.11 eV, to poor.

SPECIFIC II-VI/II-VI HETEROJUNCTIONS

The simplest approach to making II-VI LEDs would be to stay within the II-VI class of materials, not the least reason being to avoid cross-doping effects. Within this class, the conventional view is that the only p-type materials are the tellurides, as was suggested in Fig. 3. There are, however, some choices to be made within the tellurides. The three most studied of these materials appear in Fig. 5, which displays E_v values reported experimentally and theoretically.

The situation for HgTe is becoming clearer, especially at room temperature, where values near 0.35eV are reported by several investigators (KCK[15], DHF[10], SS[20], WZ[23] and CMB[9]). At low temperature, a much smaller value was reported (GBV[14]) and some recent evidence (CMB[9]) suggests that E_v falls with temperature toward the low value. Indeed a theory to explain this temperature dependence is currently being advanced[26]. Even so, in the present application HgTe does not offer E_v values lower than can be obtained with easier-to-use materials and is thus not advantageous.

Two predictions for ZnTe (OR[18] and AN[7]) suggest a striking advantage. Aside from these predictions, however, the E_v position appears, on average, to be about the same as for CdTe. Under these circumstances, the smaller lattice parameter of ZnTe makes this material preferable to pair with most HJ partners of interest.

In passing, some less common tellurides should be mentioned. MnTe, though the bulk material has a different crystal structure, can be made to grow in the zincblende structure in very thin layers[27]. It has been calculated[5] in the "frozen d-shell approximation" to have $E_v = -0.75eV$, and, allowing for d-band contributions (WZ[23]), to have $E_v = -0.25eV$. Even the latter estimate is advantageous to the present devices, however. But experimental evidence[28] has so far not particularly supported these

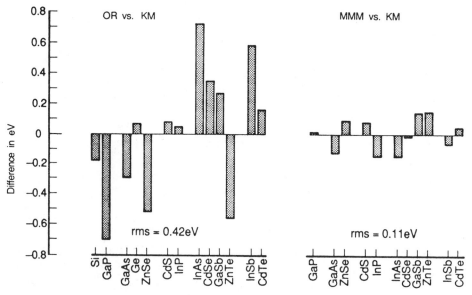

FIG. 4.   Deviations between E_v values of Katnani and Margaritondo(KM) and those of O. von Ross(OR) and McCaldin, McGill and Mead(MMM), respectively. Zero of energy scale chosen so that sum of deviations is zero in each comparison. "rms" is root-mean-square deviation.

estimates. Other possibilities are BeTe[29] and MgTe. Both appear in the zincblende structure, but are hygroscopic. However, MgCdTe alloys have been made with light-emitting pn junctions and withstood ambient conditions[30]. It would be interesting to know where such alloys would fall on a diagram like Fig. 5. For the present, however, ZnTe is the best p-type II-VI for our discussion.

We turn next to the selenides as n-type partners for p-ZnTe. CdSe has the advantage of good lattice match, but ZnSe offers a much wider bandgap. Both these

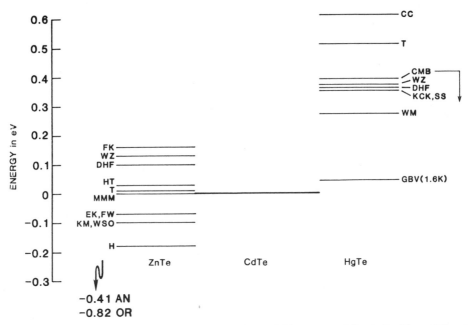

FIG. 5. Experimental and theoretical values of E_v reported for tellurides of Zn, Cd and Hg. Zero of energy is arbitrarily set at E_v(CdTe). Temperature is near 300K, except for experiments denoted 1.6K and variable temperature case labelled with arrow.

materials appear in Fig. 6, which displays predicted energy positions for both the valence and conduction bands. This figure is designed to display a "rule of thumb" that the barrier for useful current in a LED should not be more than about 0.4eV. With this barrier height, the Richardson equation indicates a thermionic current over the barrier of about 1 Amp/cm² at room temperature. Current densities smaller that this would hardly be competitive for a LED device. The current arrows in the figure are scaled to represent 0.4eV and stipling denotes energies for E_c or E_v at which a greater barrier height occurs. For simplicity we assume, slightly optimistically, that such heavy doping is possible that the quasi-Fermi energies can be brought to the band edges.

Two of the many predictions shown in Fig. 6 will be highlighted to facilitate discussion, namely, H[3] and HT[4]. The former prediction indicates that all 4 of the possible injection currents in the figure are strongly blocked. Similar conclusions are reached by FW[13] and EK[11]. Yet other predictions, which do not treat all of the 4 possible barriers, reach only pessimistic conclusions for the cases they do treat.

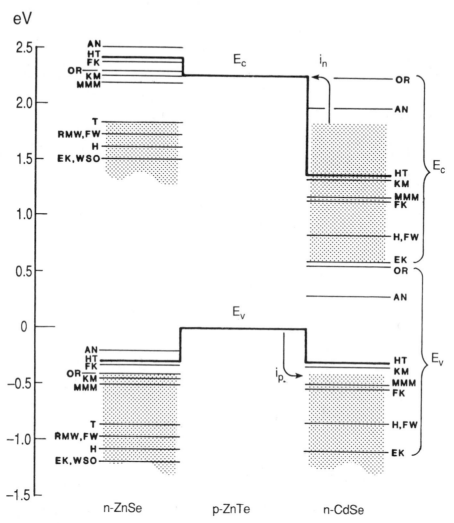

FIG. 6. Energy relationships reported for ZnSe, ZnTe and CdSe. Zero of energy is arbitrarily set at E_v(ZnTe). Stipling denotes energy ranges which do not permit substantial forward injection currents, as discussed in text. Arrows next to current symbols indicate 0.4eV range discussed in text. Heavier lines illustrate the particular predictions of Harrison and Tersoff, which appeared earlier in Fig. 3.

On the other hand, HT[4] is more optimistic and is illustrated with heavier lines in the figure. For the ZnTe/CdSe HJ, the current i_p is favored, while i_n is blocked; i.e. injection into the smaller gap CdSe is predicted. Although the band gap of CdSe at 1.7eV is not quite sufficient to offer visible light emission, this HJ could provide an "existence theorem" for these systems.

Turning now to the more interesting ZnSe/ZnTe HJ, HT[4] finds small enough offsets to allow both currents to flow, though this type I HJ tends to favor injection of electrons into the smaller gap ZnTe. If electroluminescence were produced in the red in this way, it would likely have the advantage of being more understandable, and perhaps more efficient, than the Schottky-contact type light emission[31] heretofore produced in this material. Finally, we consider the possibility of significant hole injection, i_p, into ZnSe. The very substantial barrier to holes represented by $\Delta E_v = 0.29eV$ could in principle be balanced by a comparable Fermi potential in the ZnSe, i.e. by limiting the electron concentration in the ZnSe to $\sim 10^{14} cm^{-3}$. This rather limited prospect would be improved substantially, of course, if one accepts the one prediction more optimistic than HT[4], namely, AN[7].

Plots similar to Fig. 6, can be made for the sulfides, as well. However, these materials give rise to substantially larger offsets in almost every instance and are likely to be less favorable for HJ LEDs.

SPECIFIC II-VI/III-V HETEROJUNCTIONS

Plots like Fig. 3 suggest that more favorable HJ can occur when III-V substrates are allowed. Specifically, the ZnTe/AlSb HJ mentioned earlier appears in the figure as type II but with no offset barriers. This optimistic prediction by HT[4] is not shared by all the other predictions. Nevertheless the other predictions are still fairly optimistic, as is pointed out elsewhere[32]. Thus, on the basis of the literature review of offsets considered here, this HJ would appear to be perhaps the most favorable for obtaining the desired injection currents. Recent experiments[33] on the analogous ZnTe/GaSb HJ, however, obtained the unexpected result of finding substantially higher ΔE_c than reported by any of the many theorists or experimentalists making predictions. Thus, an experiment on this specific II-VI/III-V HJ seems to indicate some difficulty in the literature on offsets.

The prototype II-VI/III-V HJ has been ZnSe/GaAs, much studied a few years ago. The offset $\Delta E_v = 0.96eV$, reported by Kowalczyk et al[34] and constituting one of the Kroemer[1] selected values, was consistent with many theoretical predictions. The experimenters did notice variability as large as 0.15eV with processing conditions, but nothing so large as the present discrepancy in the ZnTe/GaSb system. Thus, the feasibility of II-VI/III-V HJ for making II-VI LEDs is in need of some elucidation at this time.

DISCUSSION AND CONCLUSIONS

Heterojunctions involving II-VI materials are the subject of many photoluminescent studies today, particularly in the form of superlattices. One purpose of these studies is to assess feasibility for efficient light emission, particularly in the visible and hopefully well into the blue. An encouraging result is the frequent observation that the efficiency of light emission in the superlattices is markedly higher than in the corresponding alloys. While the photon pumping used in the studies is particularly useful in the laboratory, it is not attractive for eventual wider applications where electrical pumping, as in commercial III-V LEDs, is highly desirable. In the present paper, the difficulties to be expected in obtaining injection currents electrically has been estimated, using the extensive literature on band offsets. While superlattices were not specifically discussed, they are subject to much the same constraints as single HJ, as one can see by imagining a superlattice inserted into the middle of the HJs sketched in Figs. 1 and 2.

Based on the prior literature, and as can be visualized from Fig. 3, a II-VI/III-V HJ appeared to be most favorable for obtaining injection currents needed by LEDs. However, recent experiments suggest a quite different result for this particular HJ, ZnTe/AlSb, than is anticipated from the literature. The more time-tested approach of making II-VI/II-VI HJ, on the other hand, is subject to a very divided prognosis, according to the offset literature. The more favorable of the offset predictions does indicate, however, that ZnSe/ZnTe HJ should be capable of producing substantial electroluminescence in ZnTe.

REFERENCES

1. H. Kroemer, as quoted in Table II of W.A. Harrison and J. Tersoff , *J. Vac. Sci. Technol.* **B4**, 1068 (1986).
2. G. Margaritondo , *Phys. Rev.* **B31**, 2526 (1985).
3. Walter A. Harrison , *J. Vac. Sci. Technol.* **14**, 1016 (1977).
4. W. A. Harrison and J. Tersoff , *J. Vac. Sci. Technol.* **B4**, 1068 (1986).
5. J. Tersoff , *Phys. Rev. Lett.* **56**, 2755 (1986).
6. R. H. Miles, J. O. McCaldin and T. C. McGill , *J. Crystal Growth* **85**, 188 (1987).
7. M. J. Adam and Allen Nussbaum, as quoted in Table II of ref. 16.
8. Manuel Cardona and Niels E. Christensen , *Phys. Rev.* **B35**, 6182 (1987).
9. D. H. Chow, J. O. McCaldin, A. R. Bonnefoi, T. C. McGill, I. K. Sou and J. P. Faurie , *Appl. Phys. Lett.* **51**, 2230 (1987).
10. Tran Minh Duc, C. Hsu and J. P. Faurie , *Phys. Rev. Lett.* **58**, 1127 (1987).
11. Edgar A. Kraut , *J. Vac. Sci. Technol.* **B2**, 486 (1984).
12. W. R. Frensley and H. Kroemer , *Phys. Rev.* **B16**, 2642 (1977).
13. J. L. Freeouf and J. M. Woodall , *Appl. Phys. Lett.* **39**, 727 (1981).
14. Y. Guldner, G. Bastard, J. P. Vieren, M. Voos, J. P. Faurie and A. Million , *Phys Rev. Lett.* **51**, 907 (1983).
15. Steven P. Kowalczyk, J. T. Cheung, E. A. Kraut and R. W. Grant , *Phys. Rev. Lett.* **56**, 1605 (1986).
16. A. D. Katnani and G. Margaritondo , *Phys. Rev. B* **28**, 1944 (1983).

17. J. O. McCaldin, T. C. McGill and C. A. Mead , *Phys. Rev. Lett.* **36**, 56 (1976).
18. Oldwig von Ross, as quoted in Table II of ref. 16.
19. Y. Rajakarunanayake, R. H. Miles, G. Y. Wu and T. C. McGill, *J. Vac. Sci. Technol.* (to be published, 1988).
20. C. K. Shih and W. E. Spicer , *J. Vac. Sci. Technol.* **B5**, 1231 (1987).
21. Chris G. van de Walle and Richard M. Martin , *J. Vac. Sci. Technol.* **B5**, 1225 (1987).
22. Chris G. van de Walle, Khalid Shahzad and Diego J. Olego, *J. Vac. Sci. Technol.* (to be published, 1988).
23. Su-Huai Wei and Alex Zunger , *J. Crystal Growth* **86**, 1 (1988).
24. A. G. Milnes , *Solid State Electron* **29**, 99 (1986).
25. Robert S. Bauer, Peter Zurcher, and Henry W. Sang, Jr. , *Appl. Phys. Lett.* **43**, 663 (1983).
26. J. A. Van Vechten and K. J. Malloy (unpublished).
27. See, for example, R. N. Bicknell, N. C. Giles and J. F. Schetzina , *Appl. Phys. Lett.* **50**, 691 (1987).
28. S-K. Chang, A. V. Nurmikko, Ji-Wei Wu, L. A. Kolodziejski and R. L. Gunshor , *Conference on Modulated Semiconductor Structures III* , Montpellier, France, 1987 (to be published).
29. W. M. Yim, J. P. Dismukes, E. J. Stofko and R. J. Paff , *J. Phys. Chem. Solids* **33**, 501 (1972).
30. Ryoichi Yamamoto and Kohji Itoh , *Jap. J. Appl. Phys.* **8**, 341, (1969).
31. D. P. Bortfeld and H. P. Kleinknecht , *J. Appl. Phys.* **39**, 6104 (1968).
32. J. O. McCaldin and T. C. McGill, *J. Vac. Sci. Technol.* B (to be published, 1988).
33. W. G. Wilke and K. Horn, *J. Vac. Sci. Technol.* B (to be published, 1988).
34. Steven P. Kowalczyk, E. A. Kraut, J. R. Waldrop and R. W. Grant , *J. Vac. Sci. Technol.* **21**, 482 (1982).

II-VI HETEROSTRUCTURES AND MULTI-QUANTUM WELLS

T. C. McGill, R. H. Miles *, Y. Rajakarunanayake, and J. O. McCaldin

T. J. Watson, Sr., Laboratory of Applied Physics
California Institute of Technology
Pasadena, California 91125

Abstract

The ability to fabricate epitaxial layers of II-VI semiconductors has made it possible to circumvent the doping difficulties of II-VI semiconductors and to fabricate visible light emitters. Heteroepitaxial structures are one of the promising approaches. The more interesting cases involve heavy strain and require particular values of the band offsets. We review recent work on CdTe/ZnTe and ZnSe/ZnTe multiquantum well structures.

I. Introduction

Research into producing light emitters from semiconductors from column II of the periodic table and column VI dates to the 1930's.[1] While these materials have a great propensity for emitting light, difficulties have been experienced in controlling the electrical properties of the bulk materials to attain both p-type and n-type.[2] However, recently the application of wide bandgap II-VI semiconductors in electronic devices has been the subject of research because of the new and novel ways that they can be fabricated. Layers of II-VI semiconductors can now be deposited at low temperatures using molecular beam epitaxy (MBE)[3] and metal-organic chemical vapor deposition (MOCVD).[4]

These new techniques of MBE and MOCVD are ideal for the fabrication of heterojunctions and multiple quantum well structures such as those shown in Fig. 1. Heterostructures like these are already widely used in III-V semiconductors.[5] In the case of the II-VI's one might hope to overcome some of the limitations on band gap and doping control (i. e. typically the larger bandgap II-VI's have only been usefully doped n-type or p-type but not both) by fabricating similar heterostructures. A

---

* Current address: Hughes Research Laboratories, Malibu, California 90265

EPITAXIAL STRUCTURES

HETEROSTRUCTURE

THIN GROWN LAYER

SUBSTRATE

MULTI-QUANTUM WELL STRUCTURE

MATERIAL A

MATERIAL B

BUFFER

SUBSTRATE

FIG. 1.   Schematic illustration of a superlattice. The superlattice consists of a repeated layer structure grown on a substrate, typically with an intervening buffer layer.

number of different II-VI combinations have been fabricated over the last couple of years. In this paper, we will concentrate on two of the most widely studied systems, CdTe/ZnTe[6] and ZnTe/ZnSe.[7] Both of these systems have been shown to emit light and an optically pumped laser has been demonstrated in the CdTe/ZnTe system.[8]

II. Basic Considerations

Four major considerations govern the use of heterostructures in making light emitting devices: first, the band gap of the materials making up the heterostructure; second, the mismatch between the lattice constants of the two constituent materials; third, the type of doping that can be usefully attained in a given semiconductor;

BAND OFFSETS AT HETEROJUNCTIONS

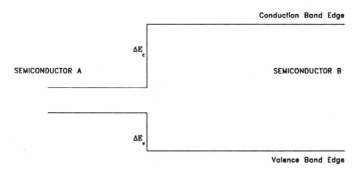

FIG. 2. Schematic diagram indicating the definition of the valence band and conduction band offsets across a heterojunction.

fourth, the offset in the band edges in going from one material to the other. This fourth property is illustrated in Fig. 2 where we have drawn the conduction and valence band edges in the two semiconductors as a function of the position normal to the heterojunction. The steps in energy between the conduction band edges and the valence band edges are called the conduction band offset ΔE_c and the valence band offset ΔE_v, respectively. The article by J. O. McCaldin in this volume provides a review of the current understanding and role of the valence band offsets.

All of these concepts can be summarized in a single plot of energy versus lattice constant. In Fig. 3, we have presented such a plot for a number of the relevant III-V and II-VI semiconductors.[6] This figure shows the positions of the conduction and valence band edges as a function of lattice parameter for a number of semiconductors. One can see that the lattice parameters fall into a few clusters within each of which the match is rather good but between which the match is poor. An example of a good lattice match is the ZnTe-CdSe couple. On the other hand, the CdTe-ZnTe and ZnSe-ZnTe couples are examples of poor lattice match. Lattice match is a very important consideration in determining what can be grown. Perfect

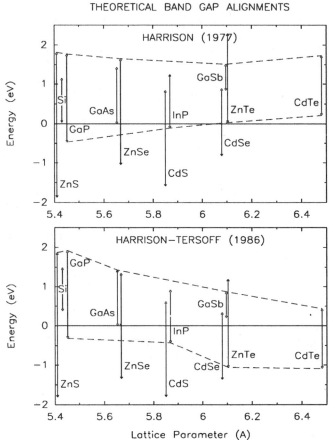

FIG. 3. Important semiconductor properties versus lattice constant. For each vertical line, the upper arrow represents the conduction band edge; similarily, the lower arrow represents the valence band edge. The zero on the energy scale arbitrarily represents the valence band edge of GaAs. In the upper figure we have used the values from Ref. 21. In the lower figure we have used the values from Ref. 23. Of the materials shown, those which can be doped usefully n-type lie at or below the upper dashed line. Similarly, those which can be doped usefully p-type have lower arrows which lie at or above the lower dashed lines.

lattice match means that epitaxial layers can be prepared that are of any thickness. However, if the lattices do not match, then strain energy builds up in an epitaxial structure until at some thickness, the so-called critical thickness, dislocations are introduced to release the strain.[10] While there are a number of simple theories of the so-called critical thickness,[11-15] these theories are not adequate to account for the metastable structures that can be grown. This phenomenon is still a subject of active research.[16-18]

Doping characteristics are also indicated in Fig. 3. For the materials shown here, the upper dashed line connects conduction band edges of materials in which n-type doping is readily achieved; all conduction band edges lying below this line represent materials which are easily doped n-type. For example, GaSb can be made usefully n-type while ZnTe can not. The lower dashed line represents a similar limit for p-type usefulness.[2,19]

Finally, we come to the key parameter characterizing ideal heterojunctions: the band offsets $(\Delta E_v, \Delta E_c)$. As illustrated in Fig. 2, these parameters gauge the relative positions of the valence band edges and conduction band edges as one goes across the interface. While these parameters are essential to understanding and predicting the properties of heterojunctions, their values are not well known. Theories of band offsets are at best qualitative since a truly first principles theory of the charge rearrangement at the interface is difficult to carry out. Among the most notable theories for the II-VI heterojunctions are the so-called common anion rule,[20] based on the concept of meaningful relative positions of the valence band edges as gauged by the tight binding atomic term values due to Harrison,[21] and the recent use of the concept of a neutral charge position, originated by Tersoff[22] and extended by Harrison and Tersoff.[23] In Fig. 3 we present band offset predictions based on two of the most widely quoted theories.[21,23] We have referenced all of the band edges to the valence band edge of GaAs (which we have arbitrarily taken to be our zero of energy).

There have been some experimental measurements of the band offsets by x-ray photoemission spectroscopy, notably by Grant and coworkers[24] and by Margaritondo and co-investigators.[25] Currently, these theories and experiments do not include the role of strain in band offsets. However, strain has been included in some detailed calculations of the band offsets.[26] In reality, strain could play a large part in determining the value of the band offset in the case of a heavily strained heterojunction. For example, the 6% strain in the CdTe-ZnTe system can move the valence band edges around by as much as a few 100 meV.

Two cases which show real promise are CdTe/ZnTe and ZnTe/ZnSe. In both cases, one of the constituents can be doped p-type while the other can be doped n-type. On the other hand, as noted above, these cases both correspond to substantial lattice mismatch. In this paper we will concentrate on these two systems.

III. Properties of CdTe-ZnTe Multi-Quantum Well Structures

Multi-quantum well structures of CdTe-ZnTe have recently been fabricated by Faurie and coworkers[27], and Feldman and coworkers[28]. We have made photoluminescence measurements on the Faurie superlattices[6] and have studied their structural properties using x-ray diffraction[29]. The photoluminescence spectrum (PLS) for one of these superlattices is shown in Fig. 4 along with the spectrum

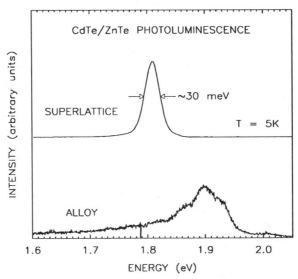

FIG. 4. The photoluminescence spectra from a CdTe-ZnTe superlattice and a $Cd_{0.37}Zn_{0.63}Te$ alloy. The superlattice consists of repeated periods each of which consists nominally of layers of CdTe 31Å thick and ZnTe 23Å thick.

from a corresponding ZnCdTe alloy. The important feature to note is that the PLS shows a prominent peak at an energy substantially below that of the band gap of the corresponding alloy, as has been noted in comparing other superlattices with the corresponding alloys.

These superlattices are heavily strained as a consequence of the 6% lattice mismatch between CdTe and ZnTe. To characterize the degree to which strain can be accommodated elastically in this system, superlattices have been grown on a number of different buffer layers, including CdTe, ZnCdTe (with approximately the lattice constant of the free-standing superlattice) and ZnTe. At issue is the defect structure between the superlattices and the substrates as well as the structure of the layers making up the superlattice. The individual layers of the superlattice are sufficiently thin that according to the criterion due to Matthews and Blakeslee[12,13] these layers should not develop misfit dislocations. Strain should only result in a tetragonal distortion of these layers, the magnitude of which can be calculated from the in-plane lattice constant through simple elasticity theory. However, the question remains as to how the strain is accommodated between the substrate and the superlattice. The results of our x-ray analysis[29] plus measurements of the lattice constants of the individual layers during growth[30] are consistent with the basic conclusion that the first layers of the superlattice develop a number of misfit dislocations during growth. These defects adjust the in-plane lattice constant of the superlattice layers from that of the substrate to that of a free standing superlattice.

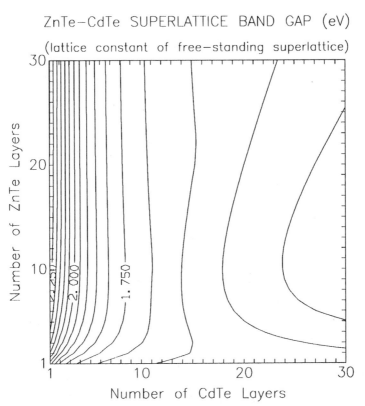

ZnTe−CdTe SUPERLATTICE BAND GAP (eV)

(lattice constant of free−standing superlattice)

Number of ZnTe Layers

Number of CdTe Layers

FIG. 5. Contour plot of the band gap of CdTe-ZnTe superlattices as a function of the number of CdTe and ZnTe monolayers within a single superlattice period. The valence band offset is assumed to be zero. The lattice constants of the layers in the superlattice are assumed to take on the values that one would obtain in a free standing superlattice. The contour interval is 50meV. The calculations were carried out using a simple two band $\vec{k} \cdot \vec{p}$ method that included strain. More details can be found in Ref. 6.

Band gaps calculated for this superlattice are relatively insensitive to the value of the band offset. Faurie[31] and Katnani[32] each report a value of $\Delta E_v \approx 0.1 eV$, although they differ on the sign. All of the theories suggest that $\Delta E_v \approx 0$ for the case of CdTe-ZnTe. Based on these offsets and the assumption of a free-standing strained structure, one can predict the band gap of CdTe-ZnTe superlattices as a function of the thicknesses of the individual layers. Figure 5 is a contour plot of the band gap of the superlattices as a function of the thickness of the individual layers making up the superlattice. The figure indicates that band gaps all the way from the near infrared to the green can be obtained in this system at low temperatures. The bowing which occurs for thick layers of CdTe and ZnTe is due to the large role strain is playing in modifying the positions of the valence band edges.

IV. Properities of ZnSe-ZnTe Multi-Quantum Wells

Superlattices of ZnSe and ZnTe have been widely studied in Japan.[7, 33, 34, 35] This system appears to be a good candidate for making green and blue light emitters. ZnSe can be doped n-type and ZnTe p-type, making electrical injection a possibility. To obtain the shorter wavelength emission afforded by ZnSe, the injected current must consist predominantly of holes originating in the ZnTe. Mn doping of the ZnSe near the interface may suffice to block the electron current originating from the ZnSe, based on recent measurements of the ZnSe/MnSe offsets.[36] A more substantial problem, however, may be the size of the valence band offset blocking hole injection.

In Fig. 6 we have indicated several predicted positions of the valence band edge of ZnTe with respect to the valence band edge of ZnSe. As one can see from this figure, the range of values is exceedingly large. Some of the values, such as the most recent predictions by Harrison and Tersoff[23] and the measurements by Katnani and coworkers[32], indicate that thermal injection of holes into the ZnSe from the ZnTe may be possible. However, the three largest predicted offsets in the figure would suggest that this is impossible.

Studies of the photoluminescence of superlattices give the band gap of the superlattices. In the case of ZnSe-ZnTe, calculations of the band gaps of the superlattices suggest that changes in the values of the band offsets result in large variations in the superlattices properties.[37] In Figs. 7 and 8, we have illustrated this point by presenting the energy gap as a function of number of layers of ZnTe and ZnSe for two different values of the band offset. In Fig. 7, we have taken the valence band offset to be 0.975 eV, a value nearer the larger valence band offsets. In Fig. 8, we have taken the valence band offset to be 0.430 eV, a value nearer the smaller offsets. As can be seen from these contour plots, the bandgap of the superlattice is very sensitive to the value of the band offset.

We have compared calculated superlattice band gaps with measurements carried out by Kobayashi and co-workers[7]. We find that their measured superlattice band gaps suggest that the band offset is near 1 eV, in agreement with the three upper predictions of Fig. 6. Recently a first principles calculation of the band offset has also yielded a value in agreement with the larger value.[26] This value of the offset would, as indicated earlier, make hole injection into ZnSe from ZnTe impossible.

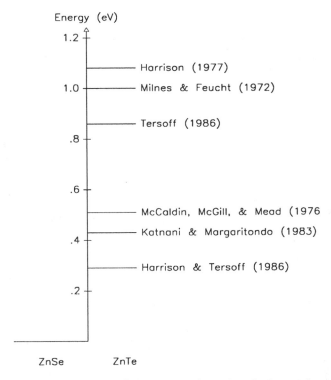

ZnTe/ZnSe VALENCE BAND OFFSET

FIG. 6. The relative position of the ZnTe valence band edge with respect to that of ZnSe, according to a number of different theories. The value labeled Milnes and Feucht (1972) was obtained by using the electron affinity difference to give the conduction band offset (after A. G. Milnes and D. L. Feucht, *Heterojunctions and Metal-Semiconductor Junctions* (Academic Press, New York, 1972).) The value labeled Harrison (1977) was obtained from Ref. 21. The value labeled Tersoff (1986) was obtained from Ref. 22. The value labeled McCaldin, McGill and Mead (1976) was obtained from Ref. 20. The value labeled Katnani and Margaritondo (1983) was obtained from Ref. 25. The value labeled Harrison and Tersoff (1986) was obtained from Ref. 23.

Thus, radiative recombination of holes and electrons at this heterojunction would have to be obtained from a process different from that of standard LEDs or lasers (e.g. by recombination *across* the junction, tunneling injection, etc.).

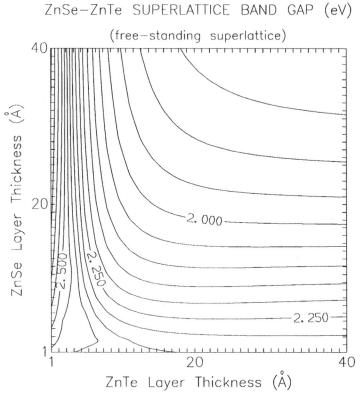

FIG. 7. Contour plot of the band gap of ZnSe-ZnTe superlattices as a function of the number of ZnSe and ZnTe monolayers within a single superlattice period. The valence band offset is assumed to be 0.975 eV. The lattice constants of the layers in the superlattice are assumed to take on the values that one would obtain in a free standing superlattice. The contour interval is 50meV. The calculations were carried out using a simple two band $\vec{k} \cdot \vec{p}$ method that included strain. More details can be found in Ref. 37.

V. Conclusions

We have examined some recent results on wide band gap II-VI heterostructures. The current situation is such that it is difficult to project whether these new structures will be useful in making optoelectronic devices. Some of the major questions include: Can we use the new fabrication techniques to obtain layers with different doping than obtained previously in bulk growth?[38] What are the values of the band offsets? What is the nature of the large mismatched layered structures?

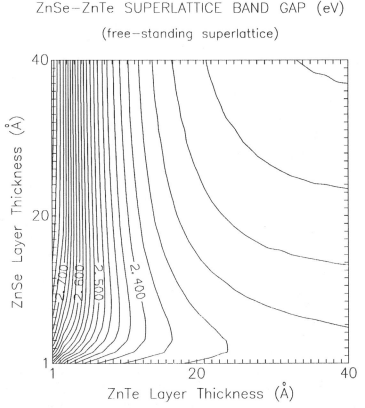

ZnSe–ZnTe SUPERLATTICE BAND GAP (eV)

(free–standing superlattice)

FIG. 8. Contour plot of the band gap of ZnSe-ZnTe superlattices as a function of the number of ZnSe and ZnTe monolayers within a single superlattice period. The valence band offset is assumed to be 0.430 eV. The lattice constants of the layers in the superlattice are assumed to take on the values that one would obtain in a free standing superlattice. The contour interval is 50meV. The calculations were carried out using a simple two band $\vec{k} \cdot \vec{p}$ method that included strain. More details can be found in Ref. 37.

What is the structural stability of these layered structures? While very few of the questions have definitive answers at the present, the field shows real promise for producing a II-VI visible light emitter.

VI. Acknowledgment

We would like to acknowledge the support of the Defense Advanced Research Projects Agency under contract No. N000014-K-86-0841. We would like to acknowledge useful discussions with J. P. Faurie, G. Y. Wu, H. Ehrenreich, and R. A. Reynolds.

References

1. G. Destriau, *J. Chim. Phys.* **33**, 587 (1936).

2. See for example, F. A. Kroger, *The Chemistry of Imperfect Crystals* (North Holland Publishing Company, Amsterdam, 1973) 2nd Edition Vol. II. Chap. 16 p728ff.

3. See for example, T. Yao in *The Technology and Physics of Molecular Beam Epitaxy*, Edited by E. H. C. Parker (Plenum Press, New York, 1985) Chapter 10.

4. See for example,H. Mitsuhashi, I. Mitsuishi, and H. Kukimoto, *J. Cryst. Growth* **77**, 219 (1986).

5. See for example, Henry Kressel and J. K. Butler *Semiconductor Lasers and Heterojunction LEDs.* (Academic Press, New York, 1977).

6. R. H. Miles, G. Y. Wu, M. B. Johnson, T. C. McGill, J. P. Faurie, and S. Sivananthan, *Appl. Phys. Lett.* **48**, 1383 (1986).

7. M. Kobayashi, N. Mino, H. Katagiri, R. Kimura, M. Koagai, and K. Takahashi, *Appl. Phys. Lett.* **48**, 296 (1986).

8. A. M. Glass, K. Tai, R. B. Bylsma, R. D. Feldman, D. H. Olson, and R. F. Austin, *Appl. Phys. Lett.* **53**, 834 (1988).

9. R. H. Miles, J. O. McCaldin, and T. C. McGill, *J. Crystal Growth* **85**, 188 (1987).

10. J. H. Van der Merwe, *J. Appl. Phys.* **34**, 123 (1963).

11. C. A. B. Ball and J. H. Van der Merwe, in *Dislocations in Solids* edited by F. R. N. Nararro (North Holland, Amsterdam,1983) Vol. 6.

12. J. W. Matthews, in *Epitaxial Growth*, edited by J. W. Matthews (Academic Press, New York, 1968), Part B.

13. J. W. Matthews and A. E. Blakeslee, *J. Cryst. Growth* **27**, 118 (1974); **29**, 273 (1975); **32**, 265 (1976).

14. R. People and J. C. Bean, *Appl. Phys. Lett.* **47**, 322 (1985); **49**, 229 (1986).

15. J. C. Bean, in *Silicon Molecular Beam Epitaxy*, edited by E. Kasper and J. C. Bean (Chemical Rubber, Boca Raton, Florida, 1987).

16. R. H. Miles, "Structural and Optical Properties of Strained-layer Superlattices" (Thesis, California Institute of Technology, August, 1988).

17. B. W. Dodson and J. Y. Tsao, *Appl. Phys. Lett.* **51**, 1325 (1987); **52**, 852(E) (1986).

18. B. W. Dodson, *Phys. Rev. B* **35**, 5558 (1987).

19. M. Aven and J. S. Prener, *Physics and Chemistry of II-VI Compounds*(North-Holland, amsterdam, 1967).

20. J. O. McCaldin, T. C. McGill and C. A. Mead, *Phys. Rev. Lett.* **36**, 56 (1976).

21. W. A. Harrison, *J. Vac. Sci. Technol.* **14**, 1016 (1977).

22. J. Tersoff, *Phys. Rev. Lett.* **56**, 2755 (1986).

23. W. A. Harrison and J. Tersoff, *J. Vac. Sci. Technol.* **B4**, 1068 (1986).

24. J. R. Waldrop, R. W. Grant, S. P. Kowalczyk, and E. A. Kraut, *J. Vac. Sci. Technol.* **A3**, 835 (1985).

25. A. D. Katnani and G. Margaritondo, *J. Appl. Phys.* **54**, 2522 (1983).

26. C. Van der Waal and R. M. Martin, *J. Vac. Sci. Technol.* **B6**, 1354 (1988).

27. G. Monfroy, S. Sivananthan, X. Chu, J. P. Faurie, R. D. Knox and J. L. Staudenmann, *Appl. Phys. Lett.* **49**, 152 (1986).

28. J. Menendez, A. Pinczuk, J. P. Valladares, R. D. Feldman, and R. F. Austin, *Appl. Phys. Lett.* **50**, 1101 (1987).

29. R. H. Miles, T. C. McGill, S. Sivananthan, X. Chu and J. P. Faurie, *J. Vac. Sci. Technol.* **B5**, 1263 (1987).

30. G. Monfroy, X. Chu, M. Lange, and J. P. Faurie (unpublished results).

31. Tran Minh Duc and J. P. Faurie, *Phys. Rev. Lett.* **58**, 1127 (1987).

32. A. D. Katnani and G. Margaritondo, *J. Appl. Phys.* **54**, 2522 (1983).

33. M. Kobayashi, N. Mino, H. Katagiri, R. Kimura, M. Konagai, and K. Takahashi, *J. Appl. Phys.* **60**, 773 (1986).

34. M. Kobayashi, R. Kimura, M.Konagai, K. Tahahashi, *J. Cryst. Growth* **81**, 495 (1987).

35. The recent experimental work in this field is reviewed in the article by Professor M. Konagai.

36. H. Asonen, J. Lilja, A. Vuoristo, M. Ishiko, and M. Pessa, *Appl. Phys. Lett.* **50**, 733 (1987).

37. Y. Rajakarunanayake, R. H. Miles, G. Y. Wu and T. C. McGill, *J. Vac. Sci. Technol.* **B6**, 1354 (1988).

38. T. Yasuda, I. Mitsuishi, and H. Kukimoto, *Appl. Phys. Lett.* **52**, 57 (1987).

A REVIEW OF THE GROWTH OF 3D II-VI COMPOUNDS

B. J. Fitzpatrick

Philips Laboratories
North American Philips Corporation
345 Scarborough Road
Briarcliff Manor, New York 10510

ABSTRACT

II-VI materials would be preferable as substrates for the growth of II-VI 2D structures. However, growth of the II-VI bulk materials is difficult, mainly because of the low stacking fault energy. In addition, several of the cubic compounds have phase transitions. Melt growth, always a preferable technique, has shown some recent advances, apparently, due to the addition of alloying elements which improve the structural quality. Vapor growth has also had some interesting developments, showing significant increases in rate, due, apparently, to the use of hydrogen as a transport agent. Other techniques have not kept pace with developments in melt and vapor growth.

INTRODUCTION

II-VI materials are a natural choice as substrates for the growth of II-VI epitaxial structures. This is due to several factors: chemical compatibility, transparency and homoepitaxy. The primary advantage of chemical compatibility is the fact that, if autodoping from the substrate takes place during epitaxial growth, the autodoping element is at least isoelectronic and thus does not alter the carrier concentrations directly. Transparency is important for devices such as longitudinal electron-beam-pumped lasers, photodetectors and waveguides. The last category (waveguides) also benefits from the lower refractive index available in II-VI materials. Despite these potential advantages, the use of II-VI substrates for epitaxy has been very limited. This is due to the difficulties inherent in the growth of these materials. The growth problems include the high vapor pressure and reactivity with container materials and encapsulants. However, the most important problem is the low stacking fault energy.[1] This is a consequence of the low difference in energy between the various stacking sequences in tetrahedrally coordinated compounds. Another consequence of this is the potential occurrence of phase transitions at high temperatures. Definite phase transitions have been established for ZnS and ZnSe. Compounds such as CdS and CdSe appear to have stable hexagonal phases from room temperature to the melting point, which greatly facilitates crystal growth. In addition, Jones[2] has pointed out that there is much more likelihood of single crystal growth in hexagonal systems because the inherent anisotropy favors growth in one direction strongly. Despite the favorable properties of the hexagonal compounds, there has not been as much interest in them as in ZnSe and ZnS, probably because these latter compounds have much higher band gaps which are potentially more useful for various device applications. These compounds suffer from severe twinning problems; ZnS, in particular, has the distinction of being one of the model materials for polytypism, which is due to the above mentioned phenomenon, i.e., the low difference in energy between the stacking sequences. These problems have stimulated research into almost all growth methods possible for II-VI compounds.

Although recently the major interest in physics has shifted to 2D materials, 3D materials have shown some advances in materials preparation which have resulted in interesting phenomena. The efficient exciton localization caused by tellurium in $ZnSe$[3] and the cyclotron resonance in highly-purified vapor-grown ZnSe are examples of spectroscopic results which may find parallels in 2D structures. The enhancement of the second harmonic generation coefficient by manganese incorporation into $CdTe$[5] may stimulate investigation of frequency doubling in manganese-doped 2D structures. Low exciton line widths (0.3 meV) and long diffusion lengths (2.0 μm) have been observed in $ZnSe$[6]. A low electron-beam laser threshold (2.0 A/cm^2 at 28 kV) has been achieved in CdS.[6]

Bulk growth has been reviewed recently[7]; in this paper more emphasis will be put on those compounds which have potential for visible light emission. Three broad categories of growth methods--melt, solution and vapor--will be discussed. Within these categories, specific techniques which have had some recent progress or which show good future prospects will be discussed in more detail.

MELT GROWTH

Gradient Freeze (Tammann) and Bridgman Techniques

The speed of melt growth (up to mm/hour) is important not only for economic reasons, but also because it allows more rapid experimentation. Of course, the complexity of melt growth apparatus, especially that necessary for the high vapor pressure compounds, somewhat dilutes this advantage. Often, massive pressure chambers have been used, which impede the easy use of the rotation and translation mechanisms of Bridgman and zone melting growth. An average pressure for the growth of the higher vapor pressure compounds has been 100 atm. Thus, the gradient freeze or Tammann technique predominated in the early days of high pressure growth.[8,9,10] Direct resistance heating using graphite elements can be used. This is very efficient, but the control of gradients is dependent on the detailed design of the element, and graphite reacts with sulfur at high temperatures. Later workers[11-14] preferred the Bridgman technique. The pressure problem may be minimized by the use of a self-sealing technique.[15] Since, in any high-pressure Bridgman technique, a drive shaft must be moved through the base of the pressure chamber in any event, it is just as easy to introduce rotation which damps out asymmetries in the thermal profile. In principle, this should lead to both better crystal structure and a more favorable ingot shape (a long cylinder). In practice, growth of these materials is so dominated by other factors, such as phase transitions and high vapor pressures, that these early Bridgman attempts at improving the sophistication of the apparatus did not have a dramatic effect on structural quality. Even in the case of CdTe, which is a relatively low vapor pressure, low melting point compound, progress toward fully single crystalline material on the basis of classical adjustment of growth parameters has been difficult. However, some recent attempts at growth under lower gradient conditions have shown reductions in structural defects. Even in this case, the best results shown were not for pure CdTe, but for $Cd_{0.96}Zn_{0.04}Te$.[16] Although no report of a phase transition in CdTe has been published, it is possible that the addition of zinc increases the stability of the cubic phase. The more conventional explanation is that zinc "hardens" the lattice, possibly by stabilizing the Cd-Te bond.[17] Similar effects have been noted in III-V compounds; the explanations are still controversial. In the case of ZnSe, "rod-like low angle grain boundaries" were eliminated by growth in a low gradient.[18] Twins were still present, however, despite the fact that, because of supercooling, the growth took place below the phase transition. The addition of relatively small amounts (8×10^{18}/cc) of manganese to ZnSe seemed to suppress twin formation.[19] Low levels of impurities can have a marked effect on structural defects; for example, dislocations in InP can be markedly reduced by the addition of zinc or sulfur at similar levels.[20]

Horizontal Bridgman techniques have shown good results in the CdTe case.[21] The success of this method appears to be due to the large melt free surface: the material from the bottom of the ingot (touching the boat) is of much poorer quality than that of the top.

Czochralski Technique

Since it is apparent that the interaction of the melt with the crucible wall is a source of defects[21,31], the Czochralski technique has an obvious appeal, since contact with crucible walls is eliminated. However, this technique has had only moderate success with CdTe[22,23], and Cd(Te,Se).[24] Some authors[23] report no reaction with the boric oxide encapsulant, while others[25] note extensive solubility of the CdTe. Since these reactivity problems are seen with the lowest melting, lowest vapor pressure compound, it is unlikely that boric oxide will be successful with the higher-band-gap compounds, although the authors in Reference 26 have shown that the rate of decomposition of CdSe at 1300°C is similar to that of GaAs. Since the rate of decomposition of the material in contact with the boric oxide is probably the major limiting factor in growth, it seems that CdSe could be grown by this technique. However, no report of crystal growth followed this work. ZnSe apparently mixes with B_2O_3.[27]

Zone Melting

It is surprising that zone melting has not proven more popular for the visible band gap compounds, although it has been used from the beginning of work on CdTe[28], and has been used for ZnSe.[29] The appeal of zone melting for the more volatile, higher melting point materials lies in the fact that only a small portion of the charge is molten at any one time. This reduces both the rate of corrosion of the crucible and the rate of evaporation of the melt.

A self-sealing and self-releasing (SSSR) technique[30] has been used for growth at low applied pressures. At the start of a run, some material evaporates from the charge at the bottom and condenses at the closure (usually a screw cap) at the top. After the crucible has been lowered through an induction coil to traverse the zone, the coil (now at the top) vaporizes the seal. This has allowed the growth of CdS, which has a vapor pressure of 3.8 atm, with an applied pressure of only 7 atm.[31] Another feature of zone melting methods is their ability to make ternary alloys of relatively uniform composition, i.e. zone leveling can be used.

VAPOR GROWTH

These methods have yielded crystals of excellent structural quality, but at generally very low rates, especially for cubic compounds of moderate vapor pressure. CdS, with its high vapor pressure and stable hexagonal structure, can be grown very well by this method without the use of a transport agent.[32] The problem with the other compounds is that attempts to increase the rate of growth lead to the same problems of stacking disorder that plague melt growth, even if the temperature is far below a phase transition. Only very low temperature (< 850°C), very slow iodine transport has resulted in twin-free growth[33] of ZnSe, for example. This temperature is far below the phase transition (1425°C). Crystals grown at 1005°C, which show excellent optical properties (cyclotron resonance and free exciton luminescence), do show high twin densities in some areas.[34] Thus it appears that just being below a phase transition does not easily lead to twin-free materials, even in very high purity conditions at low rates. Only transport with elemental iodine has been shown to give twin-free material; the rate of growth, however, is very low, < 0.7 g/day. Recently, using sublimation, some authors have shown quite high growth rates for ZnSe (1 mm/day)[35] and CdTe (2 to 15 mm/day).[36,37,38] Structural properties were not discussed in the ZnSe case, but for CdTe it seemed that lower temperature and lower rates resulted in material in which was more likely to be uniformly twinned, i.e. all the twins running in the same direction. Hydrogen has often been used as a transport agent or carrier gas in vapor epitaxy. However, it has found little use in 3D vapor growth. Vohl[39] used it "usually" as a "carrier gas", apparently not considering it of great importance. Recently, however, Akhekyan et al.[40] have used carefully-calibrated argon-hydrogen mixtures to obtain very high growth rates of ZnSe and closely related alloys. Korostelin et al.,[41] in an extension of this work showed good rates of growth of cubic ZnS with only 1-4% hexagonality at a temperature of 980°C, only 40°C below the classical value of the phase transition temperature. Another unusual modification was the use of a seed below the source (on the bottom in a vertical ampoule). This has been avoided in the past, because of the fear that particles of the source material would fall on the substrate. (In this method, the substrate sits on the bottom.) Putting the seed on the bottom assures simple, low strain mounting. Since the bottom is colder than the top, convection is suppressed, and diffusive transport should predominate. Ordinarily, this would be expected to slow the rate of growth because the gas flow would be slower. The very large diameter of the ampoule (the seed itself is 50 mm diameter) may have some influence. The crystals are only 15-20 mm high, which is somewhat awkward for wafer fabrication; this could probably be improved by lengthening the growth time, currently 3 days.

SOLUTION GROWTH

Traveling Heater Method

The most popular form of solution growth has been the traveling heater method. This is essentially a form of liquid phase epitaxy (LPE), in which a solvent zone is passed through a bar of feed material. The crucible is usually a long vertical tube of quartz on which a layer of carbon is deposited. The seed is placed in the bottom of the crucible; it is covered with a relatively small amount of solvent, and the feed material is placed on top. The crucible is lowered slowly through the heater; the rate must be kept low enough so that the movement of the zone does not fall behind that of the heater. This is so because the rate of dissolution of the source material and the rate of deposition on the seed (which must be equal) are by nature rather slow, and determine the rate of travel of the zone. This method combines the purity advantages of both LPE and zone refining. Most of the work has concentrated on the growth of tellurides from tellurium, since Te is a near-ideal solvent[42,43]. Other solvents - indium metal and lead chloride - have been used for ZnSe.[44] Both of these solvents had precipitation problems; however, the growth from lead chloride resulted in large (~ 1 cm area) single crystalline, apparently twin-free growth. Rates, however, were low (1 mm/day at 900°C). It is interesting to note that the vapor pressure of lead chloride is 400 mm at 893°C (zinc chloride is, of course, even higher). Lead fluoride has been used as a solvent for ZnO.[45]

Other Methods

Directional cooling of solutions has also given very good results for tellurides. A Bridgman-gradient freeze configuration, featuring a sophisticated profile adjustment system, has been used to grow very high quality tellurides.[46] A cold finger technique[47] has been used to produce highly perfect CdTe, but the height of the ingot is small (~ 1.5 cm). In this method, cold gas is directed at the center of the bottom of the crucible. The crystal "grows out" into the solution, minimizing contact with the walls. Solvent evaporation[48] has been used for CdTe (Cd is evaporated). A Bridgman-like ampoule is used. Nishizawa has grown ZnSe from Se solution at 1050°C by the "temperature difference under controlled vapor pressure" method.[49] The apparatus is similar to that used above, or to the vertical modified Bridgman method, i.e. a crucible in a growth zone with a gradient, and a separate reservoir for Zn. The zinc is distilled into the ZnSe-Se solution, forcing precipitation of ZnSe.

Hydrothermal growth has been successful in making crystals of laser-grade ZnO[50] and ZnS.[51] A disadvantage of this method is that it seems to be necessary to use LiOH in the solution, leading to lithium doping of the crystals.

Classical flux growth has been unsuccessful in producing anything more than small platelets. However, the use of halide solvents mentioned above for the travelling heater method is probably a better adaptation of the principles of flux growth to II-VI compounds.

SUMMARY AND CONCLUSIONS

II-VI compounds in bulk form have advantages, primarily as substrates, for the growth of 2D structures. Their application has been hindered by the difficulty of growing large single crystals, even though the quality of small samples, especially with regard to luminescence[4], may often be very high. Structural techniques (x-ray[19] and TEM [52]) support this conclusion. The cubic materials generally are afflicted with twinning problems; however, if the material is cut perpendicular to the <111> direction around which twinning occurs, the growing epitaxial layer "sees" a twin-free {111} plane, since the twin planes do not intersect the surface[53]. Twin planes in themselves should not be harmful, since simple rotational twins do not have any broken bonds. If twin planes must be eliminated, however, compositional changes have shown some promise for improvement. Growth of the hexagonal compounds is generally easier: twinning is not generally seen in hexagonal materials. Thus the use of hexagonal compounds should be considered where possible.

ACKNOWLEDGMENTS

The author wishes to acknowledge the superb assistance of P. Harnack and the continuous support of R. Bhargava in his own work. The review of the manuscript by S. Colak, J. Petruzzello, K. Shahzad and M. Shone is appreciated.

REFERENCES

1. S. Takeuchi, K. Suzuki, K. Maeda and H. Iwanaga, Philosophical Magazine A, 50:171 (1984).

2. K. A. Jones, J. Cryst. Growth 19:33 (1973).

3. A. Reznitsky, S. Permogorov, S. Verbin, A. Naumov, Yu. Korostelin, V. Novozhilov, and S. Prokoviev, Solid State Commun. 52:13 (1984).

4. T. Ohyama, E. Otsuka, T. Yoshida, M. Isshiki, and K. Igaki, Surface Science 170:491 (1986).

5. L. Kowalczyk, J. Cryst. Growth 72:389 (1985).

6. B. Fitzpatrick, J. Khurgin, P. M. Harnack and D. de Leeuw, Proceedings of the International Electron Devices Meeting, IEDM-86, p. 630.

7. B. Fitzpatrick, ibid. 86:106 (1988).

8. A. Addamiano and M. Aven, J. Appl. Phys. 31:36 (1960).

9. K. K. Dubenskiy, V. A. Sokolov and G. A. Ananin, Sov. J. Opt. Tech. 36:118 (1969).

10. S. Shionoya, Y. Kobayashi and T. Koda, J. Phys. Soc. Japan 20:2046 (1965).

11. M. Kozielski, J. Cryst. Growth 1:293 (1967).

12. I. Kikuma and M. Furukoshi, ibid. 44:467 (1978).

13. M. P. Kulakov, I. B. Savtchenko and A. V. Fadeev, ibid. 52:609 (1981).

14. A. Scharmann and D. Schwabe, ibid. 38:8 (1977).

15. U. Debska, W. Giriat, H. R. Harrison and D. R. Yoder-short, ibid. 70:339 (1984).

16. S. Sen, W. H. Konkel, S. J. Tighe, L. G. Bland, S. R. Sharma and R. E. Taylor, ibid. 86:111 (1988); B. Pellicari, ibid. 86:146 (1988).

17. K. Guergouri, R. Triboulet, A. Tromson-carli and Y. Marfaing, ibid. 86:61 (1988).

18. I. Kikuma, M. Sekine and M. Furukoski, ibid. 75:609 (1986).

19. M. Shone, B. Greenberg and M. Kaczenski, ibid. 86:132 (1988).

20. P. Roksnoer and M. Rijbroek-van den Boom, ibid. 66:317 (1984).

21. A. A. Khan, W. P. Allred, B. Dean, S. Hooper, J. E. Hawkey and C. J. Johnson, J. Electr. Matls. 15:181 (1986).

22. H. M. Hobgood, R. N. Thomas, B. W. Swanson, R. M. Ware and I. Grant, J. Cryst. Growth 85:510 (1987).

23. J. B. Mullin and B. W. Straughan, Rev. Phys. Appl. 12:123 (1977).

24. N. Klausutis, J. A. Adamski, C. V. Collins, M. Hunt, H. Lipson and J. R. Weiner, J. Electr. Matls. 4:625 (1975).

25. G. S. Meiling and R. Leombruno, J. Cryst. Growth 3:300 (1968).

26. G. S. Dubrovikov and A. L. Marbakh, Inorg. Matls. 11:32 (1975).

27. A. G. Fischer, in "Crystal Growth (2nd ed.)," ed. B. R. Pamplin, Pergamon, Oxford (1980), p. 380

28. D. de Nobel, Philips Res. Repts. 14:361 (1959).

29. M. P. Kulakov and A.V. Fadeev, Inorg. Matls. 17:1156 (1981).

30. B. J. Fitzpatrick, T. F. McGee III and P. M. Harnack, J. Cryst. Growth 78:242 (1986).

31. B. Fitzpatrick, P. M. Harnack and S. Cherin, Philips Journal of Research 41:452 (1986).

32. G. H. Dierssen and T. Gabor, J. Cryst. Growth 43:572 (1978).

33. E. Kaldis, in "Crystal Growth: Theory and Techniques", ed. C. H. L. Goodman, Plenum Press, New York (1974) V.1, p. 49.

34. T. Ohyama, E. Otsuka, T. Yoshida, M. Isshiki and K. Igaki, Surface Science 170:491 (1986).

35. T. Taguchi, T. Kusao and A. Hiraki, ibid. 72:46 (1985).

36. C. Geibel, H. Maier and R. Schmitt, J. Cryst. Growth 86:386 (1988).

37. K. Durose and G. J. Russell, ibid. 86:471 (1988).

38. R. Triboulet and Y. Marfaing, ibid. 51:89 (1981).

39. P. Vohl, Mat. Res. Bull. 4:689 (1969).

40. A. M. Akhekyan, V. I. Koslovskii, Yu. V. Korostelin, A. S. Nasibov, Y. M. Popov and P. V. Shapkin, Sov. J. Quant. Elect. 15:737 (1985).

41. Yu. V. Korostelin, V. I. Kozlovsky, A. S. Nasibov, Ya. K. Skasyrsky and P. V. Shapkin, to be published.

42. R. O. Bell, N. Hemmat and F. Wald, Phys. Stat. Solidi (a) 1:375 (1970).

43. R. Schoenholz, R. Dian and R. Nitsche, J. Cryst. Growth 72:72 (1985).

44. R. Triboulet, J. Cryst. Growth 59:172 (1982).

45. G. A. Wolff and H. E. LaBelle, Jr., J. Am. Ceram. Soc., 48:441 (1965).

46. B. Schaub, J. Gallet, A. Brunet-Jailly and P. Pellicari, Rev. Phys. Appl. 12:147 (1977).

47. K. Zanio, J. Elect. Mat. 3:327 (1974).

48. A. W. Vere, V. Steward, C. A. Jones, D. J. Williams and N. Shaw, J. Cryst. Growth 72:97 (1985).

49. J. Nishizawa, K. Itoh, Y. Okuno and F. Sakurai, J. Appl. Phys. 57:2210 (1985).

50. S. S. Demidov, G. S. Kozina, L. N. Kurbatov, I. P. Kuz'mina and Yu. V. Shaldin, Sov. J. Quantum Elect. 14:291 (1984).

51. L. N. Kurbatov, G. S. Kozina, T. A. Kostinskaya, V. S. Rudnevskii, A. N. Lobachev, V. A. Kusnetsov, I. P. Kuzmina, Y. V. Shaldin and A. A. Shternberg, Sov. J. Quantum Elect. 10:215 (1980).

52. J. S. Vermaak and J. Petruzzello, J. Electronic Materials, 12:29 (1983).

53. J. L. Schmidt and J. E. Bowers, Technical Report AFWAL-TR-80-4068, Air Force Wright Patterson Aeronautical Laboratories, Ohio 1980 (ADA086342).

THE GROWTH OF THIN LAYERS BY MOCVD OF WIDE BAND GAP II–VI COMPOUNDS

B. Cockayne and P. J. Wright

Royal Signals and Radar Establishment
St Andrews Road
Malvern
Worcs WR14 3PS, UK

INTRODUCTION

The potential device interest in the wide band gap II–VI materials, based principally on compounds of the group II elements (Zn, Cd) with the group VI elements (O, S, Se, Te) has been recognised for a long time.[1,2] These materials are known to have band gaps which are compatible with emission in the visible region of the electromagnetic spectrum and to exhibit non–linear optical effects.[3] These properties are currently being investigated in applications such as light emitting devices, optical wave guides and optical switches.

All the binary and ternary wide gap II–VI compounds listed in Table 1 have now been grown in layer form by metalorganic chemical vapour deposition (MOCVD). Hence, this process, together with molecular beam epitaxy (MBE) and the fusion of the two processes in metal organic molecular beam epitaxy (MOMBE), is now well established as a route for growing thin layer heterostructures of wide–band gap materials and, in recent years, this general field has been well reviewed.[4,5] It is now clear that the low temperature growth characteristics of MOCVD can be harnessed to produce high quality single layers of ZnSe and multilayers of ZnS/ZnSe and similar progress has been made recently with equivalent Cd–based compounds. However, the basic difficulties associated with (a) heterostructural growth due to lack of suitable substrates, (b) gas phase prereaction and (c) obtaining both chemically and structurally abrupt interfaces, are still not totally controlled. These problems become compounded when alloy systems, mixed structural systems and multilayers (including low dimensional structures) are considered. This paper reviews these factors for a number of II–VI compounds and also presents data on the growth of low dimensional structures based on the ZnS/ZnSe and CdS/CdSe systems.

GROWTH APPARATUS

The types of apparatus used for the growth of II–VI compounds are described fully in recent reviews.[4,6] Fig. 1 shows a schematic diagram of an apparatus suitable for the growth of low dimensional structures. In principle, gaseous reactants containing the appropriate group II and group VI species, plus any required dopant species, are transported in a suitable carrier gas, often H_2, to a heated substrate where reaction occurs and the compound is deposited. Growth at both atmospheric and lower pressures has been utilised in conjunction with rf, resistance or optical heating and, in some instances, with photon assistance.

The growth of the very thin layers demanded by low–dimensional structures such as strained–layer superlattices (SLS), down to 10Å thickness or less, requires extremely

Table 1. Wide band gap II–VI compounds grown by MOCVD

Compound	Growth Temperature	Comments	Reference
ZnO	300 – 500 °C	Polycrystalline layers	4,10,35
ZnS	150 – 400 °C	Single crystal layers SLS with ZnSe	4, 26
ZnSe	200 – 500 °C	Single crystal layers SLS with ZnS	4,12,26, 30
ZnTe	380 °C	Single crystal layers SLS with CdTe	27, 34
ZnS_xSe_{1-x}	250 – 400 °C	Single crystal layers Alloy composition control is a potential problem	18
CdS	300 – 450 °C	Single crystal layers SLS with CdSe	14, 32
CdSe	300 – 450 °C	Single crystal layers SLS with CdS	14
CdS_xSe_{1-x}	300 – 500 °C	Single crystal layers Alloy composition control is a potential problem	28
$Zn_xCd_{1-x}S$	200 – 500 °C	Single crystal layers Alloy composition controlled	19
$Zn_xCd_{1-x}Se$	300 °C	Single crystal layers	17
$Zn_{1-x}Mn_xS$ and $Zn_{1-x}Mn_xSe$	400 – 550 °C	Single crystal layers X limited to $\leqslant 0.1$ by low volatility of Mn precursor	29

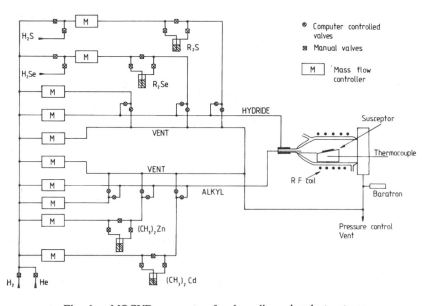

Fig. 1. MOCVD apparatus for low–dimensional structures

accurate and reproducible control of both the gas–phase composition and the growth rate. Additionally, it is important to balance the gas switching in order to avoid transient effects which, because of the very short time intervals, 5sec or less, needed to grow some layers can cause unwanted interfacial compositional variation. In general, the requisite control can be achieved only with a rapid transit gas handling system using computer control for operation of the gas–switching valves.

SUBSTRATES

High quality II–VI substrates with good structural, electrical and optical properties are essentially non–existent in useful sizes and, hence, the opportunity for the growth of lattice matched homoepitaxial structures is somewhat limited. Consequently, most of the growth reported is heteroepitaxial in character and the types of structure which can be grown are usually dictated by the substrates available. These are normally elemental semiconductor or III–V compound semiconductor single crystals. Under these constraints, factors such as interfacial mismatch, constituent atom interdiffusion and compatibility of crystal structure demand detailed investigation and control.

Fig. 2 shows the band gap energies for the main binary and ternary compounds as a function of their lattice constant, together with the lattice constants of the cubic structured substrates commonly used;[7] an indication of the colour of the light corresponding to the band gaps of the various materials is also given.

No precise lattice matching exists for any of the binary compounds with commonly available substrates, although in some instances, eg ZnSe on GaAs, the mismatch is small (0.25%). Lattice matching can be achieved by using ternary compositions, eg $ZnS_{0.06}Se_{0.94}$ on GaAs but this requires precise control of the gas phase concentrations and can pose problems due to non–linear relationships between gas phase ratios and solid compositions.

Mismatch problems are greatly enhanced when the preferred crystal structure for the epitaxial layer differs from that of the substrate, for example, hexagonal CdS on cubic GaAs. In this instance, improved matching can be obtained on (111A) orientated GaAs substrates.

Successful cleaning and removal of work damage for the substrate are further prerequisites for heteroepitaxy. Oxide removal from substrates such as Si can pose problems whilst III–V substrates can etch non–uniformly, on orientations other than (100), during damage removal.

The difficulties in finding effective solutions to these problems in most systems is one reason why the growth of deliberately mismatched and strained layer structures, either with or without buffer layers, is now receiving much attention. However, if the strain is sufficient to induce defects such as dislocations, stacking faults and/or twins, interdiffusion of atoms across the interfaces can be enhanced and further decrease the stability of these complex structures.[8,9]

PREREACTION

Typical examples of the major types of chemical reaction used thus far in MOCVD for the growth of II–VI compounds are listed in Fig. 3. The most common type utilised is that between the dimethyl group II precursor, dimethylzinc or dimethylcadmium and a group VI hydride, hydrogen sulphide or hydrogen selenide. Such reactions have undoubtedly been used to produce material of good quality but irreproducible and non–uniform layer properties frequently arise from prereaction of the precursor materials upstream from the susceptor. Whilst careful design of the reactor geometry, and particularly the gas mixing nozzle, can minimise this problem, these parameters do not eliminate the prereaction. As a consequence, considerable effort has been devoted to the study of alternative precursors and reactions in order to block or inhibit prereaction.

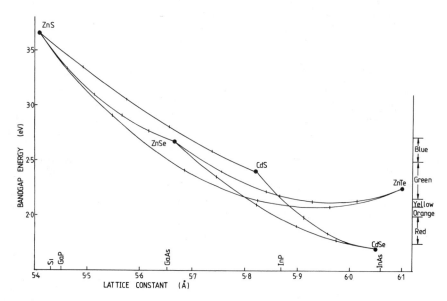

Fig. 2.   Band gap energies for main binary and ternary II–VI compounds

REAGENTS		COMMENTS	REFERENCE
$(CH_3)_2Zn$ + $(C_2H_5)_2Zn$	H_2S H_2Se	High quality layers Prereaction	26
	RSH	Prereaction stopped	12
	R_2Se	Higher growth temperature No prereaction	30,31
	 ⬡S	As R_2Se	10
$(CH_3)_2Cd$ +	H_2S H_2Se	High quality layers Prereaction	14
	R_2S	Higher growth temperature	32
$(CH_3)_2Zn$ (1,4dioxan)	H_2S H_2Se	Reduced prereaction	33

Fig. 3.   Major types of chemical reaction used for the MOCVD of II–VI compounds

One approach has been to use modified group VI precursors such as alkyl or heterocyclic compounds in reaction with the dimethyl group II precursors.[10] These readily prevent prereaction and improve uniformity but invoke the penalty of an increased growth temperature, approximately 500°C, compared to 275–300°C for the corresponding growth with group VI hydrides. However, recent work using photon–assistance[11] suggests that a growth temperature of 300°C is feasible using these reactions. A further group VI precursor, methylselenol, also shows promise for the growth of ZnSe at low temperatures.[12]

An alternative approach has been to use modified zinc and cadmium adducts. The use of various chelating ligands such as 1,4–dioxan adducted to dimethylzinc, have produced compounds which inhibit prereaction with the group VI hydrides. Adducts of this type have now provided high quality ZnSe layers at growth temperatures in the range 300–400°C. This is marginally higher than the optimum range for the conventional dimethylzinc and hydrogen selenide reaction, but the room temperature photoluminescence of such a layer grown at 350°C exhibits strong band–edge luminescence and an absence of any deep centre related emission as shown in Fig. 4 and confirms the high quality of the material produced.

The purity of the organometallic compounds of the group II metals has improved to a point where the metallic impurity levels are in the range 10–100 parts in 10^9. There is still a need to improve the purity of the group VI precursors, particularly the gaseous hydrides, where total impurity concentrations around 100ppm are specified commercially. However, many of these impurities appear to be inert in low temperature growth processes such as MOCVD.

There exists substantial evidence to suggest that once pure precursors are obtained, pure layers can be grown by MOCVD. However, this evidence is largely confined to analysis of ZnSe layers where conventional electrical and photoluminescence data can be obtained. The absence of deep centre related emission in ZnSe layers, noted above, provides clear evidence of purity and is confirmed both by the low residual carrier concentrations, routinely $<10^{15} cm^{-3}$ and the high room temperature mobilities, approaching the predicted limit of $550 cm^2 V^{-1} s^{-1}$, derived from Van der Pauw electrical measurements.

HETEROSTRUCTURES

The growth of heterostructures involves not only those formed between II–VI layers and substrates but also II–VI/II–VI combinations. The latter are often designed from theoretical predictions associated with "ideal heterostructures" where adverse features such as strain, dislocations, stacking faults, twins and interdiffusion do not occur. However, heterostructural growth must attempt to approach the ideal.

Excluding deliberately graded structures, chemically and structurally abrupt interfaces are a prime requirement for heterostructures. The importance in controlling pressure transients associated with gas switching has already been stressed, since these can readily cause unwanted changes in interface composition. An additional way of improving interface abruptness can be to introduce a pause in growth between one layer and the next so that unwanted species are swept clear. This applies to both lattice matched and mismatched structures.

A graded interface can be used to take up mismatch between two layers. In principle, the grading profile can be made to any predetermined shape but it is quite likely that, as the grading proceeds, the growth rate will alter due to the different concentrations of constituents so that a linear grading may be difficult to achieve. An alternative procedure for accommodating mismatch is the mark–space technique. Here a thin layer of compound A, below the critical thickness for the generation of misfit defects, is grown on to compound B. This is followed by a layer of B. The next layer, of A, is slightly thicker than the first layer and the next B layer is slightly thinner. This process continues keeping the thickness (A+B) constant, until B reduces

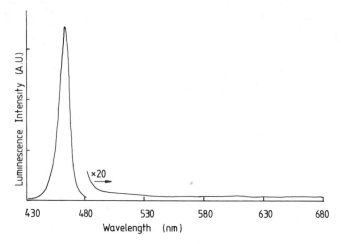

Fig. 4.   Photoluminescence spectrum (RT) of a ZnSe layer grown at 350°C
using an adduct group II source

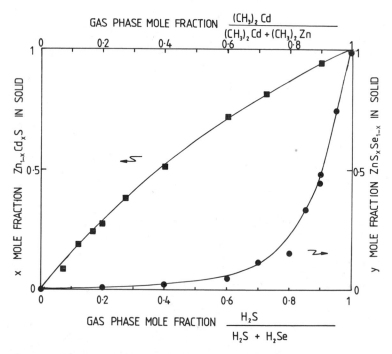

Fig. 5.   The relationship between gas phase and solid compositions for
the alloy systems $Zn_{1-x}Cd_xS$ and ZnS_ySe_{1-y}

to zero and compound A alone is grown. These techniques can be particularly useful for the growth of buffer layers onto lattice mismatched substrates.

Even with appropriate gas switching control, other problems remain. As shown in Fig. 5, the gas phase reactant and the resulting solid compositions are not always related in a simple linear manner. For the system ZnS_ySe_{1-y}, this relationship is far from linear and it is obvious that slight fluctuations in $[H_2Se]$ can lead to large solid compositional changes, especially at interfaces , unless the gas switching is perfect. The system $Cd_xZn_{1-x}S$ is nearly linear and should, therefore, be more tolerant of minor changes in gas phase composition. Hence reactant species which provide an approximately linear relationship between the gas phase and solid compositions are obviously to be preferred for abrupt interfaces. Fortunately, as Table 1 shows, the temperature ranges for the growth of these compounds are similar. Thus, in almost all cases, it is possible to use one particular temperature throughout the entire growth sequence.

A dominant problem in II–VI heterostructural growth, particularly in comparison to the more widely studied III–V compounds, is the different crystal structures, principally hexagonal or cubic, which can form for the same compound and between compounds. This means that factors such as growth temperature, substrate orientations, crystal structure of the previous layer, the degree of mismatch and strain can exert a strong influence on the structure adopted by the epitaxial layer.

The systems studied extensively are ZnS and ZnSe, which can be grown readily and with high quality in cubic form on to (100) GaAs.[13] However, on (111) GaAs the layers are heavily twinned. Correspondingly, CdS grows with the hexagonal structure on to (111A) GaAs whilst CdSe grows as a mixture of both hexagonal and cubic phases on to (100) GaAs and (0001) CdS buffer layers; the cubic phase of CdSe predominates on (111A) GaAs.[14] It is therefore clearly possible to modify the crystal structure of the epitaxial layer by a suitable choice of substrate and orientation. Thus, entirely cubic ZnS/ZnSe superlattices have been grown on to (100) GaAs whilst entirely hexagonal CdS/CdSe superlattices have been grown on to (111A) GaAs.[15]

The interaction of mismatch and thin layers may well allow metastable structures to be grown, where the epitaxial layer is sufficiently strained to match the previous layer but not strained to the point where strain relief induces the formation of line and related defects. Strain at an interface arises from two components, the lattice parameter mismatch and the differences in thermal expansion coefficient between the dissimilar layers. At the modest growth temperatures used for most growth by MOCVD of II–VI compounds, typically 250–400°C, the lattice parameter effect predominates. Metastable structures must, of course, remain stable with time and use.

The structural quality of heterostructures is conveniently assessed by double crystal X–ray diffraction. For wide gap II–VI compounds, ZnSe grown on to GaAs has again been the main research vehicle utilising this technique. With GaAs as the reference material, the layer quality has been shown to be good with full width half maximum values of 100sec being obtained, using the 400 reflection in conjunction with Cu radiation.

One factor which limits both the chemical and stability of heterostructures and which interacts strongly with structural stability is interdiffusion of constituent atoms.[16] Out–diffusion of substrate atoms is known to be enhanced by the presence of defects, eg Ga from GaAs into ZnS and ZnSe layers. Efforts to minimise such effects thus far have concentrated upon preserving the chemical integrity of the GaAs substrate prior to growth, the growth of lattice matched buffer layers and growth at low temperatures. However, in principle, both the group II and group VI atoms can interdiffuse. Secondary ion mass spectrometry has shown that this does not seem to be a major problem for S and Se at temperatures less than 500°C but for Zn and Cd, some interdiffusion has been detected at temperatures around this value.[17]

Low dimensional structures between binary II–VI compounds must always be a strained–layer type of heterostructure. Lattice matching is only possible in these systems by using binary–ternary eg ZnSe–ZnSTe, ternary–ternary, or even quaternary combinations of compounds. For growth by MOCVD this presents major problems in the control of alloy composition and associated defects,[18,19] as discussed earlier, and hence initial studies of superlattice structures have been constrained to binary combinations. Most studies have concentrated on ZnS/ZnSe and CdS/CdSe although effort has been devoted to ZnSe/ZnTe structures[20] with the aim of producing a p–type layer complementary to n–type ZnSe.

Typically, ZnSe–ZnS strained–layer superlattices and ZnSe–ZnS$_{0.1}$Se$_{0.9}$ super–lattices have been grown successfully with layer thicknesses down to 50Å using a low pressure growth system (100 Torr) and growth temperatures of approximately 400°C.[21] Reactions between either dimethyl zinc, dimethyl selenide and dimethyl sulphide or dimethylzinc, dimethylselenide and hydrogen selenide or dimethylzinc, hydrogen selenide and hydrogen sulphide have been employed. Two major important factors have emerged from this work. These are, firstly, that a strained–layer superlattice can be utilised to control misfit dislocations and that superlattices of sufficient quality for short–wavelength optical waveguides can be grown.[22,23]

More recently, CdS–CdSe superlattices have also been grown successfully by MOCVD on to a buffer layer of CdS grown on to a (111A) GaAs substrate. These layers were grown at 450°C and atmospheric pressure using dimethylcadmium, hydrogen sulphide and hydrogen selenide.[15] Layer thicknesses investigated were typically in the range 25–400Å. Fig. 6 shows a cross–sectional transmission electron image of such a multilayer consisting of hexagonal CdS (thin) and CdSe (thick) strata. The layers show no twinning and are reasonably planar, although undulations in thickness are present at some locations, particularly at the interface with the CdS buffer layer. Despite these undulations, the initial low–temperature luminescence spectra show strong infrared emission and provide preliminary evidence of a type II superlattice.[15]

DOPING

Controlled doping of II–VI layers is required in many device applications; for instance, luminescence centres are required in display applications and the ability to dope both p and n is needed to exploit semiconducting properties.

Manganese is the most widely used luminescence centre and a suitable precursor, tricarbonyl manganese, has been used to incorporate Mn into ZnS layers.[24] Bright electroluminescent panels have been made as a result. The same precursor has also been used to produce Zn$_{1-x}$Mn$_x$Se layers with $0 < x \leq 0.1$. Incorporation of other luminescence centres, particularly rare–earth ions, awaits the development of suitable volatile precursors.

N–type doping poses little problem for those II–VI compounds which are not insulators. The use of organometallic compounds such as the trialkyl compounds of Al, Ga or In, has permitted controlled doping of ZnSe in the range 10^{16} to 10^{18} atoms cm^{-3}. Indeed, the main problem with n–type doping of ZnSe is in limiting the doping level because the vapour pressures of the triethyl derivatives of Al, Ga and In are high and these atoms are also incorporated efficiently. It is thus all too easy to produce very highly doped material so that the photoluminescence becomes dominated by deep centre emission. P–type doping of compounds other than ZnTe is still only just being investigated. Again, the major problem is a lack of suitable precursors for the likely dopants, although the successful use of Li$_3$N to produce p–type ZnSe has been reported and a p–n junction demonstrated.[25]

It is clear that some doping of II–VI layers is possible but controlled doping of low dimensional structures essentially remains a new research area.

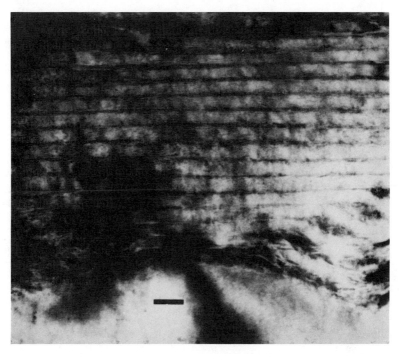

Fig. 6:   Cross—sectional transmission electron image (bright field, 0001 CdS and CdSe reflections) of a SLS of CdS (50Å) and CdSe (110Å) on a CdS buffer layer (bar marker = 28nm)

DISCUSSION AND CONCLUSIONS

It is now evident that single crystal layers of individual binary compounds such as ZnSe, ZnS, CdSe and CdS can be grown by MOCVD with sufficient control of structural and chemical properties to allow initial optical and semiconductor devices to be manufactured.   Despite the constraints discussed herein, the science and technology has allowed this initial promise to be extended into low dimensional structures such as strained—layer superlattices of ZnSe/ZnS and CdSe/CdS.   Thus both cubic and hexagonal superlattices have been grown and both have shown either useful or potentially useful device properties.

The ideal conditions for superlattice growth by MOCVD would be to have precursor materials which exhibit a linear relationship between the gas phase and solid compositions across the alloy system comprising the superlattice and which are sufficiently pure to show no evidence of deep centre related emission.   Ideally, the deposition reaction should also take place at a temperature sufficiently low to prevent or to inhibit further impurity incorporation but sufficiently high to promote sound crystallinity.   There should also be an absence of prereaction to promote uniformity and interface abruptness.   Under such conditions, control of defects and strain becomes principally a matter of controlling layer thickness.   Inevitably, few precursor compounds are currently available which approach this ideal. It is therefore encouraging that some superlattices can be grown with the materials available. This observation, coupled with the increased emphasis now being place on development of precursors with appropriate properties and the continuous advancement of MOCVD apparatus for III–V superlattices, provides a very positive platform on which to build further for the growth, assessment and exploitation of II–VI single and multiple heterostructures.

ACKNOWLEDGEMENTS

The authors wish to thank Dr A G Cullis, G M Williams and P W Smith (RSRE, Malvern) for providing the electron micrograph. The authors also wish to acknowledge the cooperation of colleagues from Hull University (Dr J J Davies, Dr E J Nicholls, M P Halsall and T J Gregory), Strathclyde University (Prof B Henderson, Dr K P O'Donnell and P J Parbrook) and Epichem Ltd (Dr A C Jones and E D Orrell).

REFERENCES

1. Mullin, J.B., Irvine, S.J.C. and Ashen, D.J., J.Crystal Growth, 55:92 (1981)
2. Proceedings of the International Conference on II–VI Compounds, Durham, UK (1982). Published as J Crystal Growth 59:1–439 (1982)
3. Miller, A, Staromlynska, J, Muirhead, I T, Lewis, K.L., Craig, D. and Steward, G, J.Crystal Growth 86:859 (1988)
4. Cockayne, B and Wright, P.J., J. Crystal Growth 68:223 (1984)
5. Yao, T, J. Crystal Growth 72:31 (1985)
6. Proceedings of the Third International Conference on II–VI Compounds, Monterey, CA, USA. Published as J.Crystal Growth, 86:1–947 (1987)
7. Ido, T, J. Electronic Mats 9:869 (1980)
8. Shaw, D, J. Crystal Growth 86:778 (1988)
9. Ohmi, K, Suemune, I, Kanda, T, Kan, Y, Yamanishi, M, Nishiyama, F and Hasai, H, J. Crystal Growth 86:467 (1988)
10. Wright, P.J., Griffiths, R J M and Cockayne B, J Crystal Growth 66:26 (1984)
11. Fujita, S, Tanabe, A, Sakamoto, T, Isemura, M and Fujita, S, Jap. J.Appl.Phys. 26:L2000 (1987)
12. Fujita, S, Sakamoto, T, Isemura M and Fujita, S, J.Crystal Growth 87:581 (1988)
13. Williams, J.O., Crawford, E.S, Jenkins, J.Ll., Ng, T.L., Patterson, A.M., Scott, M.D., Cockayne, B. and Wright, P.J., J.Mats.Sci. Letts. 3:189 (1984)
14. Halsall, M.P., Davies, J.J., Nicholls, J.E., Cockayne, B, Wright, P.J. and Russell, G.J., J. Crystal Growth – accepted for publication
15. Halsall, M.P., Nicholls, J.E., Davies, J.J., Cockayne, B., Wright, P.J. and Cullis, A.G., J.Semicon.Sci. and Tech. – accepted for publication
16. Maung, N. and Williams, J.O., J.Crystal Growth 86:629 (1988)
17. Wright, P.J., Cockayne, B., Williams, A.J. and Blackmore, G.W., J.Crystal Growth 79:357 (1986)
18. Stutius, W., J.Elec.Mats. 10:95 (1981)
19. Wright, P.J., Cockayne, B. and Williams, A.J., J.Crystal Growth 72:23 (1985)
20. Konagai, M., Kobayashi, M., Kimura, R. and Takahashi, K., J.Crystal Growth 86:290 (1988)
21. Yokagawa, T, Ogura, M and Kajiwara, T., Appl.Phys.Lett. 49:1702 (1986)
22. Fujita, S, Matsuda, Y and Sasaki, A., Appl.Phys.Lett. 47:955 (1985)
23. Yokogawa, T., Ogura, M. and Kajiwara, T., Appl.Phys.Lett. 52:120 (1988)
24. Wright, P.J., Cockayne, B., Cattell, A.F., Dean, P.J., Pitt, A.D. and Blackmore, G.W., J. Crystal Growth 59:155 (1982)
25. Yasuda, T., Mitsuishi, I. and Kukimoto, H., Appl.Phys.Lett. 52:57 (1988)
26. Yoshikawa, A., Yamaga, S., Tanaka, K. and Kasai, H., J.Crystal Growth 72:13 (1985)
27. Kisker, D.W., Fuoss, P.H., Krajewski, J.J., Amirthary, P.M., Nakahara, S. and Menendez, J., J.Crystal Growth 86:210 (1988)
28. Halsall, M.P., Wright, P.J. and Cockayne, B., – unpublished data

29. Gregory, T.J., Nicholls, J.E., Davies, J.J., Williams, J.O., Maung, N., Cockayne, B. and Wright, P.J.   Presented at '2nd European Workshop on MOVPE', St Andrews, UK, 19–22 June 1988 – to be published.
30. Sritharan, S, and Jones, K.A., J.Crystal Growth 66:231 (1984)
31. Mitsuhashi, H., Mitsuishi, I. and Kukimoto, H., Jap.J.Appl.Phys. 24:L864 (1985)
32. Kuznetzov, P.I., Shemet, V.V., Odin, I.N. and Novoseleva, A.V., Izv.Akad.Nauk. SSSR, Neorgan.Mater. 17:791 (1981)
33. Cockayne, B., Wright, P.J., Armstrong, A.J., Jones, A.C. and Orrell, E.D.,, J.Crystal Growth – to be published
34. Shtrikman, H., Raizman, A., Oron, M. and Eger, D., J.Crystal Growth 88:522 (1988)
35. Lau, C.K., Tiku, S.K. and Lakin, M.M., J.Electrochem.Soc. 127:1843 (1980)

OPTICAL STUDIES OF ZnXTe(X=Mn,Hg) ALLOYS

Philippe Lemasson[a] and Chau Nguyen Van Huong[b]

[a]Laboratoire de Photochimie Solaire/CNRS
2, Rue Henry Dunant
94320 Thiais, France

[b]Laboratoire d'Electrochimie Interfaciale/CNRS
1, Place Aristide Briand
92195 Meudon Principal Cedex, France

INTRODUCTION

ZnTe is a II-VI compound with an energy gap of ca. 2.3 eV at room temperature (RT) which crystallizes in the zinc-blende structure and presents a direct fundamental transition. Many different ternary and quaternary solid solutions based upon ZnTe are prepared by alloying with another II-VI compound like CdTe, CdSe, HgTe... or with a VI-VII material like MnSe or MnTe. In the present paper, we shall focus particularly on the ternary alloys $Hg_{1-x}Zn_xTe$ (HZT), especially for high Zn concentration ($0.5 < x < 1$) and $Zn_{1-x}Mn_xTe$ (ZMT) in the whole composition range. These alloys have received attention only in the recent past, especially HZT.

The growth of narrow gap HZT by liquid phase epitaxy[1] (LPE) and molecular beam epitaxy[2] (MBE) has been shown to be quite feasible and recently, good quality bulk HZT was obtained by the travelling heater method[3] (THM) in spite of important difficulties correlated with thermodynamic considerations[4]. Up to now, consequently, the number of experimental investigations of the physical properties of these alloys is small[5-8].

The growth of ZMT crystals is much easier and the upper limit for the solubility of MnTe in ZnTe is close to 80%. ZMT ($0 \leq x \leq 0.8$) belongs to the class of diluted magnetic semiconductors (DMS) in which the group II element of a II-VI compound is randomly replaced by Mn^{2+} ions. Among the DMSs, ZMT appears to be one of the less known examples of this class of semiconductor alloys. The magnetic properties[9], the optical properties[10] as well as the magneto-optical properties[11] have received attention.

Both HZT and ZMT crystallize with the zinc-blende structure in the whole composition range and the lattice parameter of the alloy follows Vegard's law[6,12]. The fundamental optical transition is direct and the corresponding energy E_0 ranges between 0 and 2.8 eV.

In the present paper, we propose an optical study of HZT and ZMT in which simultaneous photocurrent spectroscopy (PCS) and electroreflectance (ER) are performed in the electrolyte configuration. All the measurements are done at room temperature.

EXPERIMENTAL

HZT and ZMT samples were kindly donated by Dr. R. Triboulet (Labora-
toire de Physique des Solides, CNRS, Meudon, France). They are grown using
the THM and the Bridgmann technique, respectively. In both cases, the alloy
samples used as electrodes have a surface area of ca. 10 mm^2. Prior to electrode mounting, an ohmic contact is provided at the rear face and a gold
wire soldered to the contact. Further details have been given previously[5].
Electrodes are glued in epoxy resin in order that the front face only is
in contact with the electrolyte. The electrode surface is then mechanical-
ly polished to a mirror finish and before each experiment, the electrodes
are etched with a 1% bromine in methanol solution and transferred to the
electrolytic cell without contact with air.

In order to avoid possible reduction of dissolved oxygen, the reaction
products of which can decompose the electrode, electrolytes are deaerated
by oxy-free and dry nitrogen bubbling. In each case, the experimental conditions are carefully controlled in order that the electrode surface remains
unperturbed. Details concerning the electrochemical arrangement and experiments as well as the ER and PCS measurements have been given elsewhere[5,10g].

RESULTS

Electroreflectance

HZT. Since the energy range available experimentally extended from
1.2 to 3.5 eV, ER experiments in the fundamental gap region were possible
for Zn-rich alloys only (x > 0.6) and in the high energy range (around
the E_1 and $E_1 + \Delta_1$ transitions) for the whole composition range. Significant spectra are reported in fig.1. Such spectra can be fairly analyzed
using the three point method of Aspnes[13].

It should be mentioned that in spite of the absence of particular
treatment of the electrode surface, it is rather easy to obtain featured
ER spectra, especially in comparison with what is found for HgCdTe alloys[14].
The E_0, E_1 and $E_1 + \Delta_1$ energy values versus alloy composition are reported
in fig. 2. We notice that $E_0(x)$ is almost linear for $0.5 < x < 1$, in reasonably good agreement with previous results[7], whereas $E_1(x)$ and $(E_1 + \Delta_1)(x)$
obey an empirical parabolic law. We find:

$$E_1(x) = (2.15 + 0.76x + 0.69x^2) \text{ eV}$$

and

$$(E_1 + \Delta_1)(x) = (2.85 + 0.59x + 0.78x^2) \text{ eV}$$

A study with temperature should bring further information concerning
the physical signification of the various empirical parameters (coefficients of the x and x^2 terms) involved in such equations. Unfortunately, our
experimental set-up does not allow such an investigation due to the utili-
zation of an aqueous electrolyte for the fabrication of the Schottky barriers. However, the versatility of the electrolyte technique remains attrac-
ting, in particular for the investigation of new materials such as HZT, for
which it is important to obtain basic information about the optical properties in correlation with the conditions of crystal growth.

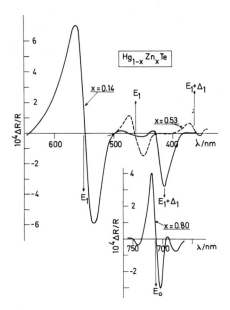

Fig. 1. Typical spectra of $Hg_{1-x}Zn_xTe$ alloys near the interband transitions E_1, $E_1 + \Delta_1$ (x = 0.14 and 0.53) and near the fundamental transition E_0 (x = 0.8).

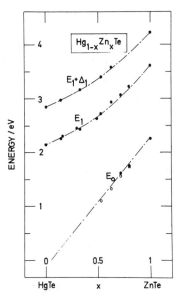

Fig. 2. Variations of the E_0, E_1 and $E_1 + \Delta_1$ energies versus the alloy composition of $Hg_{1-x}Zn_xTe$. Full circles ER data; open circles: PCS data.

ZMT. Typical ER spectra obtained for different alloy compositions are presented in fig.3. The main oscillation corresponds to the fundamental gap E_0 and can be fairly analyzed using the same method as for HZT (see above). The corresponding E_0 values deduced from this analysis are plotted in fig.4 versus the alloy composition. E_0 values are determined with good accuracy ($\pm$ 5 meV) and the linear plot corresponds to the empirical relation:

$$E_0(x) = (2.28 + 0.53x) \text{ eV}$$

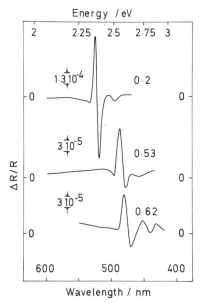

Fig. 3. ER spectra of ZnMnTe for different x values.

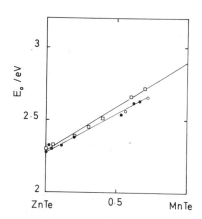

Fig. 4. Variation of the energy gap E_0 of ZMT versus x. Full circles: ER; open circles: PCS. For comparison are reported the results of Brun del Re et al.[10a] (open squares).

These results are in reasonably good agreement with previously published data[10a,10e] even if probably more reliable due to the precision of the present technique.

Photocurrent spectroscopy

In order to gain information not available by ER, photocurrent spectroscopy measurements are done at constant DC voltage both for HZT and ZMT. It should be noted that PCS results presented below are not normalized to take into account the wavelength dependence of the intensity of the light source, but this dependence is smooth and fairly flat throughout the wavelengths of interest.

HZT. Photocurrent spectroscopy is possible with the experimental set-up only for alloys with x > 0.5. The main spectra are shown in fig.5. The difference in shape for the various PCS is striking.

Whereas for x = 0.53, the onset of the photocurrent is abrupt, it is not so sharp and extends to a large energy range for x = 0.63, the sharpness of the onset increasing again for x = 0.72 and x = 0.80. Furthermore,

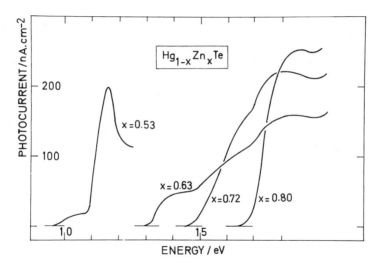

Fig. 5. Photocurrent versus photon energy for $Hg_{1-x}Zn_xTe$ alloys
with x = 0.53, 0.63, 0.72 and 0.80.

the photocurrent obtained on the plateau for x = 0.63 is smaller than for
the others. These observations tend to indicate that the sample with
x = 0.63 is not so homogeneous as the others. Thus a quantitative analysis of the corresponding spectrum must be carried out with care.

ZMT. Measurements are done, like for HZT alloys, at constant band
bending which corresponds to fixed DC voltage. The more significant spectra
are presented in fig. 6. We notice that, together with the increase of the
manganese concentration, the position of the steep increase in the photocurrent corresponding to the fundamental gap increases in energy.

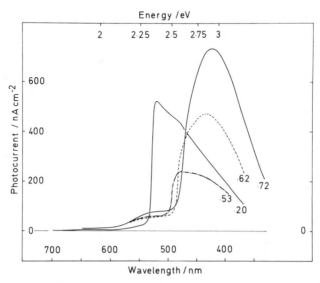

Fig. 6. Photocurrent spectroscopy of ZnMnTe
alloys for different compositions. Note the
increase of the shoulder around 600nm ($\sim$2.1
eV) together with the increase of Mn.

Further, a shoulder develops at photon energies lower than E_0, the beginning of this shoulder remaining fixed at ca. 2.05 eV. For manganese concentrations such as $0 \leq x \leq 0.72$, the relationship between E_0 and x appears to be almost linear as already pointed out in the ER experiments.

DISCUSSION

The electroreflectance experiments enable us to determine accurately the energy value of the various band-band transitions of a semiconductor. Further, the direct determination of the broadening parameter Γ of the experimental spectra gives some insight into the crystalline quality of the alloys. For ZMT, Γ lies in the range 30-40 meV, a value which is comparable to results obtained with bulk ZnTe and CdTe[15] thus indicating the good quality of the samples throughout the whole composition range.

The situation appears to be more complex for HZT. The oscillation due to the E_0 transition (x = 0.72 and x = 0.80) presents a broadening factor of ca. 30 meV, similar to those obtained with ZMT and we can conclude to a rather low defect density of these alloys[16]. Analysis of the oscillations corresponding to the E_1 transitions ($0 < x < 1$) yields Γ values which are much larger (100-130 meV) and comparable to what is obtained with the best mercury cadmium telluride samples, thus indicating that an improvement of the crystal growth should lead to even better results.

The possibility of correlating defect density with a physical parameter such as Urbach's slope was previously suggested[17]. It was shown that it was possible by analyzing the photocurrent spectra to determine Urbach's slope and to connect it qualitatively with the defect density. The larger the Urbach slope, the better the crystalline quality. According to Urbach's rule, the optical absorption coefficient in the absorption tail of a semiconductor obeys the analytical relation:

$$\alpha = \alpha_0 \exp\{ S(E_0 - h\nu)\}$$

where α is the absorption coefficient, α_0 a constant, E_0 the fundamental transition energy, $h\nu$ the photon energy and S the Urbach slope. It is easy to demonstrate[17] that in the absorption tail, the photocurrent can be expressed as:

$$i_{ph} = K\alpha_0 (W + L) \exp\{ S(E_0 - h\nu)\}$$

where K is a constant, W the space charge width and L the minority carrier diffusion length. By plotting $\ln i_{ph}$ versus $h\nu$, the S values are easily obtained. Such an analysis leads to S~ 100 eV^{-1}, 38 eV^{-1} and 54 eV^{-1} for x = 0.53, 0.72 and 0.80 respectively, indicating that the defect density is lower for x = 0.53 and larger for x = 0.72, the agreeement with ER experiments being fairly good.

Due to the existence of a shoulder on the photocurrent spectra obtained with ZMT, such an analysis appears to be impossible. The band-band transition $\Gamma_{8v} \rightarrow \Gamma_{6c}$ corresponding to E_0 is measured directly by the first oscillation on the ER spectra and is visualized by the steep increase observed in PCS. However, the existence of a shoulder at lower energy values which develops as the manganese concentration increases, cannot be attributed to a band-band transition. Absorption spectroscopy leads to curves $\alpha(h\nu)$ whose shape is similar to our PCS representations[10c,10e,10f]. The main problem in interpreting our results comes from the fact that in absorption, this shoulder can be attributed to the $^6A_1 \rightarrow {}^4T_1$ transition. This transition corresponds to the Mn^{2+} level as it appears to be general in Mn-based II-VI DMSs[18] and the upper level does not provide any possibility for the conduction of electrons. PCS currents correspond to electron-hole pairs generated inside the space charge region of the semiconductor, holes flowing through the solid to the ohmic contact (p-type conductivity), electrons being collected at the interface where they flow to the empty levels available in the electrolyte. In the case of currents existing at photon energies close to 2.2 eV, the interpretation of the phenomenon should be done with care. In the present investigations, it seems possible to assume that the observed feature can be attributed to a competition between intrinsic absorption and the intra-manganese transitions like in the case of CdMnTe with large Mn concentration[19].

The steep increase in PCS corresponding to the E_0 transition can be analyzed more quantitatively both for HZT and ZMT alloys. For an allowed direct transition, an assumption supported here by the existence of ER spectra in this energy range, the absorption coefficient is given by[20]:

$$\alpha(h\nu) = A \ (h\nu - E_0)^{1/2}$$

where A is a constant related with the hole and electron effective masses. Thus the square of the absorption coefficient is linear with photon energy and the extrapolated value allows E_0 to be determined, keeping in mind that at RT the deduced value is only approximate. In some limited cases, this treatment can be applied to photocurrent spectra. Actually, when the illumination level is weak and for large band bending, the photocurrent can be expressed as:

$$i_{ph} = K\alpha \ (W + L)$$

where K, W and L have the same meaning as before. For a given material, at constant voltage, i_{ph} is thus proportional to α.

In the case of HZT and ZMT, for photon energies higher than E_0, i_{ph}^2 is linear with $h\nu$. For HZT alloys, E_0 values found by this technique are 1.72, 1.55, 1.32 and 1.09 eV for x = 0.8, 0.72, 0.63 and 0.53 respectively. As long as the agreement is rather good with energies determined by ER, we can assume that E_0 values determined from PCS are reliable (fig.2). The same consideration holds for ZMT alloys (fig.4).

CONCLUSION

We have presented an optical investigation of mercury zinc telluride and zinc manganese telluride alloys. Electroreflectance and photocurrent spectroscopy are shown to bring complementary information about the optical transitions as well as the crystalline quality of the samples. Such an information is particularly important for HZT, an alloy grown only recently for which experimental data are lacking. The well-defined ER spectra show that generally good crystalline quality can be obtained. Both for HZT and ZMT, the energy values of the interband transitions determined by analysis of ER and PCS are reliable and lead to precise energy-composition diagrams.

REFERENCES

1. A. Sher, D. Eger and A. Zemel, Mercury zinc telluride, a new narrow gap semiconductor, Appl. Phys. Letters 46:59 (1985).
2. S. Sivananthan, X. Chu, M. Boukerche and J.P. Faurie, Growth of Hg_{1-x} Zn_xTe by molecular beam epitaxy on GaAs (100) substrate, Appl. Phys. Letters 47:1291 (1985).
3. R. Triboulet, A. Lasbley, B. Toulouse and R. Granger, Growth and characterization of bulk HgZnTe crystals, J. Crystal Growth 79:695 (1986).

4. A. Laugier, Thermodynamics and phase diagram calculations in II-VI and IV-VI ternary systems using an associated solution model, Rev. Physique Appl. 8:259 (1973).

5. C. Nguyen Van Huong, R. Triboulet and P. Lemasson, Electrochemical and electrooptical investigation of cadmium mercury telluride and zinc mercury telluride, SPIE, Materials Technologies for IR Detectors, 659:65 (1986); Electrochemical and electrooptical properties of $Hg_{1-x}Zn_xTe$, J. Crystal Growth 86:570 (1988).

6. B. Toulouse, R. Granger, S. Rolland and R. Triboulet, Band gap in $Hg_{1-x}Zn_xTe$ solid solutions, J. Physique (Paris) 48:247 (1987).

7. R. Granger, A. Lasbley, S. Rolland, C.M. Pelletier and R. Triboulet, Carrier concentration and transport in $Hg_{1-x}Zn_xTe$ for x near 0.15, J. Crystal Growth 86:682 (1988).

8. H. Mariette, R. Triboulet and Y. Marfaing, Alloy-trapped excitons in a new II-VI semiconductor solid solution $Hg_{1-x}Zn_xTe$, J. Crystal Growth 86:558 (1988).

9. (a). T. M. Holden, G. Dolling, V. F. Sears, J. K. Furdyna and W. Giriat, Spin correlations in $Zn_{1-x}Mn_xTe$ alloys, Phys. Rev.B 26:5074 (1982).

   (b). A. Manoogian, B. W. Chan, R. Brun del Re, T. Donofrio and J. C. Woolley, Electron spin resonance in $Cd_xZn_yMn_zTe$ alloys, J. Appl. Phys. 53:8934 (1982).

   (c). S. P. McAlister, J. K. Furdyna and W. Giriat, Magnetic susceptibility and spin-glass transition in $Zn_{1-x}Mn_xTe$, Phys. Rev.B 29:1310 (1984).

   (d). A. Wittlin, R. Triboulet and R. R. Gałazka, EPR studies of $Zn_{1-x}Mn_xTe$, J. Crystal Growth 72:380 (1985).

   (e). Y. Shapira, S. Foner, P. Becla, D. N. Domingues, M. J. Naughton and J. S. Brooks, Nearest-neighbour exchange constant and Mn distribution in $Zn_{1-x}Mn_xTe$ from high-field magnetization step and low-field susceptiblity, Phys. Rev.B 33:356 (1986).

10. (a). R. Brun del Re, T. Donofrio, J. Avon, J. Majid and J. C. Woolley, Lattice parameter and optical-energy gap values for $Cd_xZn_yMn_zTe$ alloys, Nuovo Cimento 2D:1911 (1983).

   (b). W. Giriat and J. Stankiewicz, Absorption edge in $Mn_xZn_{1-x}Te$, Phys. Status Solidi (b) 124:K53 (1984).

   (c). J. E. Morales Toro, W. M. Becker, B. I. Wang, U. Debska and J. W. Richardson, Identification of new absorption bands in $Zn_{1-x}Mn_xTe$, Solid State Commun. 52:41 (1984).

   (d). T. Donofrio, G. Lamarche and J.C. Woolley, Temperature effects on the optical energy gap values of $Cd_xZn_yMn_zTe$ alloys, J. Appl. Phys. 57:1932 (1985).

   (e). S. Ves, K. Strössner, W. Gebhardt and M. Cardona, Absorption edge of $Zn_{1-x}Mn_xTe$ under hydrostatic pressure, Phys. Rev.B 33:4077 (1986).

   (f). K. Hochberger, H. H. Otto and W. Gebhardt, Pressure and temperature dependence of localized $d^5 \rightarrow d^{5*}$ excitations in $Zn_{1-x}Mn_xTe$, Solid State Commun. 62:11 (1987).

   (g). P. Lemasson, C. Nguyen Van Huong, A. Benhida, J. P. Lascaray and R. Triboulet, Optical investigation of the diluted magnetic semiconductor $Zn_{1-x}Mn_xTe$, J. Crystal Growth 86:564 (1988).

11. (a). A. Twardowski, P. Swiderski, M. von Ortenberg and R. Pauthenet, Magnetoabsorption and magnetization of $Zn_{1-x}Mn_xTe$ mixed crystals, Solid State Commun. 50:509 (1984).

   (b). A. Golnik, A. Twardowski and J. A. Gaj, Interband magneto-absorption splitting as a function of magnetization in semimagnetic semiconductors, J. Crystal Growth 72:376 (1985).

   (c). G. Barilero, C. Rigaux, M. Menant, Nguyen Huy Hau and W. Giriat, Magnetization and magnetoreflectance in $Zn_{1-x}Mn_xTe$, Phys. Rev.B 32:5144 (1985).

   (d). M. C. D. Deruelle, J. P. Lascaray, D. Coquillat and R. Triboulet Magnetooptical measurements in $Zn_{1-x}Mn_xTe$ with high manganese concentration, Phys. Status Solidi (b) 135:227 (1986).

(e). J. P. Lascaray, M. C. D. Deruelle and D. Coquillat, Magnetization and magnetoreflectivity measurements in $Zn_{1-x}Mn_xTe$ with
$0.25 \leqq x \leqq 0.71$, Phys. Rev.B 35:675 (1987).

12. J. K. Furdyna, W. Giriat, D. F. Mitchell and G. I. Spiroule, The dependence of the lattice parameter and density of $Zn_{1-x}Mn_xTe$ on composition, J. Solid State Chem. 46:349 (1983).

13. D. E. Aspnes, Third derivative modulation spectroscopy with low field
electroreflectance, Surface Sci. 37:418 (1973).

14. C. Nguyen Van Huong, C. Hinnen, R. Triboulet and P. Lemasson, The semi-
conductor-electrolyte interface: surface and bulk properties of Hg_{1-x}
Cd_xTe ($0.6 \leqq x \leqq 1$), J. Crystal Growth 72:419 (1985).

15. P. Lemasson, C. Nguyen Van Huong, X. Chu, S. Sivananthan and J. P.
Faurie, Electrochemical and electrooptical study of the strained-
layer superlattice system: CdTe-ZnTe, J. Electrochem. Soc. 135:1282
(1988).

16. P. M. Raccah, J. N. Garland, Z. Zhang, U. Lee, D. Z. Xue, L. L. Abels,
S. Ugur and W. Willinsky, Comparative study of defects in semiconductors by electrolyte electroreflectance and spectroscopic ellipsome-
try, Phys. Rev. Letters 53:1958 (1984).

17. P. Lemasson, A. F. Boutry and R. Triboulet, The semiconductor-electro-
lyte junction: physical parameters determination by photocurrent spectroscopy throughout the $Cd_{1-x}Zn_xTe$ alloy series, J. Appl. Phys. 55:
592 (1984).

18. Y. R. Lee, A. K. Ramdas and R. L. Aggarwal, Origin of the Mn^{2+} optical
transition in Mn-based II-VI diluted magnetic semiconductors, Phys.
Rev.B 33:7383 (1986).

19. J. P. Lascaray, J. Calas, F. E. Darazi, M. Averous, R. Triboulet and
E. Janik, Photoconductivity experiments in $Cd_{1-x}Mn_xTe$ and comparison
with optical absorption for $0.1 < x < 0.7$ and $77 < T < 300$ K, J.
Crystal Growth 72:393 (1985).

20. J. I. Pankove, "Optical processes in semiconductors", Dover, New York
(1975).

ELECTRICAL AND STRUCTURAL PROPERTIES OF

WIDE BANDGAP II-VI SEMICONDUCTING COMPOUNDS

A. W. Brinkman

Applied Physics, SEAS
University of Durham
Durham, DH1 3LE   U.K.

1. Introduction

The II-VI semiconductors are composed of equimolar proportions of a group IIb element (Zn,Cd,Hg) and a group IVa element (S,Se,Te). They crystallise in either the cubic zincblende (sphalerite) or the hexagonal wurtzite structures, in which the metal ion is surrounded tetrahedrally by four chalcogen ions. Many of the important properties derive from the tetrahedral bonding and the valency. Band structure calculations, have established that the zinc and cadmium compounds are all direct semiconductors. Thus the need to conserve crystal momentum does not inhibit band-to-band radiative recombination and the II-VI semiconducting compounds are efficient emitters and detectors of light. It was as phosphors for CRT displays that these materials first found application.

The zinc compounds have the largest bandgaps with values of 2.25eV, 2.67eV and 3.66eV for ZnTe, ZnSe and ZnS respectively. CdS with a bandgap of 2.32 is the only other II-VI compound for which the bandgap energy is commensurate with visible light. The three zinc compounds crystallise with the sphalerite structure, while CdS adopts the wurtzite modification. Some of the II-VI compounds form mixed crystals of $A_x B_{1-x} C$ (A,B = Zn,Cd) and $M F_{1-x} G_x$ (F,G = S,Se,Te) types. The bandgap is found to vary in a quadratic manner with composition (x) in accordance with the empirical formula:

$$Eg(x) = xEg[AC] + (1-x)Eg[BC] - x(1-x)b$$

where b is a constant, termed the bowing parameter, and is a measure of the non-linearity of the band gap dependence. The lattice parameters vary linearly with composition for mixed anion compounds and for $Cd_{1-x} Zn_x Te$, but for $Cd_{1-x} Zn_x S$ and $Cd_{1-x} Zn_x Se$ there is a change from the wurtzite to the sphalerite phase at some intermediate composition. The ternary alloys are currently of much technological interest because they permit tuning of the bandgap and of the lattice parameters for matching to substrates.

The present paper will be concerned primarily with a consideration of structural aspects of the epitaxial growth of ZnSe and ZnS and the mixed anion compound $ZnS_x Se_{1-x}$ and with the electrical properties of the mixed cation system $Zn_x Cd_{1-x} S$. The former are of interest as materials for electroluminescent devices, the latter as a thin film solar cell material. For a comprehensive survey of the wideband gap II-VI semicon-

ducting compounds, the reader is referred to the very full review by
Hartmann et al[11].

2. Structural Properties of Binary II-VI Compounds

2.1 The Role of Dislocations

There is increasing evidence from various studies[2,3,4] that in
heteroepitaxial systems, the defect content of the layers is distributed
anistropically. Given the polar nature of the wide-gap II-VI
semiconductors with their sphalerite structure, differences in growth
rates on {111}A and {$\overline{11}$1}B planes are to be expected. When combined
with the known differences in the velocities of different dislocation
systems, the observed anisotropy in defect distribution seems
inevitable. Yet, this is seldom recognised explicitly in structural
studies, except in its macroscopic manifestations through etching
behaviour. The relative motion of dislocation networks lies at the
heart of this anisotropy and a knowledge of these processes is an
essential prerequisite to the understanding of how these defects can
arise.

The perfect dislocation a/2 <110> consists of a pair of extra {110}
half planes of atoms and has corresponding {111} glide planes, as
illustrated in fig. 1a. Close packing requirements mean that it is
energetically more favourable for the dislocation to proceed via two
sequential movements in complementary <211> directions. For example the
dislocation

$$a/2 \ <10\overline{1}> \ \rightarrow \ a/6 \ <2\overline{11}> \ + \ a/6 \ <\overline{2}11>$$

reaction is equivalent to a single a/2 [10$\overline{1}$] displacement and is energetically favourable. A consequence of this, is that a perfect a/2
<110> dislocation, consisting of a pair of extra half planes, will have
a tendency to break-up into two partial dislocations, each composed of a
single plane, which move independently through the lattice, as illustrated in fig. 1b. The region between the two partial dislocations will
as a result suffer slip along the relevant <211> direction and become a
stacking fault (fig. 1b).

The propagation of such partial dislocations on adjacent {111}
planes will give rise to microtwins[5], while the successive passage of
partial dislocations across adjacent pairs of {111} planes will result
in thicker twins[6]. The latter may also be formed through the interaction of twinning dislocations with other dislocations[6]. The a/2
<110> perfect dislocation takes two forms, depending on whether the
extra plane terminates at the slip plane with a line of group II atoms
(α dislocation) or a line of group VI atoms (β dislocation). An
immediate consequence of this, is that the two types of dislocation will
not, in general, move with the same velocity nor necessarily exhibit the
same stacking fault energy. In practice, the differential motion of α
and β dislocations in the sphalerite structure is strongly influenced by
factors such as the relative size of the anion and the cation, and
dopant species and concentration.

2.2 Epitaxial Growth of ZnSe on GaAs

GaAs is the most commonly used substrate for the epitaxial growth
of binary II-VI semiconductors. It is particularly suited to the growth
of ZnSe in that it has the closest lattice parameter of all the binary
systems. Even so there is a 0.27% mismatch, nearly twice that of the

AlAs/GaAs system (0.14%). The thermal expansion properties are not as close, and there is a difference of about 33% in thermal expansion coefficients, as compared with ∿19% for the AlAs/GaAs structure.

The general growth of heteroepitaxial systems has been well documented[7,8] and the effects of lattice mismatch on the structure of the epitaxial layer are comparatively well understood. A model for the nucleation and subsequent development of epitaxial growth has been formulated by Matthews[7]. The initial stages of growth are characterised by the complete registration of atomic layers across the interface, so that the epitaxial layer is constrained to adopt the lattice spacing and structure of the substrate. In consequence the layer will be in elastic strain. The associated strain energy density will increase with thickness until some critical thickness is reached, when misfit dislocations will form relieving some of the strain and lowering the total energy of the system. The lattice parameter of the layer then relaxes rapidly to a value close to that of the bulk crystal, although there may be some residual elastic strain. Matthews calculated the critical thickness by assuming that the total sum of the misfit strain energy and the misfit dislocation energy should be a minimum. For the ZnSe/GaAs regime this works out to be about 84nm[9]. An alternative approach is to equate the strain energy density with the misfit dislocation energy as suggested by People and Bean[10]. The critical thickness depends largely on the lattice mismatch and, not surprisingly, is greatest in those epitaxial structures that are most closely lattice matched.

The essential correctness of this approach for the ZnSe/GaAs regime has been verified in x-ray diffraction studies of ZnSe layers grown by both MOVPE and MBE on (100) GaAs[11,12,13,14]. In these studies, the critical thickness for ZnSe on GaAs was found to be ∿150 nm. Others have found a considerable reduction in the lattice parameter for ZnSe thicknesses in the vicinity of 200 nm[15]. For thicknesses less than the critical thickness the lattice parameter normal to the substrate is significantly greater than that of bulk ZnSe. Clearly the layer is in 2-dimensional compressive strain so that the lattice spacing in the interface matches that of GaAs. TEM observations[14] of these layers have shown that these films exhibited stacking faults lying on {111} planes and bounded by partial dislocations of the Frank type with Burger's vectors a/3 <111>. It was also observed that these faults occurred in pairs and lay preferentially on two {111} planes i.e. there was an anisotropic distribution of these stacking faults.

When the thickness of the ZnSe layer exceeds about 150 nm, the lattice parameter normal to the substrate plane reduces significantly, indicating that the strain is being relieved by misfit dislocation formation. The relaxation becoming more or less complete when the layer thickness reaches ∿ 1μm[13]. These layers are generally characterised by the formation of complete misfit dislocations with Burger's vector a/2 <110> and inclined at 60° to the dislocation line (and are thus usually referred to as being of the 60° type. The misfit dislocation network is observed to develop as the film thickness increases while stacking fault densities are observed to decrease. Yao et al[13] and Petruzello et al[14], also found that when the ZnSe layer thickness exceeded ∿ 1μm the lattice parameter normal to the substrate plane fell below the bulk crystal value, indicating that the layer was now in two-dimensional tensile strain. This was attributed to thermal stress. In principle, the misfit strain is relieved at the growth temperature (250 - 410°C) by dislocation formation and on cooling to room temperature, differential thermal contraction results in additional strain. Since the expansion coefficient for ZnSe (6.8×10^{-6} K^{-1} at 300K) is greater than that of GaAs (5.7×10^{-6} at 300K), the ZnSe layer will be in tensile strain.

Petruzello and co-workers[14] carried out a detailed TEM study of the misfit dislocations in their layers. Following the approach of Matthews[7] they identified three possible sources for the misfit dislocations. These were threading dislocations present in the sub-strate, Frank type partial dislocations and Frank-Read surface sources. Frank-Read surface sources were thought to be the predominant mechanism, since the great majority of the observed dislocations were of the 60° type, which glide along {111} planes from the surface to the interface. The partial dislocations of the Frank type were present in the thinner films, as a result of the stacking faults. They develop because of the low stacking fault energy of ZnSe (13.6mJ m^{-2} compared with 55 mJ m^{-2} for GaAs)[16] to relieve the compressive stress. The degree of relaxation will depend on the length of dislocation line that intersects the interface which for thin layers is necessarily small. Petruzello et al suggest that these evolve into perfect dislocations, possibly through the reaction

$$a/3 \ [111] \rightarrow a/2 \ [110] + a/6 \ [112]$$

where the first term represents a perfect dislocation and the second is a partial of the Shockley type which should escape from the lattice by gliding to the surface.

2.3 The Structure of Epitaxial ZnS on GaAs

In contrast to ZnSe there is little reported work on the growth of ZnS. In principle, ZnS may be grown on GaP[17,18] which is a good lattice match, but in practice GaAs is often[18,19] used as the sub-strate, because of its superior quality and availability. There is a considerable lattice mismatch of −4.3% between ZnS and GaAs and the ZnS will be in bi-axial tensile strain. Consequently, the relief of lattice strain through the formation of misfit dislocations etc. takes place in very much thinner layers than for ZnSe, and ZnS layers grown on GaAs are characterised by much higher densities of dislocations and micro-twins[18] as a result.

The distribution of these defects in ZnS/GaAs structures is found to exhibit substantial in-plane anisotropy. This was first studied by Stutius and Ponce[2] and has since been studied in detail by Brown et al[3,4] in ZnSe/ZnS/GaAs structures prepared by MOVPE for electro-luminescent studies[21,22]. These structures were found to have ridged morphologies in which the ridges lay predominantly along one <110> direction. This is illustrated in fig. 2 which shows a SEM micrograph from a ZnSe/ZnS/GaAs structure in which the thickness of the ZnS layer was ∿ 200nm and that of the ZnSe about 4μm. When examined by RHEED, these surfaces yielded two different patterns depending on whether the beam was incident along the [110] axis or the orthogonal [1$\bar{1}$0] axis as shown in fig. 3a and 3b. These orientations were established uniquely from etch pit studies on the reverse side of the GaAs substrates. It can be clearly seen in fig. 3b where the electron beam was parallel to the [1$\bar{1}$0] direction that there are extra spots at 1/3 a <111> positions. These indicate the presence in the layer of a high density of microtwins lying along {111} planes inclined at 54° to the interface. No such spots were apparent in the orthogonal [110] pattern, fig. 3a. These observations indicate that there is a strong anisotropy in the distribution of defects in the ZnSe overlayer and by inference in the ZnS layer as well.

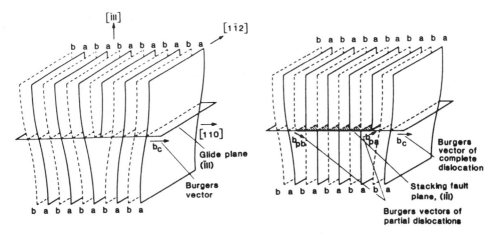

Fig.1. a) Perfect dislocation.   b) Partial dislocations.

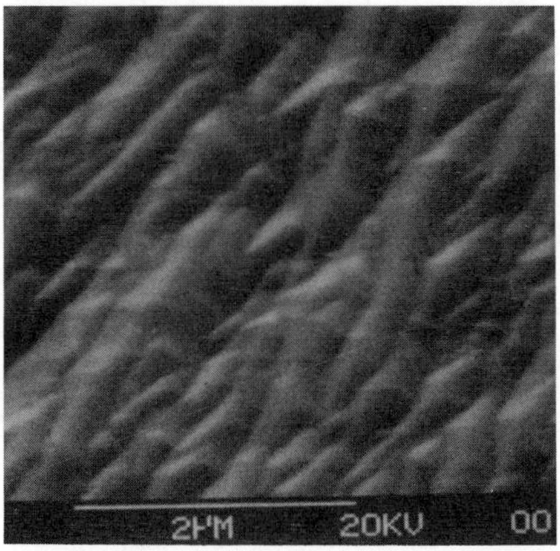

Fig.2. SEM ZnSe/ZnS/GaAs structure.

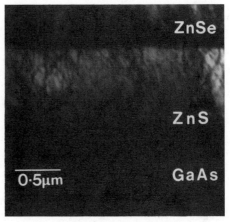

Fig.3a. RHEED [110] zone axis

Fig.3b. RHEED [1$\bar{1}$0] zone axis.   Fig.4a. XTEM [110] projection.

This was confirmed by extensive cross-sectional TEM studies. Fig. 4a shows a cross section through a ZnSe/ZnS/GaAs structure in the [110] projection, and fig. 4b shows the orthogonal [1$\bar{1}$0] projection. The latter is characterised by a large number of linear defects inclined at about 55° to the interface. The associated diffraction pattern contained extra spots in 1/3 a <111> position (as for the RHEED patterns) confirming these defects as microtwins. Most of them appear to be generated at the GaAs interface and continue through the ZnS/ZnSe interface, giving rise to the observed RHEED patterns. A network of misfit disloctions is also generated at the ZnSe/ZnS interface and is propagated through the layer. In contrast there is a notable absence of microtwins in the [110] projection (fig. 4a) and only misfit dislocations are found to initiate at the two interfaces. The microtwins, which were observed edge on in the [1$\bar{1}$0] projection are inclined to the electron beam in the [110] projection, and so give rise to features in the micrographs which exhibit parallel fringe contrast.

The anisotropic defect distribution and its relation to surface morphology is well illustrated in the low magnification cross sectional TEM micrograph of fig. 5. This sample consists of two sections cut along orthogonal <110> directions and glued face-to-face[3]. The top part of the micrograph shows the [110] projection and contains only dislocations. The lower part of the micrograph corresponds to the [1$\bar{1}$0] projection and shows the microtwins (edge on) running through the sample to the surface, which in this case shows the surface facets very clearly. Using the microtwins as a reference, the angles of intersection of the surface facets were measured and found to correspond to advancing {411}A and {511}A type planes. Detailed microdiffraction studies[4] of these samples allowed the sense of advancing {111} planes for each <110> projection to be determined unambiguously.

2.4 Anisotropy of Defect Distribution

Consider the Thompson tetrahedra in fig. 6, which illustrate the two non-equivalent groups of four {111} planes, associated with the sphalerite structure. One tetrahedra is bounded by four {111}A planes and the other by four {$\bar{1}\bar{1}\bar{1}$}B planes, where A refers to group II atoms and B to group VI atoms. When the model is viewed along the [110] direction, with growth along the [001] direction, then only the [1$\bar{1}$1]B and the [$\bar{1}$11]B have components along the [001] growth direction and are said to be advancing. When the tetrahedra are viewed along the orthogonal [1$\bar{1}$0] direction, the advancing planes are the [111]A and the [$\bar{1}$$\bar{1}$1]A planes. The important point is that the advancing B planes are only observed (edge on) along the [110] and the advancing A planes are only observed along the [1$\bar{1}$0] direction. Thus any planar defect structure associated exclusively with A planes alone, say, will be seen in [1$\bar{1}$0] projection only, and vice-versa.

Thus the microtwin structure seen only in the cross sectional TEM micrograph fig 4(a) in the [1$\bar{1}$0] projection may be associated with the {111}A planes. Epitaxial layers of ZnS on GaAs are in tensile strain, with extra half planes lying in the ZnS. Thus 60° dislocations of the α type would lie on advancing B planes while those of the β type would lie on the advancing A planes. In n-type ZnSe[23] and in n-type GaAs[24] α-dislocations are found to be much more mobile than β-dislocations, by a factor of ~100 in both instances. By inference, the same situation would be expected to apply in n-type ZnS. Thus, it is probable that the α-dislocations are swept rapidly through the epitaxial layer leaving perfect crystal behind them. In contrast the slowly moving β dislocations would be much more likely to dissociate with the leading partial being swept out to the ZnS/ZnSe interface, and hence giving rise to the microtwin formation.

The anisotropy in the defect distribution is thus seen to be a function of the differential motion of the α and β dislocations in the sphalerite structure. This in turn is affected by factors such as doping and the relative size of the cation and the anion. Since there is a large difference in the ionic radii of Zn^{2+} and S^{2-} ions, it is likely that this assymetry in the dislocation mobility is accentuated in ZnS.

3.0 Ternary II-VI Alloys

3.1 ZnS_xSe_{1-x}

ZnS_xSe_{1-x} is one of the more widely studied ternary wide bandgap II-VI alloys. The interest in this material lies primarily in the low x region where it is closely lattice matched to GaAs. The lattice parameter is normally assumed to follow Wegard's law which would give a perfect lattice match for GaAs at x=0.052[25]. At this Se rich composition the bandgap is expected to be 2.70 eV[1] not greatly different from that of ZnSe (2.67eV) and potentially the material is still useful for light emitting device applications.

ZnS_xSe_{1-x} has been grown successfully on a number of substrates including GaAs[25, 17, 18] GaP[18, 26] which is closely lattice matched at high x values and Ge[18]. The best results have been obtained with growth on GaAs[18], probably because wafers are of superior quality and the preparation of substrate surfaces is better. The growth of low-x ZnS_xSe_{1-x} of the lattice matched composition (x ≅ 0.05) on to GaAs (100) substrates gives[11] epitaxial layers that are of better quality than ZnSe. Kamata et al[11] have shown that for the lattice matched case, the crystalline quality of the layers, as determined from double crystal x-ray diffraction, increases with layer thickness, at least up to a thickness of ∿4µm and is in general superior to that of ZnSe. Kanda and co-workers[9] have established the range of critical composition for which 1µm thick films may be grown, without the development of misfit dislocations, to be 0.049 < x < 0.072. Although the stacking fault energy for ZnS is half that of ZnSe[16], Williams et al[18] have found that lattice matched ZnS_xSe_{1-x} layers are singularly free of stacking faults. This is in contrast to experience both with sub-critical thickness ZnSe layers[14] and bulk crystal ZnS_xSe_{1-x}[27] where relatively high densities of stacking faults are observed.

In their study Kanda et al[9] investigated the thermal stability of ZnSe and ZnS_xSe_{1-x} (0 < x < 0.12) by heating their samples to 600°C for periods of 5 hours, and then recording changes in the photoluminescence. They found that layers with composition outside of the critical range showed a significant increase in the Ga donor related luminescence peak after the heating treatment. This, they interpreted as indicating increased Ga diffusion along misfit dislocations. Since these were absent in material with the lattice matched composition, there was correspondingly no Ga donor related luminescence.

3.2 $Zn_xCd_{1-x}S$

$Zn_xCd_{1-x}S$ has been of considerable interest as the window material and n-type limb in Cu_2S based solar cell structures, where a composition with ∿20% Zn, is lattice matched to the Cu_2S[28]. Since these are, in general, intended to be low cost thin film structures, the material has not figured greatly in epitaxial studies, although there have been some[29]. However, very much more is known about the electrical properties, since these are of prime importance in the solar cell context.

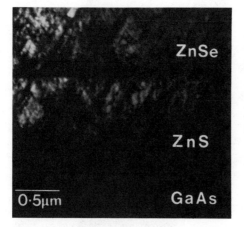

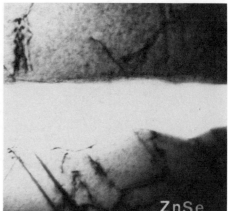

Fig.4b. XTEM $[1\bar{1}0]$ projection.   Fig.5. Orthogonal $\langle 110 \rangle$ projections.

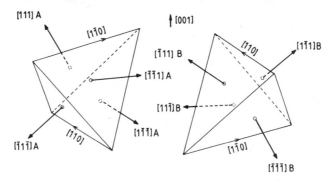

Fig 6. Thompson tetrahedra.

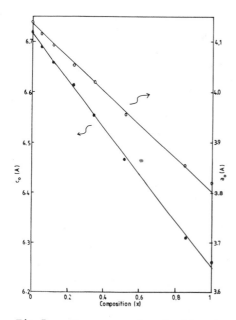

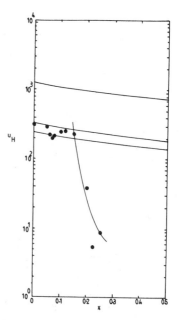

Fig.7. a & c vs. x for $Zn_xCd_{1-x}S$.   Fig.8. Mobility vs x for $Zn_xCd_{1-x}S$.

CdS normally crystallizes in the hexagonal wurtzite structure and ZnS in the sphalerite, so there must be some composition of the mixed crystal system for which the structure changes from the hexagonal to the sphalerite. The variation in structure and lattice parameter has been investigated for bulk crystals[30] and it is found that the transition does not occur until $x > 0.85$. The variation in the lattice parameter a_o and c_o with composition up to the transition is shown in fig. (7), and follows Wegard's law rather well with:

$$a = 4.136 - 0.34x$$
$$c = 6.705 - 0.46x$$

The bandgap varies quadratically and smoothly with composition with a bowing parameter of 0.61[3b] from 2.42eV for CdS to 3.66eV for ZnS. Typically measurements of conductivity, mobility and carrier concentration as a function of composition, measured at room temperature can only be made for $0 < x < 0.45$. At values of x greater than 0.45 the material becomes semi-insulating and reliable transport measurements cannot be made. The conductivity is found to fall uniformly with increasing x, as expected in line with the incease in band gap energy. The Hall mobility data is plotted against composition in fig. 8. The mobility is found to decrease only slightly from $\sim$290 $cm^2 V^{-1} s^{-1}$ at $x = 0.05$ to 235 $cm^2 V^{-1} s^{-1}$ at $x = 0.15$ but then fell dramatically at higher values of x to $\sim$6 $cm^2 V^{-1} s^{-1}$ at $x = 0.22$. The solid lines in fig 8 correspond to calculated curves obtained for (a) polar optical mode scattering[32], (b) piezoelectric scattering[32], (c) polar optical and piezoelectric modes combined according to Mathiessen's rule and (d) space-charge scattering[33]. In the low x region $x < 0.15$ the mobility appears to be well described by a combination of piezoelectric and possibly polar optical mode scattering.

At higher values of x the mobility is clearly being limited by other processes. Possible scattering processes include alloy scattering, space charge scattering and impurity ionisation scattering. Approximate calculations of mobility when limited by alloy scattering gave values for the mobility that were much too large. Space charge scattering limited mobility is given by[33];

$$\mu = \frac{3.2 \times 10^9 \, T^{-0.5}}{(m^*_e/m_o)^{0.5}(N_s A)}$$

where T is the temperature m^*_e is the electronic effective mass, μ_o the rest mass of the electron and $(N_s A)$ is the density - cross section product for the scattering centres. Although $N_s A$ was not known a curve could be fitted to the mobility data by assuming that $(N_s A)$ varied linearly with composition as assumed by Stringfellow[34] and Kaneco et al[35] in a study on GaAlAs. Adopting such a procedure resulted in the curve [d] in fig. 3.3, where values of $(N_s A)$ were of the order $10^7 \, cm^{-1}$. The temperature dependence of the mobility for samples with $x > 0.15$ suggested that at higher temperatures impurity ionization scattering was important but at lower temperatures (T < 200K) some other mechanism, possible space charge scattering was limiting the mobility.

4.0 Conclusions

The four case studies discussed above have illustrated the importance of lattice mismatch in the epitaxy of wide bandgap II-VI compounds. It is now well established that even in the comparatively well matched ZnSe/GaAs system dislocation free growth is only possible up to a thickness of about 150 nm[13,14] and that is not sufficient for device purposes, except for rather special applications[15]. Moreover, even within that thin a layer there is a high incidence of stacking faults[14]. When ZnSe layers are grown to the thickness generally required for device purpos ($\sim$ 1μm) the resulting misfit dislocation network[9] appears to provide enhanced cross diffusion from the substrate, of Ga[9], which in light emitting applications will result in yellow self-activated emission rather than the desired blue emission[36]. The implication being that while it may be possible to grow p-n structures in the laboratory[37], such structures may lack the required durability for large scale device processing.

In the hetero-epitaxy of II-VI compounds, dislocations play a fundamental role. In a system with a large mismatch such as the ZnS/GaAs case, the differential motion of α and β dislocations is shown to result in anisotropic distribution of planar defects, such as microtwins[3,4]. These studies also illustrate the importance of ensuring that the material characterisation is comprehensive and conclusive.

The use of ternary compounds to promote lattice matching has been shown to produce layers of ZnS_xSe_{1-x} of excellent quality and free of misfit dislocations at thicknesses in excess of 1μm[9,11]. These layers are also thermally stable and in that sense suitable for device processing. However the experience with $Zn_xCd_{1-x}S$ has shown that the use of ternary alloys can have serious implications for the electrical properties as additional scattering processes act to reduce carrier mobility[30,34]. Since only about 5% S needs to be included for lattice matching purposes, it seems likely that these difficulties may not be too severe.

An alternative approach to solving the lattice matching, and impurity diffusion problems associated with the use of foreign substrates, would be homoepitaxy of ZnSe on ZnSe bulk crystal substrates. The main difficulty here is due to the unavilability of high quality II-VI bulk crystal material. The low stacking fault energies of II-VI compounds lead to extensive twinning in bulk crystals making the production of large area wafers expensive. Homoepitaxy may well raise its own problems as has been found in CdTe[38] the only II-VI homoepitaxial system to have been studied.

Acknowledgements

The author wishes to thank Mr. P. D. Brown, Dr. K. Durose, Dr. G. J. Russell, Dr. M. K. Saidin and Professor J. Woods for their help with this paper.

REFERENCES

1. H. Hartmann, R. Mach & B. Selle in Current Topics in Materials Science Vol. 9, Ed. E. Kaldis, North Holland Amsterdam (1982).
2. W. Stutius and F. A. Ponce, J. Appl. Phys. 58 (1985) 1548.
3. P. D. Brown, A. P. C. Jones, G. J. Russell, J. Woods, B. Cockayne and P. J. Wright, Inst. Phys. Conf. Ser. 87 (1987) 3.
4. P. D. Brown, G. J. Russell and J. Woods, J. Appl. Phys. in press.
5. A. Lefebvre, Y. Androussi and G. Vandershaeve, Phys. Stat. Sol (a) 99 (1987) 405.
6. E. O. Hall, "Twinning and Diffusionless Transformations in Metals", Butterworth (1954), 116.
7. J. W. Matthews in "Epitaxial Growth, part B", Ed. J. W. Matthews, Academic, New York (1975) p.560.
8. P. M. Dryburgh, B. Cockayne and K. G. Barraclough (Eds.) "Advanced Crystal Growth, Part IV", Prentice Hall, New York (1987) p.289.

9. T. Kanda, I. Suemmune, K. Yamada, Y. Kau and M. Yaminishi, 4th Int. Conf. MOVPE, Hakone, Japan, 1988.
10. R. People and J. C. Bean, Appl. Phys. Lett. 47 (1985) 322.
11. A. Kamata, K. Hirahara, M. Kawachi and T. Beppu. Extended Abst. 17th Conf. Sol. State Dev. Mater. Tokyo, Japan (1985) 233.
12. J. Kleiman, R. M. Park and S. B. Qadri, J. Appl. Phys. 61 (1987) 2067.
13. T. Yao, Y. Okada, S. Matsui and K. Ishida, J. Crystal Growth 81 (1987), 518.
14. J. Petruzzello, B. L. Greenberg, D. A. Cammack and R. Dalby, J. Appl. Phys. 63 (1988) 2299.
15. G. D. Studtmann, R. L. Gunshor, L. A. Kolodziejski, M. R. Melloch, J. A. Cooper, Jr., R. F. Pierret, D. P. Munich, C. Choi and N. Otsuka, Appl. Phys. Lett 52 (1988) 1249.
16. S. Takeuchi, K. Suzuki, K. Maeda and H. Iwanaga, Phil. Mag., A50 (1984) 171.
17. S. Fujita, Y. Matsuda and A. Sasaki, J. Crystal Growth 68 (1984) 231.
18. J. O. Williams, T. L. Ng, A. C. Wright, B. Cockayne and P. J. Wright, J. Crystal Growth 68 (1984) 237.
19. T. Yao and S. Maekawa, J. Crystal Growth 53 (1981) 423.
20. S. Fujita, Y. Tomomura and A. Sasaki, J. Appl. Phys. 22 (1983) L583.
21. A. P. C. Jones, A. W. Brinkman, G. J. Russell, J. Woods, P. J. Wright and B. Cockayne, J. Crystal Growth 79 (1986) 729.
22. A. P. C. Jones, P. J. Wright, A. W. Brinkman, G. J. Russell, J. Woods & B. Cockayne, IEEE Trans. Electron. Dev., ED-34 (1987) 937.
23. Yu A. Osip'yan, V. F. Petrenko, A. V. Zaretskii and R. W. Whitworth, Adv. in Phys. 35 (1986), 115.
24. P. B. Hirsch, Phil. Mag., B52 (1985) 759.
25. W. Stutius, J. Crystal Growth, 59 (1982) 1.
26. J. Zhou, H. Goto, N. Sawasaki and J. Akasaki, Japan J. Appl. Phys. 27 (1988) 229.
27. J. R. Cutter, G. J. Russell and J. Woods, J. Crystal Growth 32 (1976) 179.
28. W. Palz, J. Besson, J. N. Duy and J. Vedel, Proc. 10th IEEE Photovoltaic Specialist Conf., Palo Alto, U.S.A. (1973) 69.
29. P. J. Wright, B. Cockayne and A. J. Williams, J. Crystal Growth 72 (1985) 23.
30. M. K. B. Saidin, Ph.D. Thesis, University of Durham (1987).
31. D. Howarth and E. H. Sondheim, Proc. Roy. Soc. A219 (1953) 53.
32. A. R. Hutson, Phys. Rev. Lett. 4 (1960) 505.
33. L. R. Weisberg, J. Appl. Phys. 33 (1962) 1817.
34. G. B. Stringfellow, J. Appl. Phys. 50 (1979) 4178.
35. K. Kaneco, M. Ayabe and N. Watanabe, Inst. Phys. Conf. Ser. 33a (1977) 216.
36. A. P. C. Jones, A. W. Brinkman, G. J. Russell and J. Woods, Semicond. Sci. Technol. 1 (1986) 41.
37. T. Yasuda, I. Mitsuishi and H. Kukimoto, Appl. Phys. Lett. 52 (1988) 57.
38. J. E. Hails, G. J. Russell, A. W. Brinkman and J. Woods, J. Appl. Phys. 60 (1986) 2624.

SOME ASPECTS OF IMPURITIES IN WIDE BAND GAP II-VI COMPOUNDS

H.-E. Gumlich

Institut für Festkörperphysik der TU Berlin
Hardenbergstr. 36, D 1000 Berlin 12

INTRODUCTION

Since the early days of research on luminescence of II-VI-compounds the impurities play an important role in our knowledge of the electronic structure of matter, of vibrational states and of energy transport in crystals. Moreover the incorporation of impurities and their special behaviour is decisive for most of the technical devices based on II-VI-compounds. Besides of the luminescence many new experimental techniques have emerged during the last decades and a wealth of new data has been generated, involving many branches of solid state physics, semiconductor technology and inorganic chemistry. Simultaneously the theorists have worked on novel theoretical techniques needed to describe localized impurities which interact with covalent bonds, including new cluster models, first principles pseudopotentials and self consistent Greens function approaches. Quite a number of reviews has been published during the last years in this field [Schulz 82 and 87, Zunger 86] dealing especially with transition metals in II-VI-compounds. The reader is refered to these publications. It is therefore not the aim of this paper to cover the field of impurities in II-VIs entirely, but to discuss some general trends and to focus the attention on phenomena, which mark the frontier of our knowledge of the interaction of the lattice with the impurities. It characterizes the situation that our knowledge of impurities giving rise to stable p- and n-doping is a more or less phenomenological one, whereas the insights into the interaction of the transition metal having incomplete 3d shell with the lattice are relatively quite advanced [Schlüter and Baraff 86].

n- AND p-CONDUCTION IN II-VI COMPOUNDS

Coming from the elemental semiconductors it is easy to conclude that the introduction of elements of a lower valence state relative to the Si or Ge, respectively gives rise to p-conduction, and vice versa the introduction of elements of a higher valence state causes n-conduction. This simple procedure does not work in II-VI compounds as far as the thermodynamical equilibrium is concerned.

At first glance it is generally admitted that the selfcompensation is to be blamed for: The II-VI-crystals fight back, when donors are introduced, they create native defects acting as acceptors and vice versa. These native defects are even present in the crystals, when no impurities were added. They can exist e.g. in form of point defects, and dislocations. The point defects belong to the thermodynamic equilibrium as point defects are vacancies, interstitial atoms and associates. The concentration of different defects depends on each other. General speaking a binary crystal MeX (Me: Metal, X: Nonmetal) tries to reach the state of the lowest Gibb's free energy $G = H - TS$ with H the Enthalpy, T the absolute temperature and S the Entropy. This is true for a given pressure and a given temperature. It implies that at all temperatures $T > 0$ there is a well defined disorder in the crystals.

The degree of disorder depends also on the energy which is necessary to create one defect: When the energy which is needed for the creation of one vacancy is large, the number of vacancies is small and vice versa. The Gibb's free energy per Mol in the equilibrium is therefore given by the pressure p, the temperature T and the composition x. Unfortunately it is quite difficult to calculate the free energy of disorder, the ionization energy and the other values, which are necessary to estimate the degree of disorder from basic principles. One is obliged to draw conclusions from experimental data. No distinction can be made between vacancies and interstitial atoms. However some simple considerations can help. As we all know we can vary the gas pressure in crystal growing when the crystals are heated. We obtain n-conducting ZnO, ZnS, ZnSe, CdO, CdS and CdSe when the crystals were fired under the vapour pressure of the nonmetal component ,provided the cation vacancies create acceptor levels near the valence band, the anion vacancies create donor levels near the conduction band. As a matter of fact ZnTe can be obtained p-conducting, but not n-conducting.

In this way the limit of n- and p-conduction is given by selfcompensation. Below this limit we should not obtain electric conductivity as we wish by doping in equilibrium. As Mandel et al. already pointed out [Mandel et al.64], the degree of selfcompensation by single charged vacancies is a function of the forbidden gap and the energy of cohesion E_{con} per gramm atome. As an equivalent of the energy of cohesion they take the half of the molar standard enthalpy of the reaction MeX = Me + X. Complete selfcompensation by simple vacancies would mean $E_g/E_{con} \gg 1$, little would occur if $E_g/E_{con} < 0.5$. The II-VI-compounds belong to the group were $E_g/E_{con} \approx 1$.

As already mentioned the degree of selfcompensation depends on the energetic distance of the level of the compensating vacancy to the neighbouring band. Assuming E_D is the energetic distance of the donor level to the conduction band, E_A is the distance of the acceptor level to the valence band the probability of selfcompensation is given by

$$W = \frac{E_g - E_A - E_D}{E_{con}}$$

This probability is getting more complicated, when the vacancies and impurities can be ionized twice, when they have double charge. In this case the position of the second ionization maximum is dominating the possible kind and number of lattice disturbances. The simplest model given by Mandel has the character of an F-center consisting of a sphere with the radius R in a surrounding dielectric. In a large distance of the cavity the effective fieldstrength is $F_{eff} = 2e/\varepsilon r2$, where e is the electric charge of an electron and ε is the effective dielectric constant. The energy within the sphere is supposed to be constant. This potential is superimposed by the potential of the trapped charge. In a first order approximation [Krumhansl and Schwarz 53] one finds that the time of stay of the charge within the trap and the ionization energy are linear functions of the radius R. Furthermore the ionization energy is a function of the dielectric constant which is dominated locally by the polarizability of the neighbours next to the vacancy and which is also a linear function of the diameter of the neighbours. The heigher this polarizibility is the smaller the ionization energy of the vacancy. Both actions of the atomic radii are opposite in view of the fact that either donors or acceptors dominate the conductivity (n- or p-conduction).

Assuming in a compound MeX the metal and therefore the Me vacancy have a large radius. This leads to a large second ionization energy of the compensating vacancy. Since the Fermi level which is situated near of the conduction band is dominated by the second ionization energy of the compensating vacancy, the crystal is n-conducting as long as the selfcompensation by single ionized vacancies is incomplet. On the other hand the strong polarizibility of the metal atom, which we assume in our example, leads to a low second ionization energy of the nonmetal vacancies forming a donor. This involves the total selfcompensation of p-conduction. The reverse is true if the radii of the nonmetal atoms are large compared with the radii of the metal atoms. In this case the p-conduction is favoured, but the n-conduction is prevented.

Of course this is a quite crude consideration, but it leads to some rules, that express the possible character of conduction by
1° the ratio of the energy of the forbidden gap Eg to the cohesion energy
2° the ratio of the atomic radii under covalent conditions.

Comparing these rules with the experimental findings in the II-VIs we see that in most cases either n- or p-conductivity can be realised. Only if $R_{Me} \approx R_x$, both types of conductivity can be obtained in the same crystal.

We should keep in mind that this is not the whole story, but describes the situation only in the thermodynamic equilibrium and if really the selfcompensation plays the dominant role what seems to be a controversial topic nowadays.

The theoretical situation for the II-VIs is quite poor relative to the elemental semi-conductors and to the III-V-compounds. On the other hand we can expect that methods applied successfully for the III-Vs may be used by now also for the II-VIs.

TRANSITION METALS

Besides the impurities, which control the p- and n-conductivity, the transition metals with incomplete 3d shell are very important. They are quite ubiquitous in the II-VIs, they have a direct influence on the lifetime of free carriers and on the luminescence intensity and moreover they give very detailed insights into the basic physics of II-VI-crystals. [Zunger 86], [Schulz 86], [Kamimura Watanabe 86.] Many experimental data are collected in the handbook of Landolt-Börnstein [82]. First we are going to discuss general trends and ideas and than we will illustrate the frontier of our experimental knowledge by some examples.

General Trends

The most important feature of the observed phenomena is the duality of the transition metals with an incomplete 3d shell. As Zunger summarized recently they show simultaneously atomically-localized and covalently-bonded characteristics [Zunger 86b]. He collects for the covalently delocalized model the following feature: 1° The rate at which ionization energies change with atomic number is about 10 times smaller than in free ions; 2° The Mott-Hubbard Coulomb repulsion energy for impurities E (o/ +) - E (o/-) is ~50 to 100 times smaller than for free ions; 3° The absolute impurity ionization potential taken as the difference between the host work function and the impurity energy with respect to the valence band is about 5-7 eV in all systems, which implies strong polarization and interaction with the media; 4° The hyperfine coupling constants in semiconductors are reduced relative to free ions or to 3d impurities in ionic compounds, what implies again strong covalency; 5° The angular momentum parts of the g-values are quenched, the spin-orbit splittings are reduced, what can be explained only by covalent effects; 6° Finally the Mössbauer isomershift of Fe in Si and III-Vs does not depend on the formal oxydation state, what he explains by a supply of s electrons of the host crystal, when an electron of the transition metal is removed. So far the arguments for a covalently delocalized model.

Arguments for a localized model which Zunger collects are the following. 1° The atomiclike multiplet structure is largely observed in solids and can be related to the structure of free atoms; 2° The ground states obviously obey Hund's rule. This means that the exchange effects; which increase with localization; are dominating, 3° The relatively small Jahn-Teller-effect suggests the dominance of localization-induced exchange interactions over symmetry-lowering elastic deformation; 4° The spin density is atomically localized, which is proved by ENDOR experiments; 5° The half shell stabilization d^5/d^4 is like that observed in free atoms; 6° The excitation and emission energies in zincblende and wurtzite structures differ only within the range of a few meV, which backs up the idea of a strong localization of the 3d wave functions.

Zungers theoretical approach to solve this duality is to describe them as atomically-localized and covalently delocalized at the same time. He starts from the well confirmed feature, that parts of the 3d shell of the transitions metals hybridize with the valence band of the host giving rise to maxima about 3.5 eV below the top of the valence band, whereas a weak shadow of the structure is observed within the band gap range. Most of the effective charge and local magnetic moment is contributed by the valence band resonance. It is important to note that symmetry considerations do not allow any coupling between the e-type orbitals and the nearest ligand atoms so the e-type orbitals remain localized. This is true for heavy atoms, when they resonate in the valence band and for light atoms in the band gap as well. By this the coexistence of localized and delocalized structures in the same crystals may be understood.

One essential point of this model is the answer of the host crystal on the ionization of 3d elements. In this case the charge around the impurity atom is reduced, but the valence band resonance wave functions respond by becoming more localized around the impurity, replacing in such way most of the charge by the lost gap level. This implies that the effective charge around the impurity atom remains nearly constant in different ionization states. It should be mentioned, however, that only direct Coulomb interactions are screened in that way, but not the magnetic moments. The band gap levels of the 3d atoms and the valence band levels have opposing contributions to the impurity charge, but both types contribute in the same direction to the local magnetic moment. This result predicts that spin densities would be localized in the central cell and will change with ionization far more than the impurity site charge density.

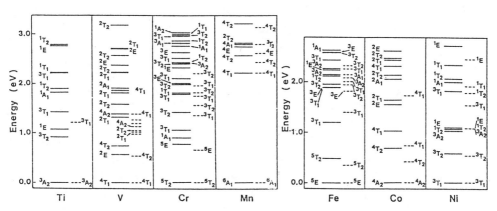

Fig. 1 Calculated (solid line) and observed (dashed line) multiplet structures for the transition metal impurities in ZnS. (Watanabe and Kamimura 1986).

Exchange interactions are hence affected far less by this screening. The survival of atomiclike groundstates, the validity of Hunds rule and the characteristic break in the trends of the ionization energies at the d^5/d^4 point are due to the small screening of exchange interactions. The model of Zunger can explain many "duality effects" of transition metals in II-VI-compounds but some predictions, however, are still to be tested.

As for chemical trends some rules seem to be valid. The most important rule seems to be the "vacuum refered binding energy-rule". As Caldas et al. pointed out [Caldas et al. 84] that if the donor and acceptor energies are referred to an intrinsic reference energy characteristic of the host, which might be the vacuum level, these "vacuum referred binding energies" are about constant for the same impurity in the different hosts. This knowledge can be used to calculate the intrinsic valence band offset between two semiconductors [Zunger 85; Langer and Heinrich 85].

Luminescence of Transition Metals

One of the most powerful tools to obtain insights into the microscopic structure of the transition metals in the II-VIs is the luminescence spectroscopy. All 3d transition metals show luminescence emission due to internal transitions within the 3d shell. The law of decay is predominantly a single exponential one, as far as the concentration of transition metals is low and no energy transfer processes are involved. The decay times reach from 0.3 µs to 1.7 ms [Goetz and Schulz 88] [Busse et al. 76]. The decay depends to a certain extend on the environment of the impurity. The spectral distribution of excitation and emission is influenced by the site of the transition metals too. Therefore, the site selecting time resolved spectroscopy is the best way to obtain some knowledge of the influence of the host on the 3d shell of the transition metals and vice versa.

In the following we concentrate as an example on the Mn^{2+} in the II-VIs. In the system $II_{1-x}Mn_xVI$ we can investigate the emission due to Mn concentrations as low as $x = 0.00001$. This allows for instance some insights into the polymorphic properties of the crystals. At higher concentrations ($0.001 < x < 0.01$) the Mn-Mn pairs produce special zerophonon lines.

At still higher concentrations ($x > 0.01$) some new luminescence features appear which obviously are due to a phase transition of the lattice.

Polymorphic Properties of $ZnS_{1-x}Mn_xS$

One of the most studied polymorphs is Zn, which can be grown with the zincblende structure (site symmetry T_d), the wurtzite structure (site symmetry C_{3v}) and with the zincblende structure with stacking faults; in this case cubic sites of T_d symmetry and several inequivalent sites of C_{3v} symmetry appear as a consequence of irregular sequences of closed -packed ZnS layers along the common $[111]_w$ axis of the cubic parts giving rise to shifts of the zerophonon lines relative to the zerophonon lines of Mn^{2+} on pure cubic Zn sites. Fig. 2 gives an example. As has been shown by

Parrot et al. [Parrot et al. 81] the Jahn-Teller effect is of primary importance to analyze the relatively large shifts of the centers of gravity of the $4T_1$, $4T_2$ and $4E$ levels of Mn^{2+} at axial sites with respect to the levels at cubic sites and to explain all observed fine structure patterns. As the luminescence emission due to the transition $4T_1$ (G) $\rightarrow$ $6A_1$ (S) of Mn can be excited by five excitation bands and their zerophonon lines ($4T_1$ (G), $4T_2$ (G), $4E$, $4A$ (G) T_2 (D) and $4E$ (D)), we can compare the shift of these lines with respect to the cubic lines. Fig. 3 summarizes the experimental findings. Having in mind that the difference of sites in cubic and wurtzite surroundings consists only in the action of a third neighbour, the shift of about 200 cm^{-1} is quite high.

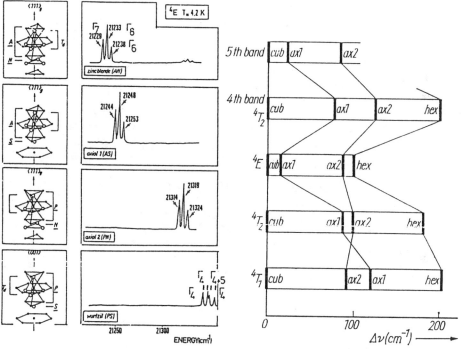

Fig. 2   Zerophonon lines of excitation of the transition $6A_1$ (S) $\rightarrow$ $4E$ (G) of Mn^{2+} in polymorphic ZnS. Emission due to the transition$4T_1$ (G) $\rightarrow$ $6A_1$ (S).

Fig. 3   Energy shift of Mn^{2+} levels in polymorphic ZnS with respect to Mn^{2+} at cubic sites (Benecke et al. 1985).

Obviously the combined action of several stacking faults introduces a variety of additional zerophonon lines, as has been shown by Busse et al. [Busse et al. 80]. It turns out that the zerophonon lines of Mn^{2+} are an effective tool for studying the polymorphic properties of wide band gap II-VI-compounds.

Mn-Mn Pairs

With increasing amount of Mn we reach the concentration range where we can expect an interaction of the Mn ions forming Mn-Mn pairs. After the pioneering work of McClure on ZnS:Mn [McClure 63] it was confirmed by time resolved spectroscopy that the zerophonon lines due to Mn-Mn pairs in $Zn_{1-x}Mn_xS$ are to be observed in the concentration range [Busse et al.76]. It should be mentioned, however, that while EPR measurements have confirmed the existence of Mn-Mn pairs in $Zn_{1-x}Mn_xS$ [Röhrig 73], no experiments are known demonstrating the appearance of these pairs by both luminescence and EPR-technique on the same crystals. Following the line of introducing sufficient Mn into ZnS, we should expect two species of Mn-Mn-pairs, because statistical considerations deliver for adequate concentrations two types of pairs for the case of 1nn and 2nn interaction range. By time resolved laser spectroscopy two types of pairs systems were discriminated. However within the framework of statistical considerations it is not possible to attribute unambiguously the systems to the 1nn or 2nn structure [Benecke et al. 85].

Quite a step forward in this field has been made very recently by Pohl [Pohl 88]. He calculated the probability distribution of impurity clusters in a zincblende lattice with an interaction range to the 16th nearest neighbours (16 nn) including singles, pairs and triples. This probability distribution is correlated to four zerophonon lines in emission and excitation, which are not due to the action of accidentally introduced impurities. Their intensity depends on the Mn concentration. These lines have the characteristic features of the $^4T_1$ (G) → $^6A_1$ (S) emission of the Mn^{2+}, namely the appearance of a thermalized line at temperatures above 4.2 K and the correct fingerprint of the $^6A_1$ (S) state in Zeeman experiments. Therefore there is no doubt that these lines are due to tetrahedrally coordinated Mn. The energetic distance relative to the zerophonon lines of isolated singles can be expres-

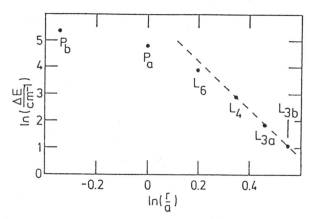

Fig. 4.   Energetic distance of the Mn2+ zerophonon lines of emission of pairs relative to the zerophonon lines of ioslated Mn2+ as a function of distances of neighbouring lattice sites, which can be attributed to the lines Pb to L6 as 1nn to 6nn. Double logarithmic scale

sed by relative series of lines with the isolated line as a limit of these series. It is interesting to note that the spin-orbit coupling of the new lines changes with respect to the value of the isolated Mn with increasing energetic distance. Therefore it is very likely that we are dealing here with the influence of Mn on the Mn atoms with different spatial distance between these atoms.

High Concentration Effects

When Mn is introduced into the II-VI compounds with $x > 0.01$ they become alloys $II_{1-x} Mn_x VI$. These semimagnetic semiconductors or diluted magnetic systems show a number of very peculiar effects which where reviewed recently by Brandt and Moshchalkov [85], by Gebhardt [87] and Goede and Heimbrodt [88]. One of the peculiar effects is the appearance of additional luminescence emission bands. While all wide band gap II-VI compounds with low amounts of Mn show the yellow emission due to the internal transition $4T_1$ (G) $\rightarrow$ $6A_1$ (S), additional low energy emission bands appear when the Mn exceeds $x = 0.01$. These long-wavelength emission bands are observed with $Zn_{1-x}Mn_xS$, $Zn_{1-x}Mn_xSe$, $Cd_{1-x}Mn_xTe$ and $Cd_{1-x}Mn_xS$. All these materials show an emission band within the infrared, some of them have also bands in the yellow and red range. Most of the data are available for $Zn_{1-x}Mn_xS$ which shows at high concentration X four emission bands. These bands are obviously connected by phonon assisted energy transport processes and depend drastically on temperature. The flux of energy can be registered by time resolved spectroscopy. The key to the understanding of this high concentration luminescence may be found in the time resolved excitation spectra. As a matter of fact the appearance of additional emission bands is accompanied by additional excitation bands in the UV and in the long-wave length part of the spectra [Benecke et al. 87]. The comparison of the additional excitation bands of the infrared emission of different II-VI-compounds shows a surprising similiarity. For example the infrared emission bands of $Zn_{1-x}Mn_xS$ and $Zn_{1-x}Mn_xSe$ are shifted 40 meV against each other. The same value is found for the shift of the additional excitation bands comparing both materials.

Some models have been proposed for this Mn emission in highly doped II-VI compounds [Vecchi et al. 81] [Becker et al. 84]. It turns out that these models include either quite questionable assumptions or cannot explain all the experimental findings. The most promising model follows the line of Goede et al. [88] who investigated the luminescence emission and absorption of MnS. Even if the numerical values do not yet fit perfectly, we assume tentatively that the infrared emission in II-VI-compounds with high amounts of Mn is due to the formation of Mn clusters leading to a phase transition into a rocksalt structure. This implies that parts of the Mn are no more tetrahedrally but octahedrally coordinated in MnS clusters. Of course this model, which still has to be confirmed by other experiments, might have some consequences in the discussion of semimagnetic semiconductors.

LITERATURE

W.M. Becker, R. Bylsma, M.M. Moriwaki, R.Y. Tao; Solid State Comm $\underline{49}$ (1984) 241

N.B. Brandt, V.V. Moshchalkov; Advances in Physics $\underline{33}$ (1984) 193

W. Busse, H.-E. Gumlich, B. Meißner, D. Theis; J.Lum. $\underline{12/13}$ (1976) 693

W. Busse, H.-E. Gumlich, W. Knaak, J. Schulze; J.Phys.Soc. Japan 49, Suppl. A, (1980) 581

C. Benecke, W. Busse, H.-E. Gumlich, U.W. Pohl; phys.stat.sol.(b) $\underline{128}$ (1985)

C. Benecke, W. Busse, H.-E. Gumlich, H.-J. Moros; phys.stat.sol.(b) $\underline{142}$ (1987)

C. Benecke, W. Busse, H.-E. Gumlich, H. Hoffmann, S. Hoffmann, A. Krost, H. Waldmann
(to be published 1988)

M. Caldas, A. Fazzio, A. Zunger; Appl.Phys.Lett. $\underline{45}$ (1984) 671

A. Fazzio, M. Caldas, A. Fazzio, A. Zunger; Phys.Rev.B $\underline{30}$ (1984) 3430

W. Gebhardt; "Excited state spectroscopy in Solids", p. III, Editrice Compositori,
Bologna 1987

N. Gemma; Journ.Phys. C $\underline{17}$ (1984) 2333

O. Goede, W. Heimbrodt; phys.stat.sol.(b) $\underline{146}$ (1988) 1

G. Goetz, H.-J. Schulz; J. Lum. $\underline{40/41}$ (1988) 415

B.S. Gourary, F.J Adrian in "Solid State Physics" 10 Ed. Seitz and Turnbull Academic
Press New York (1960) 127

H. Kamimura, S. Watanabe; 18th International Conference on the Physics of Semiconductors Stockholm, World Scientific, (1986) 979

J.A. Krumhansl, Schwartz; Phys.Rev. $\underline{89}$ (1953 1154)

Landolt Börnstein; New series Vol. 17 b, O. Madelung (ed.) Springerverlag Berlin (1982)

J.M. Langer, H. Heinrich; Phys.Rev.Lett. $\underline{55}$ (1985) 1414

G. Mandel, F.F. Morehead, P.R. Wagner; Report "II-VI-Laser Materials study, IBM
Watson Research Center, New York (1964), NR 017-903/4-5-63

D.S. McClure; J.Chem.Phys. $\underline{39}$ (1963) 2850

R. Parrot, A. Geoffroy, C. Naud, W. Busse, H.-E. Gumlich; Phys.Rev. B $\underline{23}$ (1981) 5288

U.W. Pohl; Doctoral Thesis, Technische Universität Berlin, Germany, D 83 (1988)

R. Röhrig; Doctoral Thesis, Freiburg (1973)

M. Schlüter, G.A. Baraff; 18th International Conference on the Physics of Semiconductors Stockholm,World Scientific, (1986) 793

H.-J. Schulz; Materials Chem. and Phys. $\underline{16}$ (1987) 373, Elsevier Sequoia

H.-J. Schulz; J.Crystal Grows $\underline{59}$ (1982) 85

M.P. Vecchi, W. Giriat, L. Videla; Appl.Phys.Lett. $\underline{38}$ (1981) 99

A. Zunger; 18th International Conference on the Physics of Semiconductors
Stockholm;World Scientific, (1986) p.21

A. Zunger, in H. Ehrenreich, F. Seitz and D. Turnbull (eds) Solid State Physics,
Academic Press, New York, Vol. $\underline{39}$ (1986) 275

A. Zunger; Phys.Rev.Lett. $\underline{54}$ (1985) 848

CONDUCTIVITY CONTROL OF WIDE GAP II-VI COMPOUNDS

Hiroshi Kukimoto

Imaging Science and Engineering Laboratory
Tokyo Institute of Technology
4259 Nagatsuda, Midori-ku, Yokohama 227, Japan

INTRODUCTION

The rapidly expanding field of optoelectronics, has relied on the establishment of the epitaxial growth technology for III-V compounds and related alloys. In particular, the light emitting devices, LEDs and laser diodes, which have been developed to date are made of epitaxial layers of GaAs, GaP, InP, GaAsP, GaAlAs, InGaP, InGaAsP and InGaAlP, for which the defect densities are at low level and the n- and p-type conductivity control is easily achievable. These materials have the bandgap energies ranging from 0.8 eV to 2.4 eV, and are suited for the devices in optical fiber communication systems operating at 1.3-1.5 μm, in laser printing and readout at 0.78 μm, and in displays from red to green. For the light emitting devices operating at shorter wavelengths, e.g., blue LEDs required for full-color displays to be used together with already available green and red LEDs and short-wavelength laser diodes for high density memory and printing systems, one must rely on the wider-bandgap materials. Wide gap II-VI compounds, such as ZnSe, ZnS and ZnSSe, have long been expected as candidates for the purposes.

The conductivity control of wide bandgap II-VI compounds has been a subject of great concern over the past years. For II-VI compounds, besides the impurity compensation due to impurities inadvertently incorporated in crystals, the so-called self-compensation tends to take place under thermal equilibrium conditions during the conventional high-temperature crystal growth. The situation of the self-compensation for electrons is illustrated at the upper part of Fig. 1. Suppose that two donor impurities are introduced in ZnSe, for example, and a zinc vacancy which forms as a double acceptor state is created. Then, two electrons are transferred to the acceptor states, and they do not any more contribute to the n-type conduction. If the energy to form a vacancy H_v is smaller than the energy gained by the electron transfer, roughly equal to $2E_g$, which is the case for wide gap II-VI compounds, the self-compensation would take place. The situation is also applicable to holes at doped acceptors in II-VIs. Recently, a growth technique of metalorganic vapor phase epitaxy (MOVPE), which features low temperature and non-equilibrium growth, has overcome the difficulty of conductivity control due to such self-compensation.

This paper is concerned with a fundamental principle of achieving n- and p-type controls of the wide gap II-VI compounds, based on our MOVPE growth and a preliminary result of its application for ZnSe blue p-n junction LEDs, followed by prospect of developing short wavelength lasers.

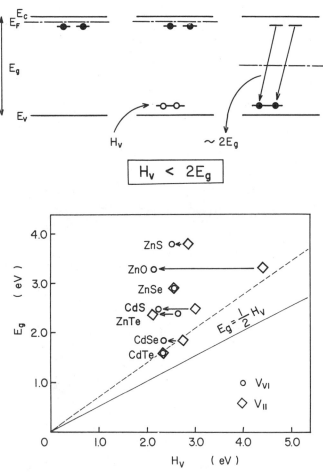

Fig. 1. A mechanism of self-compensation in II-VI compounds. The self-compensation becomes dominant if $H_v < 2E_g$.

MOVPE GROWTH OF ZnSe, ZnS AND ZnSSe

During the last decade, among wide gap II-VI compounds, ZnSe has been most extensively grown by MOVPE using dialkyl zincs (dimethyl or diethyl zinc, DMZn or DEZn) and hydrogen selenide (H_2Se).[1-6] A problem of the growth was a premature reaction taking place even at room temperature, resulting in unsatisfactory surfaces with respect to morphology and uniformity. Recently, this problem has been solved by adopting appropriate source material combinations such as dialkyl zincs (DMZn or DEZn) and dialkyl selenides (DMSe or DEZn) or adducts of dialkyl zincs with dialkyl selenides (DMZn-DMSe, DMZn-DESe, DEZn-DMSe or DEZn-DESe) and H_2Se.[7,8] These source combinations are obviously better than the conventional combination of dialkyl zinc and H_2S with respect to the reduction of premature reaction. By the use of these sources, the mass-transport limited growth which is a feature of MOVPE was also achieved. The examples of the growth rate behavior shown in Fig. 2 clearly indicate that a temperature-independent (and mass transport limited) growth rate region is present at temperatures higher than 500 °C for the growth using dialkyl zincs and dialkyl selenides. For the growth using adducts and H_2Se the temperature-independent growth rate

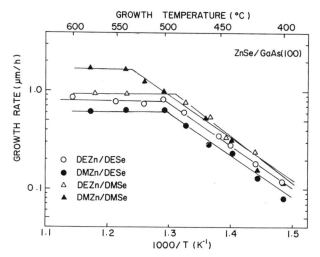

Fig. 2. Temperature dependence of growth rate for four kinds of source combinations, for the growth of ZnSe on GaAs by MOVPE at a reactor pressure of 300 Torr. Transport rates of dialkyl zincs and dialkyl selenides are 10 and 20 μmol/min, respectively.(Ref. 8)

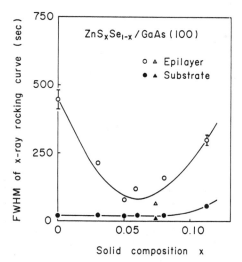

Fig. 3. Line width (full width at half maximum) of X-ray rocking curves as a function of composition x in ZnS_xSe_{1-x} epitaxial layers grown on GaAs(100) substrate.(Ref. 10)

region is located at lower temperatures, typically even at 300 °C. A similar argument about source combinations is applicable to the growth of ZnS.[9]

The further improvement in the layer quality, at the surfaces and at the epilayer-substrate interfaces, has been accomplished by growing the ZnS_xSe_{1-x} alloys with x=0.06 and x=0.83 lattice which are matched to GaAs and GaP substrates, respectively.[10,11] The line width of X-ray diffraction rocking curves indicates the improved crystalline quality for the lattice matched layers as shown in Fig. 3.

N-TYPE CONTROL

The key point to avoid the self-compensation is to reduce the growth temperature as already described. In view of this, the growth of n-type ZnSe has been carried out by using an adduct of DMZn-DESe and H_2Se as sources to reduce growth temperature as low as possible, typically down to 350 °C, and i-$C_8H_{17}Cl$, C_2H_5Br, C_2H_5I or trimethylaluminum(TMAl) as dopants. By using such appropriate dopants, high conductivity n-type ZnSe layers with carrier concentrations as high as $10^{19} cm^{-3}$ have been easily obtained for Al-, Cl-, Br-, or I-doped ZnSe layers grown by MOVPE at a typical temperature of 350 °C, as shown in Fig. 4. Above 400°C, it is difficult to achieve high conductivity; it has been evidenced for Al-doped ZnSe that the self-compensation becomes dominant, but for halogen-doped samples a decrease in impurity incorporation efficiency as well as self-compensation becomes important.

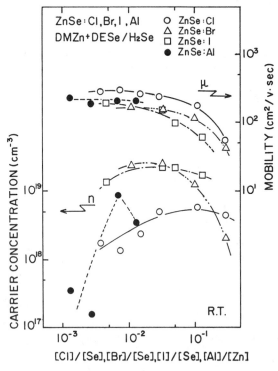

Fig. 4. Carrier concentration and mobility of n-type ZnSe doped with four kinds of dopant sources. For example, [Cl]/[Se] indicates the flow rate ratio of Cl dopant source to that of Se source.

High conductivity Al-doped ZnS layers have also been successfully grown by using an adduct of DEZn-DES and H_2S as source materials and triethylaluminum (TEAl) as the dopant,[9] as shown in Figs. 5 and 6. The electron concentration increases dramatically with temperature, reaches $8\times10^{18}cm^{-3}$ at 300°C, is independent of temperature between 300°C and 350°C, and tends to decrease above 350°C, as shown in Fig. 6. At temperatures lower than 300°C, the decomposition efficiency of TEAl increases, resulting in the increase of uncompensated Al donors with temperature. At temperatures higher than 300°C, the self-compensation start to take place, in view of the increasing intensity of self-activated photoluminescence (the so-called SA emission) which is ascribed to the transition between donors (Al) and acceptors (Zn vacancy and Al complex), resulting in a saturated and decreased carrier concentration with growth temperature.

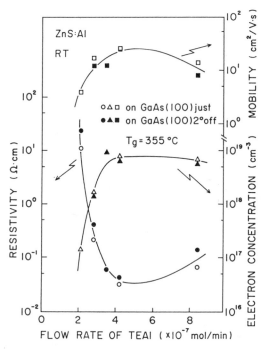

Fig. 5. Carrier concentration, mobility and resistivity of n-type ZnS doped with Al as a function of TEAl flow rate.

The growth of ZnS by using dialkyl zincs and dialkyl sulfides requires the growth temperatures as higher than 500°C. To our best knowledge it is impossible to grow high conductivity n-type ZnS at such a high growth temperature. This fact also confirms that low temperature growth is essential for achieving n-type control of wide gap II-VI compounds. The situation would also be the same for the p-type control.

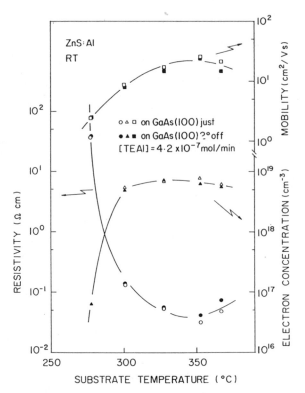

Fig. 6. Carrier concentration, mobility and resistivity of n-type ZnS doped with Al as a function of growth temperature. (Ref. 9)

P-TYPE CONTROL

Extensive studies have been performed in the past on p-type conductivity control by doping the group V elements of N, P and As,[12] but the results have given only high resistivity materials. In view of the fact that high quality epitaxial layers can be grown by MOCVD at high VI/II ratios, e.g., VI/II>10, we have tried to dope the group I_a elements. It has been shown that high conductivity p-type ZnSe layers with carrier concentrations up to $10^{18}cm^{-3}$ (a mobilty of about 40 cm^2/Vs, and a resistivity of 0.2 Ω cm) can be grown by doping Li using Li_3N as the dopant.[13] Te electrical properties of a lightly-doped p-type ZnSe layer are shown in Fig. 7. An activation energy of about 80 meV, estimated from the slope of the carrier concentration vs. 1/T straight line, is somewhat smaller than the reported Li acceptor depth of 114 meV. This suggests a decrease of hole binding energy due to the interaction among high concentration (compensated and uncompensated) acceptors.

The photoluminescence spectra of the samples grown by using Li_3N(Fig. 8), $LiN(CH_3)_2$ and cyclopentadienyl lithium (CpLi) as dopants indicates that the Li and N co-doping would play an important role in the attainment of high conductivity p type.

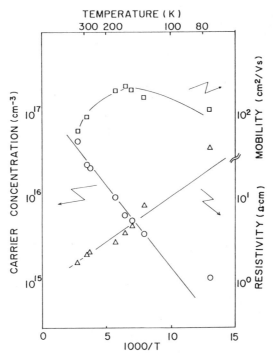

Fig. 7. Temperature dependence of carrier concentration, mobility and resistivity of p-type ZnSe.

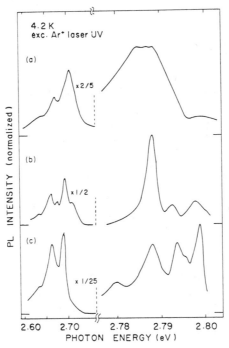

Fig. 8. Photoluminescence spectra at low temperature for (a) heavily doped ($p=3\times10^{16}cm^{-3}$), (b) lightly doped ($p<10^{15}cm^{-3}$) and (c) undoped ZnSe layers.(A part of the spectra is in Ref. 13.)

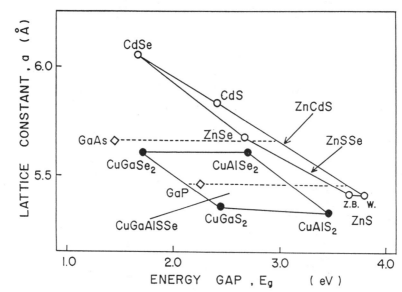

Fig. 9. Lattice parameters and bandgap energies plotted for wide gap II-VI and I-III-VI$_2$ compounds and related alloys.

APPLICATIONS AND FUTURE PROSPECTS

On the basis of these results we have recently demonstrated blue emission from ZnSe p-n diodes fabricated on the GaAs substrate[13]. The emission band has a spectral peak at 467 nm and a long tail in the longer wavelength region. The further improvement in efficiency, spectrum, and quality of junction interface is expected by optimizing growth conditions and device fabrication processes.

One of the most attractive applications of the conductivity-controlled ZnSSe system is to fabricate the laser structure in conjunction with wide gap chalcopyrite compounds, since epitaxial layers of $CuAl_xGa_{1-x}S_2$ and $CuGa(S_xSe_{1-x})_2$ including end materials of $CuAlS_2$, $CuGaS_2$ and $CuGaSe_2$ have already been grown by MOVPE on GaAs and GaP substrates.[14,15] The lattice parameter and bandgap energy plotted for wide-gap II-VI and chalcopyrite compounds, which is shown in Fig. 8, would be helpful for designing the double heterostructure lasers operating in the visible region from green to blue.

REFERENCES

1) W. Stutius, Organometallic vapor deposition of epitaxial ZnSe films on GaAs substrates, Appl. Phys. Lett. 33:656 (1978).
2) P. Blanconnier, M. Cerclet, P. Henoc, and A. M. Jean-Lois, Growth and characterization of undoped ZnSe epitaxial layers obtained by organometallic chemical vapour deposition, Thin Solid Films 55:375 (1978).
3) P. J. Wright and B. Cockayne, The organometallic chemical vapour deposition of ZnS and ZnSe at atmospheric pressure, J. Cryst. Growth 59:148 (1982).
4) F. A. Ponce, W. Stutius, and J. G. Werthen, Lattice structure at ZnSe-GaAs heterojunction interfaces prepared by organometallic chemical vapor deposition, Thin Solid Films 104:133 (1983).

5) S. Fujita, Y. Matsuda, and A. Sasaki, Growth and properties of undoped n-type ZnSe by low-temperature and low-pressure OMVPE, Jpn. J. Appl. Phys. 23:L360 (1984).

6) A. Yoshikawa, K. Tanaka, S. Yamaga, and K. Kasai, Effects of [H_2Se]/ [DMZn] molar ratio on epitaxial ZnSe films grown by low-pressure MOCVD, Jpn. J. Appl. Phys. 23:L773 (1984).

7) H. Mitsuhashi, I. Mitsuishi, M. Mizuta, and H. Kukimoto, Coherent growth of ZnSe on GaAs by MOCVD, Jpn. J. Appl. Phys. 24:L578 (1985).

8) H. Mitsuhashi, I. Mitsuishi, and H. Kukimoto, Growth kinetics in the MOVPE of ZnSe on GaAs using zinc and selenium alkyls, J. Cryst. Growth 77:219 (1986).

9) T. Yasuda, K. Hara, and H. Kukimoto, Low resistivity Al-doped ZnS grown by MOVPE, J. Cryst. Growth 77:485 (1986).

10) H. Mitsuhashi, I. Mitsuishi, and H. Kukimoto, MOCVD growth of ZnS_xSe_{1-x} epitaxial layers lattice-matched to GaAs using alkyls of Zn, S and Se, Jpn. J. Appl. Phys. 24:L864 (1985).

11) I. Mitsuishi, H. Mitsuhashi, and H. Kukimoto, MOCVD growth of ZnS_xSe_{1-x} epitaxial layers lattice-matched to GaP substrates (to be published)

12) R. N. Bhargava, Materials growth and its impact on devices from wide band gap II-VI compounds, J. Cryst. Growth 86:873 (1988), and references therein.

13) T. Yasuda, I. Mitsuishi, and H. Kukimoto, Metalorganic vapor phase epitaxy of low-resistivity p-type ZnSe, Appl. Phys. Lett. 52:57 (1988).

14) K. Hara, T. Kojima, and H. Kukimoto, Epitaxial growth of $CuGaS_2$ by metalorganic chemical vapor deposition, Jpn. J. Appl. Phys. 26:1107 (1987).

15) K. Hara, T. Shinozawa, J. Yoshino, and H. Kukimoto, MOVPE growth and characterization of I-III-VI_2 chalcopyrite compounds, J. Cryst. Growth (in press) (1988).

PHOTOASSISTED DOPING OF II-VI SEMICONDUCTOR FILMS

J.F. Schetzina, N.C. Giles, S. Hwang, and R.L. Harper

Department of Physics
North Carolina State University
Raleigh, NC

ABSTRACT

Photoassisted molecular beam epitaxy (PAMBE), in which the substrate is illuminated during film growth, is being employed in a new approach to controlled substitutional doping of II-VI compound semiconductors. Substitutional doping of these materials has been a long standing problem which has severely limited their applications potential. The PAMBE technique gives rise to dramatic changes in the electrical properties of as-grown epilayers. In particular, highly conducting n-type and p-type CdTe films have been grown using indium and antimony as n-type and p-type dopants, respectively. Double-crystal x-ray rocking curve data indicate that the doped epilayers are of high structural quality. Successful n-type doping of CdMnTe, a dilute magnetic semiconductor, with indium has also been achieved. Most recently, the photoassisted growth technique has been employed to prepare doped CdMnTe-CdTe quantum well structures and superlattices. In addition, HgCdTe films which exhibit excellent optical and electrical properties as well as exceptional structural perfection have been grown by the PAMBE technique.

INTRODUCTION

The II-VI compound semiconductors feature bandgaps ranging from the near UV to the far IR as shown in Figure 1. As a consequence, these materials could be used in a wide variety of electronic and optoelectronic devices. However, their use in device applications has been limited to date because of severe problems that have been encountered in attempts to control their electrical properties through substitutional doping with selected impurity atoms.

At North Carolina State University, we have developed a new technique for substitutional doping of II-VI semiconductor films, photoassisted MBE (PAMBE), in which the substrate is illuminated during the deposition process as shown in Figure 2. In our initial studies [1-6], an argon ion laser operating with broad band yellow-green optics (488.0-528.7 nm) was used as an illumination source. However, we have recently shown that conventional (incoherent) light sources can also be used in certain applications. The laser power density at the substrate during film deposition is approximately 150 mW/cm^2, corresponding to a photon flux density of about 4×10^{17} cm^{-2} s^{-1}. Thus, for deposition rates of 1-2 Å/s, the impinging

129

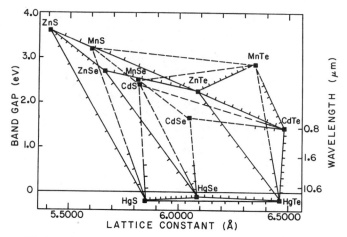

Figure 1. Bandgap versus lattice constant for selected II-VI semiconductors.

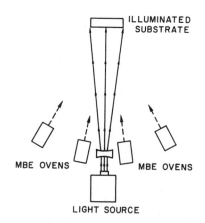

Figure 2. Photoassisted molecular beam epitaxy film growth process.

photon flux density is several orders of magnitude greater than the vapor impingement rate at the film growth surface. Using the method of Uzar et al. [7], we estimate the increase in substrate surface temperature due to the incident illumination to be less than 5 $^{\circ}$C for the laser intensities employed. Thus, it would appear that photochemical reactions, rather than purely thermal effects, play a dominant role in the PAMBE process.

The rationale behind the PAMBE technique is to provide high energy, low momentum particles (photons) at the film growth surface which can influence surface chemical reactions during the deposition process. Effects which may result from the introduction of light include enhancement of the surface mobility of constituent atoms, modification of surface bonds, conversion of surface molecules into atoms, and modification of the electrical potential of the surface through generation of photoexcited carriers.

In the present work, the photoassisted MBE process was employed to successfully grow n-type CdTe:In films [1]. The resulting as-grown layers show essentially 100% activation of the dopant species, as determined by Hall effect and SIMS measurements, when optimum growth parameters are employed. P-type CdTe:Sb films have also been successfully prepared [2]. More recently, the photoassisted MBE technique has been used to grow doped CdMnTe films and CdMnTe-CdTe superlattices [3], as well as HgCdTe films of high structural perfection. Thus, it appears that the PAMBE technique can be applied to a variety of materials to enhance their properties.

The photoassisted MBE technique gives rise to dramatic changes in the point defect character of the growing epilayers. In addition to being electrically conducting, the doped samples exhibit photoluminescence (PL) spectra which reflect these changes. The PL results provide evidence that the new growth technique substantially reduces the density of compensating defects in these doped films. Compensation effects in II-VI semiconductors play a major role in processes which greatly reduce the percentage of dopant atoms which are activated in these materials [8].

EXPERIMENTAL DETAILS

The substitutionally-doped films and superlattices were grown in an MBE system that has been described in previous publications [9, 10]. The MBE machine was modified as shown in Figure 2 to permit illumination of the substrate during the film growth process. Four MBE effusion ovens are employed for the growth of doped CdTe films. Two of the cells contain high purity polycrystalline CdTe while the other two contain, respectively, elemental In and Sb or As. The impurity content in the doped layers is controlled by varying the dopant oven temperature. For deposition of doped CdMnTe epilayers, Mn rather than a second dopant is used in one of the MBE ovens. Doped CdMnTe:In-CdTe superlattices were prepared by opening and closing the appropriate source shutters in a cyclic fashion. The HgCdTe films were prepared in a second MBE machine, designed and built at NCSU specifically for growth of Hg-based materials.

Chemimechanically polished (100) CdTe wafers, or (111) CdZnTe wafers with 4% Zn, were used as substrates. The substrates were degreased using standard solvents, etched in a weak bromine-in-methanol solution, and rinsed in methanol. Finally, just prior to loading into the MBE system, the substrates were dipped in hydrochloric acid and rinsed in deionized water. Substrate temperatures T_s ranging from 160-320 $^{\circ}$C were used for film growth.

Electrical characterization of the samples consisted of van der Pauw Hall effect measurements over the temperature range 20-300 K. Ohmic contacts to the n-type samples were obtained using an indium-based solder. Gold was used as the contact material for the p-type CdTe samples, although non-ohmic behavior was often encountered.

131

Double-crystal x-ray rocking curves for selected samples were obtained using a Blake Industries double-crystal diffractometer which employs a Phillips model 1729 x-ray generator with a Cu x-ray source. A high quality InSb wafer was used as the first crystal in the diffractometer. The instrument has a rocking curve FWHM resolution of approximately 1 arc sec. Rocking curves for PAMBE-grown samples were generally obtained using an incident x-ray beam with a 1 mm x 1 mm cross-sectional area.

Low temperature PL spectra were obtained with the samples mounted in a Janis SuperVaritemp dewar. The chopped beam from a He-Ne laser (5 W/cm^2) was used as an excitation source and the luminescence was measured with an ISA model HR-640 grating spectrometer equipped with a cooled S-1 photomultiplier tube. A lock-in amplifier was used to detect the photoluminescence signal.

CdTe FILMS

Illumination of the substrate during the growth process was found to cause a profound change in the electrical properties of CdTe epilayers. Only those films grown under illumination were conducting. All others were found to be semi-insulating. Undoped CdTe films grown by PAMBE are n-type and exhibit high electron mobilities and low carrier concentrations at low temperatures. This behavior is illustrated by the Hall data shown in Figures 3 and 4 for three different CdTe films grown by PAMBE. Note, in particular, that the electron mobility increases with decreasing temperature reaching a value as large as 6,600 cm2/v.s for film sample M291. In addition, the undoped films show a high degree of structural perfection as manifested by very narrow x-ray rocking curve FWHMs (Figure 4).

The best n-type doping results were obtained for CdTe:In samples grown at T_s = 230 $^\circ$C, although samples grown at temperatures as low as 180 $^\circ$C were highly conducting. By using In oven temperatures ranging from 375-450 $^\circ$C from run to run, n-type CdTe:In films having room temperature carrier concentrations of 2 x 10^{16} - 6 x 10^{17} cm^{-3} were obtained. Room temperature mobilities ranged from 450-800 cm^2/V·s and increased with decreasing temperature. An example of this behavior is shown in Figure 5 in which Hall data for a 0.6 μm thick CdTe:In epilayer is shown. The highest mobility of 2,400 cm^2/V·s exhibited by this film (BCTCT12) occurs at a temperature of approximately 80 K. Also shown in the figure is an x-ray rocking curve obtained for a CdTe:In film (M265) of thickness 0.4 μm. Note the exceptionally narrow FWHM(400) = 18 arc sec, indicative of high

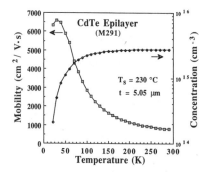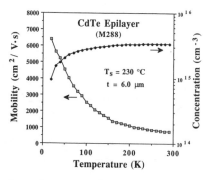

Figure 3.   Hall effect data for two CdTe films grown by PAMBE.

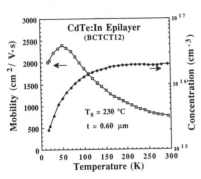

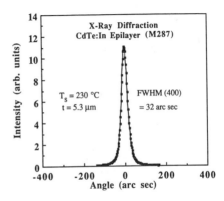

Figure 4.   Hall effect and x-ray diffraction
data for CdTe film M287.

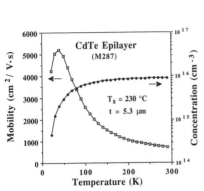

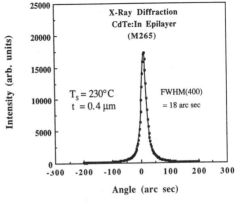

Figure 5.   Hall effect and x-ray diffraction
data for CdTe:In films

structural perfection. While it is true that the Cu K_α x-ray penetration
depth in CdTe is about 2 μm, so that the narrow rocking curve recorded for
M265 is due to diffraction from both the epilayer and substrate, the result
is important since it implies that, for thin layers, the PAMBE technique
produces epitaxial films which replicate the structural quality of the
substrate. As a consequence, PAMBE-grown CdTe:In layers of thickness
0.4-0.8 μm were recently able to be used to fabricate the first CdTe metal-
semiconductor field effect transistors (MESFET's) [11].

Low temperature PL spectra commonly observed for In-doped CdTe grown
by conventional MBE, which is often semi-insulating as grown, is dominated
by defect band emission at about 1.4 eV (Figure 6). In sharp contrast to
this, we observe spectra from conducting CdTe:In films grown by photoassisted MBE where the defect band emission is completely absent, as is shown
in Figure 6. The near edge emission region at 1.6 K reveals several sharp
bright emission lines which can be identified as free exciton recombination
(X), donor-bound (D,X) and acceptor-bound (A,X) excitonic transitions, and
donor-valence band recombination (D^o,h). On the basis of a detailed analysis of PL spectra from conducting CdTe:In films grown by PAMBE, we estimate the ionization energy of an In donor on an activated lattice site to

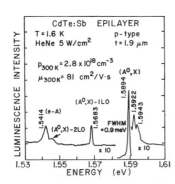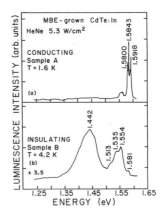

Figure 6. Photoluminescence spectra for CdTe:In grown by PAMBE and conventional MBE (left) and PAMBE-grown CdTe:Sb (right).

be E_D = 14.8 meV, in excellent agreement with the predictions of the hydrogenic donor model.

CdTe:Sb Films. Antimony has been successfully employed as a dopant to grow p-type CdTe films using the photoassisted MBE technique. At 300 K, Hall measurements yield hole mobilities of 81 $cm^2/V \cdot s$ for epitaxial CdTe:Sb layers having hole concentrations of 2.8 x 10^{18} cm^{-3}. For comparison, the highest reported Hall mobilities for holes in bulk p-type CdTe at 300 K are about 70 $cm^2/V \cdot s$ [12].

In contrast to the low temperature PL spectra for conducting CdTe:In films, which show donor-associated transitions in the near-edge region, the low temperature PL spectra from p-type CdTe:Sb epilayers are dominated by acceptor-related emission lines. Figure 6 shows a 1.6 K PL spectrum for a CdTe:Sb film which covers the near-edge energy region. The narrow dominant PL line at 1.5894 eV is due to acceptor-bound (A,X) excitonic transitions. We believe the acceptor is Sb. A phonon replica of this emission line is observed at 1.5683 eV.

CdTe:As Films. Recently, the photoassisted MBE technique has been employed at NCSU to grow highly conducting p-type CdTe films doped with arsenic. In these experiments the substrate temperature was varied between 180 °C to 230 °C. 180 °C was found to be the optimum temperature for growth of highly conducting CdTe:As films. Using this substrate temperature and an As oven temperature of 200 °C, room temperature hole concentrations as large as 6 x 10^{18} cm^{-3} have been achieved. This extremely high doping level is nearly two orders of magnitude greater than p-type doping levels normally obtainable for bulk CdTe, suggesting that a non-equilibrium mechanism associated with the photoassisted MBE technique is responsible. Hole mobilities at 300 K generally range from 65-74 cm^2/V-s in CdTe:As films grown under optimum conditions. Van der Pauw Hall data for a heavily doped CdTe:As film is shown in Figure 7. The mobility at 290 K is 74 cm^2/V-s and increases with decreasing temperature to a value of 157 cm^2/V-s at 230 K. The hole concentration of this film remains above 2 x 10^{18} cm^{-3} even at 230 K. In fact, the doping level achieved is close to the value required for degenerate p-type doping of CdTe.

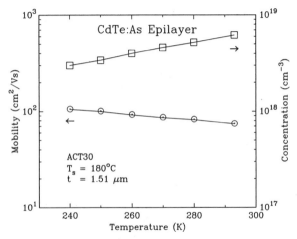

Figure 7.   Hall effect data for CdTe epilayer
ACT30 doped with arsenic.

In general, the low temperature photoluminescence from the CdTe:As films is brighter than that observed from the CdTe:Sb films discussed above. Figure 8 shows a PL spectrum for the CdTe:As epilayer (ACT30) whose electrical properties are shown in Figure 7. Note that the PL spectrum is dominated by a sharp (FWHM = 0.5 meV) bright (A^o,X) acceptor-bound exciton peak at 1.5907 eV. We associate his feature with an exciton bound to a neutral As acceptor. Application of Haynes rule [13] to this peak then yields 60 meV for the As acceptor ionization energy, which is close to the ionization energy associated with an effective-mass acceptor in CdTe (56.8 meV)[14]. In fact, recently completed temperature-dependent PL studies on a series of CdTe:As films have produced spectra in which the electron-to-acceptor (e,A) peak is present, corresponding to an acceptor ionization energy for As_{Te} of 58 meV.

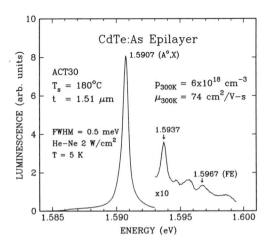

Figure 8.   Photoluminescence spectrum for CdTe:As
film grown by PAMBE.

DOPING MECHANISM

Here we suggest a mechanism which accounts for the high p-type doping levels achieved by photoassisted MBE. It has recently been reported [15], on the basis of reflection-high-energy-electron-diffraction (RHEED) studies, that the introduction of light from a He-Ne laser during CdTe film growth increases the rate of Te desorption from CdTe surfaces. The existence of these vacant Te sites thus favors the incorporation of substitutional acceptors such as Sb_{Te} and As_{Te}. This, coupled with increased atomic surface mobility brought about by the impinging photon beam, is what we believe gives rise to the highly doped p-type layers that have been produced by PAMBE.

In the absence of a p-type dopant, and at high laser illumination intensities, one might thus expect n-type layers to be produced, since the film growth surface would be rich in Cd, thus favoring the incorporation of interstitial Cd and/or Te vacancies, either of which act as an n-type dopant. We have recently observed this type of behavior in CdTe films grown under high laser illumination (greater than 100 mW/cm^2) with no dopant present.

The mechanism described above is also consistent with growth of highly activated n-type CdTe:In by PAMBE. In this case, we attribute the high degree of dopant activation (100% in some films) to suppression of the second-phase defect structure In_2Te_3, which we believe to be the p-type compensating defect in CdTe:In samples grown by conventional means. In the case of PAMBE-grown CdTe:In epilayers, we suggest that formation of In_2Te_3 is suppressed or absent because the number density of Te atoms available at the growth surface has been reduced by the incident light. As a consequence, tetrahedral bonding of In_{Cd} occurs.

ACKNOWLEDGEMENTS

This work was supported by ARO contract DAAG29-84-K-0039, DARPA/ARO contract DAAL03-86-K-0039, and NSF grant DRM83-13036. We wish to acknowledge the assistance of J. Matthews and K. Bowers for their help with substrate preparation. R.N. Bicknell and K. Harris grew two of the CdTe films included in this study.

REFERENCES

1. R.N. Bicknell, N.C. Giles, and J.F. Schetzina, Appl. Phys. Lett. 49, 1095 (1986).
2. R.N. Bicknell, N.C. Giles, and J.F. Schetzina, Appl. Phys. Lett. 49, 1735 (1986)
3. R.N. Bicknell, N.C. Giles, and J.F. Schetzina, Appl. Phys. Lett. 50, 691 (1987).
4. R.N. Bicknell, N.C. Giles, and J.F. Schetzina, J. Vac. Sci. Technol. B 5, 701 (1987).
5. N.C. Giles, R.N. Bicknell, and J.F. Schetzina, J. Vac. Sci. Technol. A 5, 3064 (1987).
6. R.N. Bicknell, N.C. Giles, and J.F. Schetzina, J. Vac. Sci. Technol. A 5, 3059 (1987).
7. C. Uzar, R. Legros, Y. Marfaing, and R. Triboulet, Appl. Phys. Lett. 45, 879 (1984).
8. Y. Marfaing, Rev. de Phys. Appl. 12, 211 (1977).
9. T.H. Myers, Yawcheng Lo, R.N. Bicknell, and J.F. Schetzina, Appl. Phys. Lett. 42, 247 (1983).
10. T.H. Myers, J.F. Schetzina, T.J. Magee, and R.D. Ormond, J. Vac. Sci. Technol. A 1, 1598 (1983).
11. D.L. Dreifus, R.M. Kolbas, K.A. Harris, R.N. Bicknell, R.L. Harper, and J.F. Schetzina, Appl. Phys. Lett. 51, 931 (1987).

12. S. Yamada, J. Phys. Soc. Jpn. 15, 1940 (1962).
13. R.E. Halsted and M. Aven, Phys. Rev. Lett. 14, 64 (1965).
14. E. Molva, J.L. Pautrat, K. Saminadayar, G. Milchberg, and N. Magnea, Phys. Rev. B 30, 3344 (1984).
15. J.D. Benson and C.J. Summers, J. Cryst. Growth 86, 354 (1988).

EXCITONIC COMPLEXES IN WIDE-GAP II-VI SEMICONDUCTORS

J. Gutowski

Institut für Festkörperphysik, Technische Universität Berlin
Hardenbergstr. 36, D-1000 Berlin (West) 12

1. INTRODUCTION

During the last years it became evident that excitonic complexes play a very important role concerning the nonlinear optical properties of wide-gap II-VI semiconductors. Thus, tremendous efforts have been made to find out and describe excited states of free and, particularly, bound-exciton complexes, their excitation and decay mechanisms, and related scattering phenomena. This paper surveys recent results regarding bound-exciton complexes and their interaction with free-exciton systems (biexcitons) in wide-gap II-VI materials at moderate to high excitation densities. Through means of resonant excitation spectroscopy it is shown that bound-exciton systems in II-VI's always possess a manifold of excited electronic states in which optical transitions often only become allowed under intense light irradiation. A comparison of spectroscopic investigations at CdS, ZnO, and ZnS is presented which allows for the development of comprehensive term schemes for neutral-impurity-exciton complexes in II-VI's. Mainly three types of excited states exist for bound-exciton complexes : (i) states which are described through electronic excitation of one electron or hole, while the other particles involved remain in their single-particle ground states, (ii) states which result from the participation of holes from lower-lying valence bands, and (iii) vibronic or rotational excited states. Measurements in magnetic fields up to 15 T allow for a well-founded assignment of optical transitions to these states. Additionally, exciton complexes give rise to resonant phononic or electronic Raman scattering processes which often break symmetry selection rules and provide further information on the levels involved. The energies and Stokes shifts of the scattered lines sensitively depend on quantities as complexes extension and binding energy.

2. NEUTRAL-IMPURITY-EXCITON COMPLEXES IN WURTZITE-TYPE II-VI MATERIALS

It is well-known that optical transitions of excitons bound to neutral impurities are the dominant feature in the spectra below the free-exciton lines in Wurtzite-type II-VI compounds. They always possess sets of excited states into which transitions often occur only for enhanced excitation densities. Acceptor-exciton complexes in CdS[1-5] and ZnO[5-7] are well-comparable examples of these systems in Wurtzite-type compounds. Excited states of these complexes have been characterized by electronic excitation of one particle involved, or by the origin of the exciton's hole from the second, lower-lying B band of the threefold split A-, B-, C-valence band structure of Wurtzite crystals.

Figure 1 shows a typical spectrum of bound-exciton luminescence of a pure CdS crystal. The lines marked with subscript 1 belong to the recombination of acceptor-exciton complexes which all possess chemically and/or structurally different neutral-acceptor centers. The lines I_1^g-I_1^l, however, do not show up in all undoped CdS crystals, and were attributed to complexes involving a more complicated acceptor center[4], e.g., a shallow double acceptor A_{Cd} on Cd site combined with a single donor D_{is} on interstitial site. While this interpretation of the center was supported by the observed giant diamagnetic shift of the lines in a magnetic field, the line I_1 - often a narrow doublet structure $I_1^{a,b}$ - is known as recombination of acceptor-exciton complexes at single Li or Na acceptors on Cd site[8,9]. The development of the spectra for increasing excitation densities (3.9 and 150 kW/cm[2], excited with 485.5 nm pulsed dye-laser radiation) demonstrates the importance of bound-exciton transitions even for high light intensities. Although biexciton-related recombination bands M and P show up, especially the line I_1 and its TA phonon sideband I_1-TA remain very dominant structures.

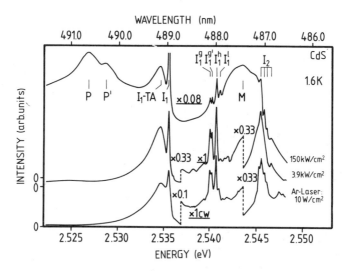

Fig. 1 Luminescence spectrum of an undoped CdS sample in the bound-exciton energy range, under excitation with the 476.5 nm line of an Ar ion laser for low and 485.5 nm pulsed dye laser light for higher intensities. For the description of the lines see text. From Ref. 4

One further proof that all I_1^i lines are due to acceptor-exciton complexes is obtained from their excitation spectra[4] for polarization E ∥ c (Fig. 2). In each excitation spectrum, four distinct resonances are seen on the low-energy side of the free-exciton transitions A_F (triplet, 2.5518 eV) and A_L (longitudinal, 2.5541 eV), two sharp lines and two broader resonances at higher energies. They are due to the fine structure of the electronic ground states of the so-called $(A°,X_B)$ complexes with one hole from the second B valence band. Due to the combination of one A ($j_z^A = 3/2$) and one B hole ($j_z^B = 1/2$), hole-hole exchange interaction leads to a sum $J_h = 1,2$, and combined with the electron ($j_z^e = 1/2$) to $J = 1/2$ (Γ_7), 3/2 (Γ_9), 3/2 (Γ_9), 5/2 (Γ_8), split by electron-hole interaction. Since the excitation of an $(A°,X_B)$ complex in one of its states is immediately followed by B-A hole conversion, the recombination of the corresponding $(A°,X_A)$ complex is enhanced for these characteristic excitation energies. Although all excitation spectra of I_1, I_1^h, and I_1^g look very similar, the resonance energies differ. It becomes obvious that the $(A°,X_B)$ states of the complex belonging to the I_1^g recombination exhibit the smallest binding energies although the $(A°,X_A)$ system correlated with this luminescence has a larger binding energy than that correlated with I_1^h. An analogous phenomenon had already been observed before at the Na and Li correlated acceptor-exciton complexes in CdS[2] and was

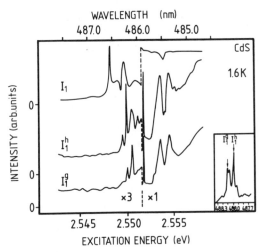

Fig. 3 Luminescence spectrum of an undoped ZnO crystal in the bound-exciton energy range under excitation with laser light 365 nm. From Ref. 7

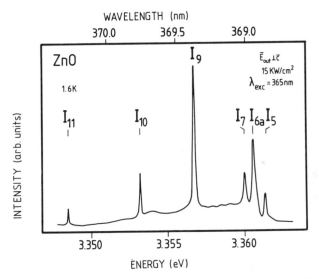

Fig. 2 Excitation spectra of the lines I_1, I_1^h, and I_1^g for polarization $\mathbf{E} \parallel \mathbf{c}$ of the incident laser light. From Ref. 4

explained by the changing interparticle distances if an $(A°,X_B)$ converts into the corresponding $(A°,X_A)$ complex[2,3]. It shows that the energies of the electronic states of a bound exciton sensitively depend on the structural nature of the system and the chemical nature of the impurity center involved.

A comparison with ZnO impressively demonstrated how general this form of excited states is for acceptor-exciton complexes in Wurtzite II-VI crystals. Fig. 3 shows the bound-exciton luminescence spectrum of an undoped ZnO sample. All lines I_5 to I_{11} are identified to belong to acceptor-exciton complexes with chemically or

141

structurally differing acceptor centers not only by coinciding Zeeman patterns[6,7] but also by striking structural similarities of their excitation spectra[7] (Fig. 4A and B). In the case of ZnO, the excitation spectra are even more complicated than for CdS. All I_{iB}^j transitions into the (A°, X_B) states are called Group II resonances, and due to the interchange of the upper two valence bands (ZnO: $A(j_z = 1/2)$, $B(j_z = 3/2)$), the strongest transitions I_{iB}^1 and I_{iB}^3 now appear for $E \perp c$ polarization. Since in ZnO the A-B valence-band distance only amounts to 5 meV[10], the resonant transitions into the (A°, X_B) levels are found very near to the respective luminescence line energies, and below those resonances which can be interpreted as excited electronic states (Group III resonances). This is in contrast to CdS where excited electronic states lie energetically below the (A°, X_B) states[1].

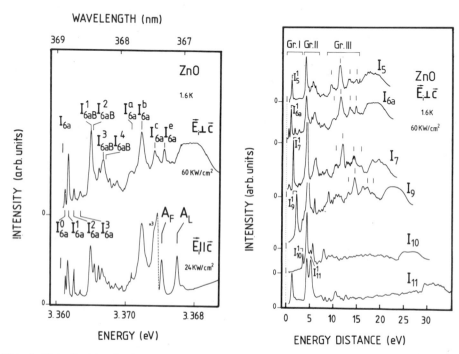

Fig. 4 A) Left side: Excitation spectrum of the acceptor-exciton recombination line I_{6a} in ZnO. In the upper spectrum, the specific resonances of $E \perp c$ polariza-tion are visible, in the lower spectrum the resonances for $E \parallel c$ are depicted which are of common nature for all acceptor-exciton lines in ZnO.
B) Right side: Comparison of the $E \perp c$ excitation spectra of I_5 to I_{11}, drawn as functions of the respective energy distances to the corresponding luminescence lines. From Ref. 7

Magnetic-field dependent studies[7] of the Group II resonances allowed for their assignment to the, like in CdS, fourfold split (A°, X_B) ground state, with a larger electron-hole exchange split than in CdS and rather similar g values of the A and B holes involved, respectively. These values obtained for (A°, X) complexes in CdS and ZnO are collected in Table 1, compared with results for (A°, X) systems in Zincblende ZnS which will be discussed in Chapter 3. Thus, bound-exciton spectroscopy provides information about the exchange interaction and the g values which are of rather fundamental significance for the excitonic properties of the II-VI materials.

Table 1. Hole-hole ΔE (hh) and electron-hole exchange split ΔE (eh) and g values of the particles involved in (A°,X) complexes with simple acceptor centers in II-VI materials

	CdS [a]	ZnO [b]	ZnS [c]
ΔE (hh)	1.47 .. 1.55 meV	1.5 meV	1.4 .. 2.0 meV
ΔE (eh)	< 40 µeV	0.2 .. 0.5 meV	unresolved
g_e	- 1.78	-1.96	- 1.88
g_h(A°)	-2.7 [d]	-3.0	-
g_{hAII}	- 3.23	-3.02	-
g_{hBII}	-4.22	-2.98	-

a) Ref. 2   b) Ref. 7   c) Ref. 21   d) Refs. 11, 12

The Group III resonances I_1^a to I_1^e of the I_i luminescence lines in ZnO have been interpreted in terms of excited electronic states where the single electron within the system is in an excited one-particle state[7]. This is comparable to excitation resonances of the I_1 lines in CdS[1] which, however, only show up for high excitation densities, and rather similar to the situation of donor-exciton systems in CdS[13-17] and CdSe[17] when the roles of the particles are simply exchanged: for the (D°,X) complexes, the single hole is in an excited orbital[13-15]. Fig. 5 shows the excitation spectrum of a (D°,X) recombination line I_2 in CdS for varying excitation densities. Those resonances lying very close to I_2 are even better resolvable in the excitation spectrum of the biexciton decay luminescence M. It has been demonstrated[17] that the biexciton recombination bands M and P in CdS and CdSe are excellently excitable via all excited electronic states of the (D°,X) complex. This was explained by a two-step process of biexciton creation: The first photon builds a (D°,X) complex, a second photon of suitable energy induces dissociation of the exciton from this system and immediate creation of the biexciton. - A comparison of the resonances I_2^a-I_2^e in CdS with the Group III resonances in ZnO shows the similarity of the structure and the intensity ratios of the resonance sets. Such excited electronic states have to be calculated based on quantum mechanics for a three-particle system within the potential of a fixed, fourth charge, i.e., two mobile holes and one mobile electron for (A°,X) complexes, or two mobile electrons and one mobile hole for (D°,X) systems. The Schrödinger equation for this system involving a product wavefunction of three single-particle functions, respectively, has been treated along the analogy from molecular physics by Puls et al.[13,14] by choosing an adequate potential to describe the attraction of the single hole for the (D°,X) system in CdS. Recently, the same calculation was applied to the (A°,X) complexes in ZnO where now the single electron is the excited particle[7]. The excited states are characterized by the radial and angular momentum quantum numbers n_r and l, respectively. Up to five stable states $I n_r,l >$ with $n_r = 0,1,2$ and $l = 0,1,2$ have been calculated to lie below the dissociation limit for the complexes. Due to the strong mass anisotropy of the holes in CdS the degeneracy of the states of equal total angular momentum is lifted, in contrast to ZnO, where m_h is nearly isotropic for the A as well as for the B valence band. For both compounds, the convincing coincidence of theoretical and experimental values becomes evident from Table 2.

The ionization energies $E_{I°}$ of the acceptor centers involved in the acceptor-exciton complexes in ZnO were determined by adjusting *one* theoretical to *one* experimental energy value, and then by simply using the Haynes rule[18] on the basis of the experimentally determined binding energies E^B(A°,X) of the (A°,X) complexes. The Haynes rule requests a simply linear dependence of E^B(I°,X) on $E_{I°}$. As is visible from the Table, the acceptor ionization energies of 60 to 114 meV (I_{11}, not listed) are relatively small compared with those known from CdS (about 160 meV[8,9]) so that the manifold of lines I_5 to I_{11} in ZnO may resemble the I_1^9-I_1^1 lines in CdS which also have been attributed to $(A°,X)$ complexes at acceptor centers of low ionization energies. In the right-hand column of Table 2, the experimentally observed and calculated excited hole states of the $(D°,X)$ complex in CdS are listed which show a promising coincidence as well. However, the term sequence regarding the $I\,0,2>$ and $I\,1,0>$ states is changed in CdS compared to ZnO. The expected oscillator strengths of the transitions satisfactorily agree with the observed intensity ratios of the resonances. For both systems, magnetic field dependent studies of the excitation resonances[7,15] strongly support the given assignments of the transitions. Splittings and g values allow for the determination of the wavefunctions involved. In conclusion, the model of a one-particle excited electronic state obviously is very suitable to describe these sets of excitation resonances of neutral-impurity-exciton complexes in Wurtzite II-VI compounds.

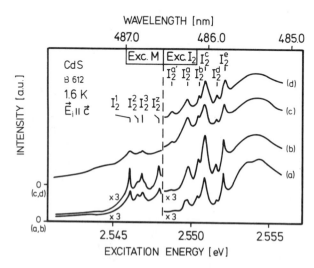

Fig. 5 Excitation spectra of the donor-bound-exciton line I_2 and the M band in CdS for different laser light densities. (a) 20 (b) 50 (c) 100 (d) 200 kW/cm[2]. From Ref. 15

The third Group I of resonances of the $(A°,X)$ recombination lines I_i which lie very close to the luminescence lines were tentatively assigned to possibly vibronic or rotational excited states of the acceptor-exciton complexes[7], as was done for a set of resonances of the I_1 $(A°,X)$ recombination line in CdS[1]. For both of them, a calculation with the molecular Kratzer potential which is possible without fit parameters yielded promising energy values very close to the experimental distances of the resonances to the respective luminescence lines[1,7]. This potential is suitable to describe a vibration/rotation of a very light against a heavy mass which correspond to the exciton and the impurity center in the neutral-impurity-exciton systems. However, due to the difficulties to distinguish between vibronic and rotational states for light particles, the assignment up to now is of rather tentative nature.

In conclusion, excited states of neutral-impurity-exciton complexes in Wurtzite II-VI materials are now well understood and assigned to electronic excitation or contribution of holes from lower-lying valence bands. Many of the transitions only occur for increased excitation densities but cover a wide energy range in the excitonic spectral regime. A serious influence on the optical properties of II-VI materials for high-excitation conditions is thus well established.

3. NEUTRAL-IMPURITY-EXCITON COMPLEXES IN ZINCBLENDE II-VI MATERIALS

In contrast to the Wurtzite materials, the band structure of Zincblende II-VI compounds results in a zero-field split of the electronic ground state of acceptor-exciton complexes. An antisymmetric combination of the wavefunctions of the two holes allows for sum angular momenta $J = 0$ and 2 already if both holes stem from the fourfold degenerate upper Γ_8 valence band[14]. In the crystal field and with the single electron coupled to the holes, a zero-field split set of $\Gamma_6(J = 1/2)$, $\Gamma_8(J = 3/2)$, and $\Gamma_7 + \Gamma_8(J = 5/2)$ of ground state levels is expected.

Table 2. Relative distances of the excitation resonances I_i^{a-e} (Group III) of the I_i luminescence lines of (A°,X) complexes in ZnO and I_i^{z-c} of the I_2 luminescence in CdS. Right-hand columns: theoretical values; left-hand columns: experimental values. The last row gives the theoretical values for the first states which lie above the free-exciton energy, respectively, and thus are not observable as excitation resonances

Lumines-cence line	ZnOᵃ: I_5		ZnOᵃ: I_6		ZnOᵃ: I_7		ZnOᵃ: I_9		CdSᵇ: I_2	
$E^B(I°,X)$ (meV) $E_{I°}$ (meV)	14.6 60.0		15.4 63.2		16.0 65.7		19.3 79.3		6.6 32.0	
Excit. res./state	th	ex	th	ex	th	ex	th	ex	th	ex
I_i^a $\|0,1>$	8.8	9.7	9.3	10.4	9.7	10.8	11.7	13.4	$\|0,1^0>$: 1.15 $\|0,1^{+1}>$:3.67	1.80 3.51
I_i^a $\|1,0>$	11.1	11.6	11.7	11.7	12.1	12.1	14.6	14.3	7.51	-
I_i^a $\|0,2>$	13.3	13.5	14.0	13.7	14.5	14.6	17.5	17.3	$\|0,2^0>$: 5.12 $\|0,2^{+1}>$:5.71	4.29 4.75
I_i^a $\|1,1>$	13.6	-	14.4	-	14.9	-	18.0	-	-	-
I_i^a $\|2,0>$	14.4	15.0	15.2	15.0	15.8	16.0	19.1	18.4	-	-
$E > E_{FE}$ (ZnO)	15.3	-	16.2	-	16.8	-	20.3	-	-	-

a) Ref. 7   b) Refs. 13,14

Cubic ZnS of high quality shows a variety of relatively sharp bound-exciton lines I_1 to I_{5b} in particular if higher excitation densities are applied[20,21] (Fig. 6). Due to the influence of hexagonal contributions in the polytypic ZnS, the cubic-related lines are somewhat broadened. For increasing excitation densities, the bound exciton lines remain the dominant structures until for more than 100 kW/cm^2 a broad-band emission begins to cover the whole excitonic energy regime. No evidence is given for biexciton or exciton-exciton-collision luminescence bands which could be compared to the M and P bands in Wurtzite II-VI compounds. Especially the line sets $I_{3a,b}$ and their LO phonon replica $I_{3a,b,c}$-LO, and the doublet $I_{5a,b}$ are due to two acceptor-exciton complexes with different impurity centers. This is not only shown by magnetic field dependent measurements of the luminescence lines but also by excitation spectroscopy. Fig. 7 presents the excitation spectra of the strongest luminescence lines, I_{3a} and I_{5a}. It becomes evident that they show resonance maxima exactly for those energies where the luminescence lines I_{3b} and I_{5b}, respectively, were found, and, additionally, other nearby maxima I_{3c} and I_{5c}, respectively.

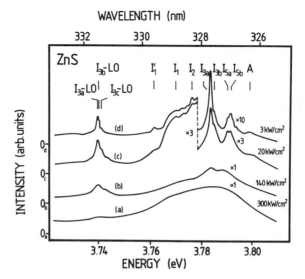

Fig. 6 Luminescence spectrum of cubic ZnS in the bound-exciton range for different excitation intensities and laser wavelength 308 nm. From Ref. 21

Since excited electronic states do not come into question to explain these close lying transitions, it is most reasonable that they are due to the fine structure of the electronic ground state. This is in accordance with theory[22] and the observations at other Zincblende structured materials[22-24]. From the intensity ratio of the transitions which is 1:2.8:2 for the resonances $I_{3a,b,c}$ in the I_{3a}-LO excitation spectrum[21], a term ordering $\Gamma_6(J = 1/2)$ lowest, $\Gamma_7 + \Gamma_8(J = 5/2)$ second, and $\Gamma_8(J = 3/2)$ highest level is most reasonable where the transition probabilities were theoretically determined to be 1:3:2. The pure, unthermalized intensity ratio is expected for absorption and excitation spectra where the transitions start with the simple neutral acceptor center. In luminescence, where the transitions start from the bound-exciton levels thermalization occurs which strongly alters the measured ratios.

Table 3 Hole-jj-splitting $\Delta E(jj)$, crystal-field splitting $\Delta E(V_c)$, and electron-hole splitting $\Delta E(eh)$ for acceptor-exciton complexes in Zincblende II-VI compounds and doped Si.

	ZnS: I_3 set [a]	ZnS: I_5 set [a]	CdTe:Cu [b]	Si:In [c]	Si:Ga [c]
$\Delta E(jj)$ (meV)	2.0	1.4	1.2	4.7	1.3
$\Delta E(V_c)$ (meV)	1.2	1.0	0.4	3.2	0.3
$\Delta E(eh)$ (meV)	-	-	0.02	-	-

a) Ref. 21   b)   Ref. 23   c) Ref. 24

Since the a-b and the a-c spacings are obviously very sensitive to the acceptor-exciton binding energy (1.4 and 2.6 meV for the I_3 set, 0.9 and 1.9 meV for the I_5 set), they should be caused by the hole-hole interaction which is strongly related to the interparticle distance. The b-c distance, however, remains nearly unaffected by the binding-energy variation (1.2 meV for the I_3 set, 1.0 meV for the I_5 set), and represents the crystal-field induced split. The hole-hole split thus corresponds to the distance between the a lines and the center of gravity of the b and c lines. As in most Zincblende compounds, the electron-hole split is too small to be observed. Only in CdTe:Cu, a split of 0.02 meV was detected[21]. Values compared for different materials are collected in Table 3, where Si as a group IV material is included since it exceptionally distinctly shows this ground state splitting.

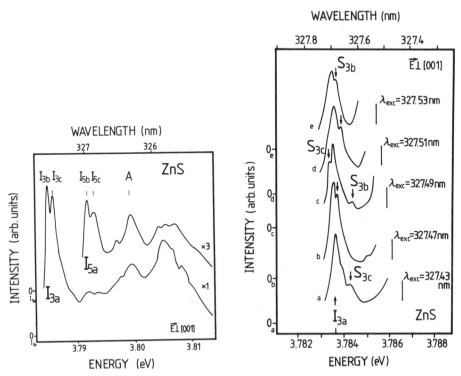

Fig. 7 Excitation spectra of the luminescence lines I_{3a} and I_{5a} in cubic ZnS. From Ref. 21

Fig. 8 Raman scattering lines S_{3b} and S_{3c} in the I_{3a} luminescence line range in cubic ZnS. From Ref. 21

Really excited electronic states of exciton-impurity complexes in Zincblende II-VI's have been reported for $(A°,X)$ systems in ZnTe[25] and CdTe[26] where excitation resonances and luminescence lines have been attributed to excited states of the electron in the field of a quasi-donor composed of the acceptor center and the two holes which remain in their single-particle ground states. However, no quantitative analysis of such states is given up to now.

4. ELECTRONIC RAMAN SCATTERING AT IMPURITY-EXCITON COMPLEXES

A very interesting and helpful phenomenon is the occurrence of electronic Raman scattering at impurity-exciton complexes in II-VI semiconductors at high excitation densities, e.g., at the just described $(A°,X)$ complexes in ZnS[21] (Fig. 8). Two scattering lines S_{3b} and S_{3c} are observable which shift with the laser wavelength and show just a resonance maximum if they coincide, subsequently, with the I_{3a} luminescence line. This is the case if the laser coincides with the I_{3b} or I_{3c} excitation energy, respectively. Thus, these scattering lines are explained as electronic RS where a virtual acceptor-exciton complex is created in one of its ground state levels discussed above. After immediate relaxation into the lowest ground state level, a recombination of the virtual complex takes place. The energy of the scattered photon is just diminished by the relaxation energy.

The Raman line's Stokes shift is not constant but varies with varying laser energy. This is a result of the change of the binding energy and thus the spatial extension of the virtually created acceptor-exciton complexes. The binding energy influences the interparticle interactions and thus the hole-hole split energy which is depicted in the Stokes shift. The shift consequently coincides with the real level distances just if the Raman lines are on resonance with the real electronic transitions[21]. A very similar process has been observed, e.g., at acceptor-exciton complexes in CdS[3] where the Stokes shift corresponds to the energy distances between the (A°, X_B) and (A°, X_A) levels which have been described in Sec. 2 (Fig. 2).

5. OUTLOOK

Bound excitons strongly influence the optical properties of II-VI materials and represent interesting electronic systems with a variety of excitable states in these compounds. Moreover, even optic switches and bistability are obtained from these systems which, however, exceeds the purpose of this contribution and will be treated in detail in a forthcoming paper[27].

REFERENCES

1. R. Baumert, I. Broser, J. Gutowski, and A. Hoffmann, Phys. Rev. B 27:6263 (1983)
2. J. Gutowski, Phys. Rev. B 31:3611 (1985)
3. J. Gutowski and I. Broser, Phys. Rev. B 31:3621 (1985)
4. J. Gutowski and A. Hoffmann, Phys. Rev. B 37:4076 (1988)
5. J. Gutowski, T. Hönig, N. Presser, and I. Broser, J. Lum. 40/41:433 (1988)
6. G. Blattner, C. Klingshirn, R. Helbig, and R. Meinl, phys. stat. sol. (b) 107:105 (1981); G. Blattner, PhD thesis, University Karlsruhe (F.R.G.) 1982
7. J. Gutowski, N. Presser, and I. Broser, submitted to Phys. Rev. B
8. C. H. Henry, K. Nassau, and J. W. Shiever, Phys. Rev. B 4:2453 (1971)
9. D. G. Thomas, R. Dingle, and J. D. Cuthbert, Proc. 7th Int. Conf. on II-VI Semiconducting Compounds, Providence (USA) 1967, W.A. Benjamin Inc. (New York 1967), p. 863
10. K. Hümmer, phys. stat. sol. (b) 56:249 (1973)
11. D. G. Thomas and J. J. Hopfield, Phys. Rev. 128:2135 (1962)
12. C. H. Henry, R. A. Faulkner, and K. Nassau, Phys. Rev. 183:798 (1969)
13. J. Puls, F. Henneberger, and J. Voigt, phys. stat. sol. (b) 119:291 (1983)
14. J. Puls, PhD thesis, Humboldt-University (GDR) 1985
15. J. Gutowski, Solid State Communications 58:583 (1986)
16. B. S. Razbirin and I. N. Yral'tsev, Fiz. Tverd. Tela 13:605 (1971) (Sov. Phys. - Solid State 13:493 (1971))
17. J. Gutowski and I. Broser, J. Phys. C 20:3771 (1987)
18. J. R. Haynes, Phys. Rev. Letters 4:361 (1960)
19. P. J. Dean and D. C. Herbert, in 'Excitons', ed. K. Cho, Springer (Berlin) 1979, p. 55, and the literature cited therein
20. J. Gutowski, J. Crystal Growth 86:528 (1988)
21. J. Gutowski, I. Broser, and G. Kudlek, submitted to Phys. Rev B
22. D. Bimberg, W. Schairer, M. Sondergeld, and T. O. Yep, J. Lum. 3:175 (1970); A. M. White, P. J. Dean, L. L. Taylor, and R. C. Clarke, J. Phys. C 5:L110 (1972); A. M. White, J. Phys. C 6:1971 (1973); A. M. White, P. J. Dean, and B. Day, J. Phys. C 7:1400 (1974)
23. E. Molva and Le Si Dang, Phys. Rev. B 32:1156 (1985)
24. K. R. Elliott, G. C. Osbourn, D. L. Smith, and T. C. McGill, Phys. Rev B 17:1808 (1978)
25. Le Si Dang, A. Nahmani, and R. Romestain, Solid State Commun. 46:743 (1983)
26. K. Cho, W. Dreybrodt, P. Hiesinger, S. Suga, and F. Willmann, Proc. 12th Int. Conf. on the Physics of Semiconductors, Stuttgart (F.R.G.) 1974, Ed. M. H. Pilkuhn, B. G. Teubner (Stuttgart 1974), p 945
27. T. Hönig, J. Gutowski, and I. Broser, to be published

OPTICAL PROPERTIES OF BULK II-VI SEMICONDUCTORS : EFFECT OF SHALLOW DONOR
STATES

E. Kartheuser

Institut de Physique, Université de Liège, Sart Tilman, B.5,
B-4000 LIEGE 1, Belgium

ABSTRACT

In the present work we discuss two experimental techniques that enable to identify shallow donor levels in II-VI semiconductors : extrinsic
oscillatory photoconductivity in n-CdTe and electric dipole spin resonance
of donor states in $Cd_{1-x} Mn_x Se$. Comparison between theory and experiment
lead to useful informations on electron lifetime and capture cross section
as well as on energy levels, spin-orbit coupling and g-factor of donors.

I. INTRODUCTION

A precise identification of impurities present in a semiconductor material is important, both in context of basic research and semiconductor
technology. Because impurities provide carriers for conduction, radiative
and non-radiative means for the recombination of electrons and holes, impurities are expected to play a key role in determining the optical and
electronic properties of these crystals. Therefore there has been a continuously growing interest in a detailed study of the impurity states in
single crystals of a semiconductor[1]. This interest has been further stimulated recently by the successfull preparation of artificially structured
solids[2].

In the present paper we concentrate on two experimental techniques :
extrinsic oscillatory photoconductivity and far infrared magnetoabsorption.

Oscillatory photoconductivity measurements have been performed in various polar semiconductors both with extrinsic[3-7] and intrinsic excitations[8-19]. Here we limit ourselves to extrinsic excitations which involve
shallow donor levels in n-CdTe[7].

Far-infrared magnetoabsorption spectra due to dipole spin resonance
(EDSR) of donor electrons have been observed in InSb[20-23] and $Cd_{1-X} Mn_X$-
Se[24,25] both in Faraday and Voigt configuration. In these crystals, electric dipole spin flip transitions in the donor ground state have been attributed to a parity-violating spin-orbit coupling which mixes states with
opposite spin orientations and parities[22,26]. The intensity of the EDSR-
lines are unusually strong because of the giant Landé g-factors of small
gap-semiconductors and magnetic semiconductors.

In section II we present a simple theoretical model for extrinsic
photoconductivity restricted to the case of weak electric fields. We apply the theoretical model to the measurements of Nicholas et al. in n-type
CdTe [7]. From the fitting of the theoretical predictions with experiment
we obtain information on the donor energy levels and average lifetime of
the electrons in the conduction band.

Section III deals with the magnetoabsorption of donor bound electrons.
Following the work of Rodriguez[27] we obtain a general expression for the
integrated absorption coefficient for spin flip transitions in the donor
ground state, valid for any crystal structure. The theory is compared to
the experimental results of Dobrowolska et al.[25] on far-infrared electric-
dipole spin resonance of donor electrons in $Cd_{1-X} Mn_X$ Se. The effect of
temperature and applied magnetic field is analysed.

II. EXTRINSIC OSCILLATORY PHOTOCONDUCTIVITY IN CdTe

II.1. Theoretical Model

The mechanism responsible for the oscillations in the photoresponse
can be viewed as follows :

Let us suppose that the photoexcitation takes place between the ground
state E_0 of the impurity center and a state relatively high in the conduction band of a polar semiconductor. The injected charge carriers will
then relax their energy by emission of longitudinal optical (LO) phonons
in a time τ_{LO} given approximately by[28]

$$\tau_{LO} \sim (\alpha\omega)^{-1} , \tag{1}$$

where α is the Fröhlich electron-LO-phonon coupling constant, and ω is
the dispersionless LO-phonon angular frequency. For example, in the case
of CdTe[29], $\omega = 3.23 \times 10^{13}$ s^{-1} and $\alpha = 0.315$, i.e., $\tau_{LO} = 9.8 \times 10^{-14}$ s.

This time is very short as compared to the lifetime of an electron
in the conduction band. Thus the condition for the observation of a

strong LO-phonon structure,[30]

$$\tau_{LO} \ll \tau \quad , \tag{2}$$

is well satisfied at low temperature.

The lifetime of the photoinjected electrons depend strongly on the phonon energy. Indeed, the photoelectrons relax their energy almost immediately by emitting a cascade of phonons. In this way they come to a state with a kinetic energy ε_k lower than $\hbar\omega$. If this kinetic energy is close to the energy of a donor level E_i plus $\hbar\omega$, the electrons are rapidly captured, and the electron concentration in the conduction band as well as the electrical conductivity are at a minimum.

On the contrary, if their kinetic energy is such that the conservation of energy forbids this resonant capture, the electron lifetime in the band is larger, and the conductivity goes through a maximum. Therefore, one expects to find dips in the conductivity at energies $n\hbar\omega$ $(n = 1,2,3,...)$ above the different impurity levels. Another possibility is a capture of electrons close to the bottom of the band without emission of LO phonons. This can occur when the electrons are captured in orbits with a large radius, for instance, the excited states of relatively shallow impurities or defects. For this case one expects to find a decrease in the conductivity for initial kinetic energies close to $n\hbar\omega$.

Taking these capture mechanisms into account we solve the Boltzmann transport equation[31] written in the steady state situation. In other words, we look for a solution of the probability of occupation of a quantum state in **k** space, $f(\mathbf{k})$, given by

$$- \frac{e\mathbf{\mathcal{E}}}{\hbar} \nabla_{\mathbf{k}} f(\mathbf{k}) + \sum_i \sum_{\mathbf{k'}} [f(\mathbf{k'})W_i(\mathbf{k'},\mathbf{k})$$
$$- f(\mathbf{k})W_i(\mathbf{k},\mathbf{k'})] + W_s(\mathbf{k}) - W_c(\mathbf{k})f(\mathbf{k}) = 0 \quad , \tag{3}$$

where the summation over i refers to the different scattering processes with density of transition probability per unit time respectively equal to $W_i(\mathbf{k'},\mathbf{k})$ and $W_i(\mathbf{k},\mathbf{k'})$. Here we have included a source term $W_s(\mathbf{k})$, which describes the photoinjection of n_e electrons per unit time and unit volume into the conduction band and a capture term $W_c(\mathbf{k})f(\mathbf{k})$, which describes the capture processes, i.e. the transitions of the electrons out of the conduction band into impurity states. We consider injection of monoenergetic photoelectrons from the donor ground state level E_0 into the conduction band, which is assumed to be isotropic and parabolic, i.e.,

$$\varepsilon_k = \hbar^2 k^2 / 2m \qquad (4)$$

where m is the electron effective mass and $\hbar k$ is its momentum with respect to the band minimum.

This yields for the source term,

$$W_s(k) = (2n_e \pi^2 / \sqrt{\varepsilon_k})(\hbar^2 / 2m)^{3/2} \delta(\varepsilon_k - \varepsilon) \qquad (5)$$

with

$$n_e = \frac{2}{(2\pi)^3} \int dk \; W_s(k) \quad . \qquad (6)$$

The electron excitation energy ε is given as a function of the incident photon energy $h\nu$ by

$$\varepsilon = h\nu - E_0 \quad . \qquad (7)$$

In the case of weak-bias electric fields E , the distribution function $f(k)$ can be developed into a power series of E , restricted to the first two terms :

$$f(k) = f_0(k) + f_1(k) \quad . \qquad (8)$$

In cubic crystals the first term is isotropic and independent of E , whereas the second one is linear in E .

At first order, Eq. (3) takes a simple form. Indeed the elastic scattering mechanisms do not contribute to the scattering terms in Eq. (3) when the electron distribution is isotropic in k space. Moreover, it is well known that the scattering by impurities or acoustic phonons is nearly elastic. These scattering mechanisms allow a fast momentum relaxation, but a slow energy relaxation. Therefore, in polar semi-conductors such as CdTe, the energy relaxation is mainly controlled by the polar interaction with the long-wavelength LO phonons and the field-independent part of Eq. (3) becomes

$$\sum_{k'} \{f_0(k')W_{LO}(k',k)-f_0(k)W_{LO}(k,k')\} + W_s(k)-W_c(k)f_0(k) = 0 \qquad (9)$$

where $W_{LO}(k,k')$ is the rate of transition from state k to state k' because of electron-LO-phonon scattering. For weak electron-LO-phonon coupling we have[28] :

$$W_{LO}(\mathbf{k},\mathbf{k'}) = \frac{\alpha}{\pi}\,\omega\,\frac{(2\pi)^3}{V}\left(\frac{\hbar}{2m\omega}\right)^{3/2}\frac{\hbar\omega}{|\mathbf{k'}-\mathbf{k}|^2}\{<N>\delta(\varepsilon_{k'}-\varepsilon_k-\hbar\omega)$$

$$+ <N+1>\delta(\varepsilon_{k'}-\varepsilon_k+\hbar\omega)\}\quad. \tag{10}$$

The first and second terms between the brackets describe, respectively, the absorption and emission of LO phonons and

$$<N> = (e^{\hbar\omega/kT} - 1)^{-1} \tag{11}$$

is the number of LO phonons present at thermal equilibrium.

For the probability of capture per unit time for an electron in a quantum state of the **k** space, we write :

$$W_c(k)f_0(k) = \frac{2}{\sqrt{\varepsilon_k}}\left(\frac{\hbar^2}{2m}\right)^{3/2}\sum_{j=0}^{2}W_{,j}(\varepsilon_k)f_0(k) \tag{12}$$

where $2W_j(\varepsilon_k)f_0(k)$ is the number of electrons captured per unit of time, unit of volume, and unit of energy. The summation over j takes account of the different transitions involved in the capture processes.

Obviously, the conservation of the carrier density requires that the number of electrons captured per unit time and unit volume be equal to that of the photoinjected ones, i.e.

$$n_e = \frac{2}{(2\pi)^3}\int d\mathbf{k}\;W_c(k)f_0(k) \tag{13}$$

which results immediately from the integration of Eq. (9) over **k** space.

The photoresponse is given by the photocurrent versus the energy of the incident photons. We calculate the photocurrent in the framework of the usual relaxation-time approximation. This approximation is justified as follows : because of the fast relaxation of energy with emission of LO phonons, the proportion of electrons with a kinetic energy larger than $\hbar\omega$ is small, and therefore, the main contribution to the conductivity comes from the electrons lying near the bottom of the band, below $\hbar\omega$. These electrons cannot emit LO phonons, and for the case of low temperature considered here, the probability of LO-phonon absorption is weak, since $\hbar\omega$ is relatively high. Therefore, the scattering mechanisms responsible for the resistivity, i.e. scattering by acoustic phonons and by impurities or defects, are essentially elastic and can be described by a relaxation-time approximation.

II.2. Comparison with experiment in n-CdTe

In the present section we apply the theoretical model discussed above to the interpretation of the measurements performed in n-CdTe.[32] In this experiment electrons are photoinjected from a donor level E_0 below the minimum of the conduction band into this conduction band. The excitation spectrum has two series of minima due to the capture of electrons with a kinetic energy respectively close to $\varepsilon_1 = \hbar\omega - |E_0|$ and $\varepsilon_2 = \hbar\omega - |E_1|$. The process is a LO-phonon-assisted capture by two donor levels lying, respectively, at E_0 and E_1 below the conduction band. In addition to this, we suppose that the carriers that lie close to the bottom of the band are captured in orbits with large radii, possibly in a cascade process.[33] As a result of this, we propose the following expression for the probability of capture for given number of photoinjected electrons :

$$W_c(\varepsilon) = \frac{2\omega}{\sqrt{\varepsilon}} (\hbar\omega)^{\frac{1}{2}} \sum_{j=0}^{2} \left(\gamma_j \ e^{-n_j \left(\frac{\varepsilon - \varepsilon_j}{\hbar\omega} \right)^2} \right) \tag{14}$$

Here the dimensionless parameters γ_0 and n_0 are used to describe the probability of capture and the width of the trapping level acting at the bottom of the band. In the same manner, γ_1 and n_1 represent the capture by the ground state E_0 of the donor level, and γ_2 and n_2 refer to the excited state E_1 .

The parameters are then adjusted to obtain the best fit with the low-temperature (T = 8K) measurements.[32] Detailed numerical calculations are given elsewhere.[34]

The results of the fitting are shown in Fig. 1. The points are obtained by interpolation between data taken from the measurements.[32] The solid line gives the theoretical results for values of the parameters γ_j, n_j and ε_j , which are displayed in Table 1.

Figure 1 shows that the theoretical model reproduces the dips appearing in the experimental curve for $\varepsilon = n\hbar\omega + \varepsilon_1$ and $n\hbar\omega + \varepsilon_2$.

From table 1 we find for the shallow donor energy levels, respectively, E_1 = 3 meV and E_0 = 12 meV. These results agree with the magneto-optical studies.[35]

In addition, the knowledge of the parameters γ_j, n_j, and ε_j enables us to calculate the probability of capture per unit time $W_c(\varepsilon)$ from Eq. (14). The total probability of trapping per unit time or, equivalently, the inverse of the lifetime τ is given by

$$\tau^{-1} = \frac{\int d\mathbf{k} \ W_c(\mathbf{k}) f(\mathbf{k})}{\int d\mathbf{k} \ f(\mathbf{k})} \tag{15}$$

since the population of the electron states is weak for electron energies above $\hbar\omega$, an approximate value of the average probability of capture can be obtained by averaging on energies up to $\hbar\omega$. We find for the average probability of capture per unit time associated with the j-th capture process the values given in the fifth column of Table 1. This leads to an average total lifetime, defined as

$$\langle\tau\rangle^{-1} = \sum_{j=0}^{2} \langle\tau_j\rangle^{-1} \quad , \tag{16}$$

equal to $\langle\tau\rangle = 2.2 \times 10^{-12}$ s.

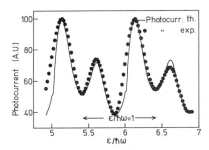

Fig. 1. Photocurrent as a function of the initial kinetic energy of the photoinjected electrons measured in units $\hbar\omega$. The experimental points are taken from Ref. 7. The solid line gives the results of the present theoretical model.

Table 1. Numerical values of the parameters obtained from fitting the theoretical model to experiment in n-CdTe. The three values of j correspond to the different capture processes

j	ε_j (meV)		n_j	$\langle\tau_j\rangle^{-1}$ (10^9 s^{-1})
0	0	1.0×10^{-3}	2.5	0.53
1	9	4.1×10^{-3}	2.6	1.3
2	18	9.5×10^{-3}	2.3	2.7

III. MAGNETOABSORPTION OF ELECTRONS BOUND TO DONORS

III.1. General formalism for the integrated absorption coefficient

We consider the absorption of electromagnetic radiation by bound electrons in the presence of an externally applied magnetic field $\mathbf{B}_0$. At low temperature the absoprtion coefficient is given in the usual manner by the expression

$$\alpha_a = n_e \sum_{i,f} \frac{\hbar\omega \ W_{if}}{I_0} \tag{17}$$

where n_e is the number of electrons per unit volume involved in the absorption process, ω and I_0 are respectively the angular frequency and intensity of the incident radiation and W_{if} is the probability per unit time that a transition from an initial electron state $|\psi_i\rangle$ to a final electron state $|\psi_f\rangle$ occurs.

Taking into account the natural line width of the absorption line at the transition energy $\hbar\omega_{fi} = \hbar(E_f - E_i)$, Wigner-Weisskopf time dependent perturbation theory[36] yields :

$$W_{if} = \frac{\Gamma}{\hbar} \frac{|\langle\psi_f|H'|\psi_i\rangle|^2}{(\omega-\omega_{fi})^2 + (\Gamma/2)^2} \tag{18}$$

where Γ is the full width of half-height of the absorption line and H' characterizes the interaction between the bound electron and the radiation field.

Consider an electromagnetic wave with electric field $\mathbf{E} = \hat{e}E$ polarized along the $\hat{e}$-direction and magnetic field $\mathbf{B}$ given by

$$\mathbf{E} = -\frac{n}{c} \dot{\mathbf{A}} \tag{19}$$

$$\mathbf{B} = \nabla \times \mathbf{A} \tag{20}$$

and let the real part

$$\mathbf{A} = \text{Re} \ \{A \ \hat{e} \ e^{i(\mathbf{q}\cdot\mathbf{r}-\omega t)}\} \tag{21}$$

be the corresponding vector potential with amplitude A and wave vector $\mathbf{q} = \frac{\omega n}{c}\cdot\hat{q}$, propagating along the $\hat{q}$-direction inside the material of index of refraction $n = \sqrt{\varepsilon_\infty}$ (c is the velocity of light).

Keeping only first order terms in A we obtain for the electron-photon interaction in Eq. (18) the following expression :

$$H' = \frac{A}{4i\hbar c} \ (\{e^{i\mathbf{qr}}, \ [e\hat{e}\mathbf{r}, \ H]\} - n \ g \ \mu_B \ e^{i\mathbf{qr}}(\hat{q}\wedge\hat{e})\sigma\hbar\omega) \tag{22}$$

where $[\mu,\nu]$ and $\{\mu,\nu\}$ denote respectively the commutator and anticommutator of the operators u and v .

156

The first term in the right-hand side of Eq. (22) is due to electric transitions, whereas the second contribution in the right-hand side of Eq. (22) gives rise to magnetic transitions resulting from the coupling of the intrinsic magnetic moment of the electron with the oscillating magnetic induction of the incident radiation. Here g is the Lande' g-factor of the bound electron, $\mu_B = e\hbar/2m_0c$ is the Bohr magneton, σ, the Pauli spin matrix and H is the Hamiltonian of a donor electron in the presence of an applied constant magnetic field directed along the $\hat{z}$-axis : $\mathbf{B}_0 : (0,0,B_0)$.

Using the effective mass theory of shallow donors[37], the envelope functions corresponding to the bound electron states are described by the effective Hamiltonian

$$H = H_0 + H_{SO} \tag{23}$$

where

$$H_0 = \frac{p^2}{2m} - \frac{e^2}{\varepsilon_0 r} + \mu_B \frac{m_0}{m} L_z + \frac{1}{8} \frac{e^2 B_0^2}{mc^2} (x^2 + y^2) + \tfrac{1}{2} g\mu_B B_0 \sigma_z \tag{24}$$

and H_{SO} is the spin-orbit coupling. Here ε_0 is the static dielectric constant of the crystal, m the effective mass of the electron and L_z the z-component of its orbital momentum.

Let H_{SO} be a small perturbation and the states $|\nu,s\rangle$ with orbital quantum number ν and spin quantum number s be solutions of the unperturbed problem :

$$H_0|\nu,s\rangle = (E_\nu + g\mu_B B_0 s)|\nu,s\rangle \quad . \tag{25}$$

We note that the transitions involved in expression (18) are related to the states $|\psi_f\rangle$ and $|\psi_i\rangle$ of the total Hamiltonian including the spin-orbit interaction H_{SO}. However we can always describe the states of H by the unitary transformation $e^{iS}|\nu,s\rangle$, where S is an operator to be determined so that the quantity $e^{-iS} H e^{iS}$ is diagonal in the representation $|\nu,s\rangle$. It follows that to first order in H_{SO}, the Hermitian operator S is given by the relation

$$H_{SO} + i[H_0,S] = 0 \tag{26}$$

and the transition-matrix element becomes

$$\langle\psi_f|H'|\psi_i\rangle = \langle\nu',s'|H' + i[H',S]|\nu,s\rangle \quad . \tag{27}$$

In the present work we concentrate on electric and magnetic-dipole transitions between the Zeeman sublevels $|0,\pm\frac{1}{2}\rangle = |0\rangle|\pm\frac{1}{2}\rangle$ associated with the ground state of H_0. This simplifies considerably the electron-photon interaction energy and thus Eq. (27). The intensity of the spin resonance line at transition energy $\hbar\omega = \hbar\omega_{fi} = g\mu_B B_0$ is then given by the integrated absorption coefficient

$$\alpha_I = \alpha_a \Gamma = 8\pi \frac{n_e}{n} \frac{e^2}{\hbar c} \omega |M|^2 \tag{28}$$

with

$$M = \frac{g\mu_B B_0}{\hbar\omega} \langle 0|[S_-,\hat{e}.\mathbf{r}]0\rangle + i\frac{n}{4}g\frac{\hbar}{m_0 c}(\hat{q}\times\hat{e})_- \tag{29}$$

for a transition from spin-down to spin-up state. Here we have used the notation $(\hat{q}\times\hat{e})_- = (\hat{q}\times\hat{e}).(\hat{x}+i\hat{y})$ and taken into account the relation Eq. (26) between the operators S and H_{SO} rewritten in terms of the Pauli matrices σ_z and $\sigma_\pm = \sigma_x \pm i\sigma_y$ as

$$S = S_+\sigma_- + S_-\sigma_+ + S_z\sigma_z \tag{30}$$

$$H_{SO} = H_+\sigma_- + H_-\sigma_+ + H_z\sigma_z \quad . \tag{31}$$

The last term in expression (29) is the intrinsic-magnetic-dipole transition which leads to the well known paramagnetic resonance. We notice from Eq. (29) that the intensity of this contribution is obtained in a straightforward manner for a given field configuration, i.e. for a given $\hat{q}$ and polarization $\hat{e}$ of the incident radiation. Usually this is the only contribution in this case because electric dipole transitions between states of the same parity are not allowed. However in crystals with inversion asymmetry the parity-violating spin-orbit interaction leads to a finite electric-dipole contribution given by the matrix element $\langle 0|[S_-,\hat{e}.\mathbf{r}]|0\rangle$ which can always be written as

$$\langle 0|[S_-,\hat{e}.\mathbf{r}]0\rangle = -i \sum_\nu \left(\frac{\langle 0|\hat{e}.\mathbf{r}|\nu\rangle\langle\nu|H_-|0\rangle}{E_\nu - E_0 + g\mu_B B_0} + \frac{\langle 0|H_-|\nu\rangle\langle\nu|\hat{e}.\mathbf{r}|0\rangle}{E_\nu - E_0 - g\mu_B B_0}\right) \tag{32}$$

and clearly shows that the spin-orbit coupling gives rise to an admixture of orbital states $|\nu\rangle$ of different parities.

We shall see in section III.2 that in semi-conductors such as Cd_{1-x}^- Mn_x Se this contribution is predominant due to the resonances appearing in Eq. (32) for shallow donor-levels and a large Lande' g-factor.

In order to eliminate the summation over the intermediate states in

Eq. (32), we follow a procedure due to Dalgarno.[38] We define a "dipole operator" $\mathbb{D}_\pm$ satisfying the condition

$$([\mathbb{D}_\pm,H_0] \mp g\mu_B B_0 \mathbb{D}_\pm)|0\rangle = \hat{e}.\mathbf{r}|0\rangle \quad . \tag{33}$$

This leads to the closed form

$$\langle 0|[S_-,\hat{e}.\mathbf{r}]\mathbb{D}\rangle = -i\langle 0|H_-\mathbb{D}_- + D_+^\dagger H_-|0\rangle \quad . \tag{34}$$

We note that the present approach allows the determination of the integrated absorption coefficient for any crystal structure, i.e. for any form of the spin-orbit coupling given by H_- , provided we know the expression for the "dipole operator".

III.2. Comparison with experiment in $Cd_{1-x}Mn_x Se$

The purpose of the present section is to compare the analytic expression of the integrated absorption coefficient deduced from the results of section III.1 with the far-infrared measurements[25] of the electric-dipole spin-resonance of donor-bound electrons in $Cd_{1-x}Mn_x Se$. The measurements show the dependence on incident photoenergy of the integrated absorption coefficient in a sample with a maganese concentration x = 0.10. The data are taken at two different temperatures, T = 4.7 K and T = 9.8 K in the Faraday geometry with circularly polarized light of positive helicity propagating along the hexagonal axis $\hat{c}$ of the crystal as seen in Figure 2.

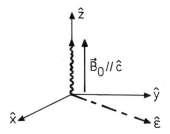

Fig. 2.   Faraday configuration : incident radiation circularly polarized in the xy-plane, propagating along the hexagonal axis $\hat{c}$ and the applied magnetic field $\mathbf{B}_0$.

We assume that for small values of the manganese concentration, the physical parameters of $Cd_{1-x}Mn_x Se$ are the same as those given for CdSe. We take[29] for the effective mass the isotropic value m = 0.13 and for the static and high frequency dielectric constants respectively $\varepsilon_0 = 9.3$ and $\varepsilon_\infty = n^2 = 6.1$. This yields an effective Rydberg $R = 20.45$ meV. We note that the intensity of the magnetic field used for the measurements never exceed 40 KG. In this case the magnetic-energy $\hbar\omega_c/2 = 1.8$ meV is much smaller than the Coulomb-energy, i.e. the ratio

$$\gamma = \frac{\hbar\omega_c}{2R} \ll 1 \tag{35}$$

and the diamagnetic contribution in Eq. (24) can be neglected. As a first approximation, we shall thus choose an envelope function with constant effective Bohr radius a, i.e.

$$\langle r|0\rangle = [\pi a^3]^{-\frac{1}{2}} e^{-r/a} \quad, \tag{36}$$

for the donor ground state, in order to find a solution for the dipole operator $\mathbb{D}_\pm$ from Eq. (33). Taking advantage of the spherical symmetry, we obtain

$$\mathbb{D}_\pm(r,\vartheta,\varphi) = 2 \sum_{m_\ell = 1,0,-1} D_\pm^{m_\ell} \hat{\beta}.r/r \quad, \tag{37}$$

where $D_\pm^{m_\ell}$ is a solution of the differential equation

$$\frac{d^2 D_\pm}{d\xi^2} + 2(\frac{1}{\xi} - 1) \frac{dD_\pm}{d\xi} - \frac{2}{\xi^2} D_\pm - \Gamma_\pm D_\pm - \xi = 0 \tag{38}$$

with

$$\hat{\beta} = (\frac{e^2}{\varepsilon_0 a^2})\hat{e} \tag{39}$$

and

$$\Gamma = \pm\gamma g \frac{m}{m_0} (1 \pm m_\ell \frac{m_0}{mg}) \quad. \tag{40}$$

Here, m_ℓ takes the values 1, 0, -1 and ξ is the dimensionless quantity $\xi = r/a$.

We now search for a physical solution of Eq. (38) in the form of a power series in ξ and Γ:

$$D_\pm = \sum_{k,\ell}^{\infty} a_{k\ell} \Gamma^\ell \xi^k \quad. \tag{41}$$

Notice that for zero magnetic field ($\Gamma = 0$), Eq. (38) leads to

$$D_{\pm} = -\xi/2 - \xi^2/4 \quad . \tag{42}$$

Moreover, in order to obtain the intensity of the spin resonance line from Eq. (34), we need to know the particular form of the spin-orbit interaction.

In a crystal, such as $Cd_{1-x} Mn_x$ Se with point symmetry group C_{6V}, the predominant contribution to the spin-orbit interaction is linear in the electron momentum and of the form[39]

$$H_{SO} = \gamma/2\hbar \; \hat{e}.(\sigma \times p) \quad , \tag{43}$$

where λ is a measure of the spin-orbit coupling strength, $\hat{e}$ is a unit vector along the hexagonal axis and σ and p are respectively the spin and momentum operators.

From expression Eq. (43) it is evident that the spin-orbit interaction mixes eigenstates of the $\hat{e}$-component of the angular momentum, hence allowing electric-dipole spin resonance due to transitions between Zeeman levels of the donor ground state. The intensity of these transitions is of course proportional to the square of the coupling constant λ . Therefore, the comparison between theory and experiment provide a means of determining the strength of the spin-orbit coupling.

Using the definition Eq. (31) and the form of the envelope function Eq. (36), we obtain

$$H_{\pm}|0\rangle = \frac{\lambda}{2a} (\pm c_z \sin \vartheta \; e^{\pm i\varphi} \mp \cos \vartheta \; c_{\pm}) \tag{44}$$

with

$$c_{\pm} = c_x \pm i c_y \quad . \tag{45}$$

In the Faraday geometry (see Fig. 2), $\hat{q} : (0,0,1)$ and $(\hat{q} \times \hat{e}) = -i\sqrt{2}$, which yields the value $\dfrac{n}{2\sqrt{2}} g \dfrac{\hbar}{m_0 c}$ for the magnetic contribution in Eq. (29).

Moreover, combining the expressions for H_-, D_- and D_+^+, a straightforward calculation of the quantum average Eq. (34) leads to the following analytic expression for the integrated absorption coefficient (in units cm^{-2}) :

$$\alpha_I = A\omega\{\varepsilon_\infty g^2 + (\frac{8 \; m_0 c}{3E_0 \hbar})^2 (\frac{27}{32} C + 1.25 \; C^2 + 2C^3 + 3.4C^4 + 5.87C^5 + 10.23C^6)^2\} \tag{46}$$

with

$$A = \frac{n_e}{2c} \left(\frac{e^2}{\hbar c}\right)\left(\frac{\hbar}{m_o c}\right)^2 \quad . \tag{47}$$

and

$$C = \left(\frac{\hbar\omega}{E_0}\right)^2\left(1 - \frac{m_0}{mg}\right)^2 \quad . \tag{48}$$

Notice that additional powers of λ and c give negligeable contributions to Eq. (45) for the excitation energies considered here.

For comparison with experiment in the Faraday geometry we use for the donor concentration[25] $n_e = 2.10^{16}$ cm^{-3}.

In the case of the diluted magnetic semi-conductor Cd_{1-x} Mn_x Se, the Lande' g-factor depends strongly on temperature and applied magnetic field due to the exchange interaction between the localized magnetic moments of Mn^{2+} and the itinerant electrons.[40] The energies of the spin-up (+) and spin-down (-) substates of the donor ground state can then be written in the form

$$E_{\pm} = E_0 \pm \tfrac{1}{2} \left(g^*\mu_B B_0 + N_s \alpha_e \langle S_z \rangle\right) \tag{49a}$$

or

$$= E_0 \pm \tfrac{1}{2} g\mu_B B_0 \tag{49b}$$

where

$$g = g^* + \frac{N_s \alpha_e \langle S_z \rangle}{\mu_B B_0} \tag{50}$$

is an effective g-factor made up of two parts : the intrinsic part and the magnetic part. The intrinsic contribution g^* is essentially given by the band-contribution. Magneto-optical studies of exciton states yield a value[41] $g^* = 0.5 \pm 0.1$ for CdSe. However, in large gap magnetic semi-conductors, the predominant part comes from the exchange energy $N_s \alpha_e \langle S_z \rangle$, depending on the effective Mn^{2+} ion-density N_s, the exchange integral α_e for the s-like Γ_6-electrons and $\langle S_z \rangle$, the thermal average of the Mn-spin in the magnetic field direction.

The quantity $<S_z>$ is directly accessible from magnetization measurements.[42] Expressing the exchange energy within the framework of the molecular field approximation, the effective g-factor can be written in the form

$$g = g^* + \frac{5}{2} \frac{N_s \alpha_e}{B^B_0} B_{5/2}(y) \tag{51}$$

with

$$y = \frac{5}{2} \frac{g_{Mn} \mu_B B_0}{k_B (T+T_0)} . \tag{52}$$

Here, $B_{5/2}(y)$ is a modified Brillouin function, $g_{Mn} = 2$ is the g-factor of the Mn^{2+} ions and T_0 is an effective temperature that reflects the antiferromagnetic interaction between Mn ions.

In order to obtain the temperature and magnetic field dependence of g , we have fitted the simple analytic expression Eq. (46) to the far-infrared observations[25] of the electric-dipole spin resonance and the spin flip Raman scattering data[43] performed on the same ingot of $Cd_{0.9} Mn_{0.1} Se$. The experimental results are taken from the position of the resonances at $\hbar\omega = g\mu_B B_0$.

The fitting procedure yields for the exchange energy $N_s \alpha = 12.7$ meV and for the effective temperature $T_0 = 3.5$ K. A variation of the effective g-factor ranging from g = 62 up to 65 at 9.8 K and from g = 123 up to g = 138 for 4.7 K has been obtained for the magnetic field intensities considered here. The value of $N_s \alpha$ is in reasonable agreement with previous work.[43] Knowing the range of values for the effective g-factor to be considered respectively at the temperatures 4.7 K and 9.8 K we are now in a position to compare the analytic expression Eq. (46) for the integrated absorption constant, with the experimental results[25] at T = 4.7 K and T = 9.8 K displayed in Figure 3, respectively by the solid and open circles.

The dashed and solid curves in Fig. 3 represent the theoretical results.[44] The best fit is obtained with the physical parameters : $\lambda = 4 \ 10^{-10}$ eV.cm, $E_0 = 18$ meV for both temperatures and for g = 62 at 9.8 K and g = 123 at 4.7 K.

The present value for the spin-orbit coupling constant λ is slightly smaller than the values of previous works.[25,26] In addition, a small increase of λ with increasing temperature has been obtained from the fitting procedure. However, because of the lack of experimental points, a definite conclusion on this modification cannot be given.

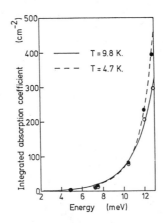

Fig. 3. Integrated absorption coefficient of $Cd_{0.9} Mn_{0.1} Se$ as a function of excitation energy $\hbar\omega$. The solid and open circles are respectively the data of Ref. 25 at 4.7 and 9.8 K, taken in the Faraday geometry. The corresponding dashed and solid curves show the best fit of the analytic expression Eq. (46).

IV. CONCLUSION

In the present work two experiments on optical transitions of wide-gap II-VI semi-conductors involving shallow donor states have been analyzed by means of simple theoretical models.

It has been shown that the oscillatory behaviour of the excitation spectrum of extrinsic photoconductivity observed in n-CdTe can be explained by emission of a cascade of LO phonons combined with capture mechanisms involving shallow donor states lying respectively at $E_0 = 12$ meV and $E_1 = 3$ meV below the conduction band. From the probability of capture an average electron lifetime $\langle\tau\rangle = 2.2 \times 10^{-12}$ s has been obtained.

The effect of temperature and magnetic field on electric dipole spin resonance of shallow donor-bound electrons in $Cd_{1-x} Mn_x Se$ has been investigated. The comparison between theory and far-infrared magnetotransmission measurements lead to giant Landé g-factors, respectively equal to $g = 62$ at 9.8 K and $g = 123$ at 4.7 K resulting from the exchange interaction between the magnetic moments of Mn^{2+} ions and the electron spin. Moreover, by fitting the experimental data and the theoretical model for the integrated absorption coefficient we obtain for the strength of the spin-orbit interaction, a value $\lambda = 4 \ 10^{-10}$ eV.cm and for the donor ground state energy a value of $E_0 = 18$ meV.

ACKNOWLEDGEMENTS

     I am indebted to R. Evrard, J.K. Furdyna and S. Rodriguez for stimulating discussions.

REFERENCES

1. A.K. Ramdas, Physica 146B+C:6(1987).
2. B.V. Shanabrook, Physica 146B:121(1987).
3. W. Engeler, H. Levinstein and C.R. Stannard, Jr., Phys.Rev.Lett. 7:62 (1961).
4. H.J. Stocker, H. Levinstein and C.R. Stannard, Jr., Phys.Rev. 150:613 (1966).
5. V.J. Mazurczyk and H.Y. Fan, Phys.Rev.Lett. 26:220(1968).
6. A.L. Mears, A.R.L. Spray and R.A. Stradling, J.Phys. C1:1412(1968).
7. R.J. Nicholas, A.C. Carter, S. Fung, R.A. Stradling, J.C. Portal and C. Houlbert, J.Phys. C13:5215(1980).
8. M.A. Habegger and H.Y. Fan, Phys.Rev.Lett. 12:99(1964).
9. H.J. Stocker, C.R. Stannard, Jr., H. Kaplan and H. Levinstein, Phys. Rev.Lett. 12:163(1964).
10. Y.S. Park and D.W. Langer, Phys.Rev.Lett. 13:392(1964).
11. D.N. Nasledov, V.V. Negreskul, S.I. Radantsan and S.V. Slobodchikov, Sov.Phys. Solid State 7:2965(1966) [Fiz.Tver.Tela. 7:3671(1965)].
12. R.E. Nabory and H.Y. Fan, Phys.Rev.Lett. 17:251(1966).
13. H.J. Stocker, H. Levinstein and C.R. Stannard, Jr., Phys.Rev. 150:613 (1966).
14. V.J. Mazurczyk, G.V. Ilmenkov and H.Y. Fan, Phys.Lett. 21:250(1966).
15. J.C. Ayache, M. Zonaghi and Y. Marfaing, Phys.Stat.Sol.(a) 2:61(1970).
16. R.E. Nabory, Phys.Rev. 178:1293(1969).
17. R.W. Shaw, Phys.Rev. B3:3283(1971).
18. A. Barbarie and E. Fortin, Solid State Comm. 14:267(1974).
19. T. Instone and L. Eaves, J.Phys. C12:L345(1979).
20. B.D. McCombe and R.J. Wagner, in Proceedings of the 11th Conference on the Physics of Semi-conductors, edited by M. Miasek (Polish Scientific Publishers, Warsaw, 1972), p. 321.
21. G. Appold, H. Pasher, R. Ebert, W. Steigenberger and K. von Ortenberg, Phys.Status Solidi B86:557(1978).
22. Y.F. Chen, M. Dobrowolska, J.K. Furdyna and S. Rodriguez, Phys.Rev. B32:890(1985).
23. Z. Barticevic, M. Dobrowolska, J.K. Furdyna, L.R. Ram Mohan and S. Rodriguez, Phys.Rev. B35:7464(1987).
24. M. Dobrowolska, H.D. Drew, J.K. Furdyna, T. Tchiguchi, A. Witowski and P.A. Wolff, Phys.Rev.Lett. 49:845(1982).
25. M. Dobrowolska, A. Witowski, J.K. Furdyna, T. Ichiguchi, H.D. Drew and P.A. Wolff, Phys.Rev. B29:6652(1984).
26. S. Gopalan, J.K. Furdyna and S. Rodriguez, Phys.Rev. B32:903(1985).
27. S. Rodriguez, Physica 146B+C:212(1987).
28. B.R. Nag, Theory of Electrical Transport in Semi-conductors (Pergamon, Oxford, 1972).
29. See, for instance, E. Kartheuser, in Polarons in Ionic Crystals and Polar Semi-conductors, edited by J.T. Devreese (North-Holland, Amsterdam, 1972), p. 727.
30. P.G. Harper, J.W. Hodby and R.A. Stradling, Rep.Prog.Phys. 36:1(1973).
31. J.M. Ziman, Electrons and Phonons (Oxford University Press, London, 1961), chap. 7, p. 209.
32. The comparison is made with the results for the sample called GE in Ref. 6. See Fig. 1 of this reference.

33. M. Lax, J.Phys.Chem.Solids 8:66(1959).
34. E. Kartheuser, J. Schmit and R. Evrard, J.Appl.Phys. 63:784(1988).
35. P.E. Simmonds, R.A. Stradling, J. Birch and C.C. Bradley, Phys.Status Solidi B64:195(1974).
36. See for instance W.H. Louisell, Radiation and Noise in Quantum Electronics (McGraw Hill, New York, 1964), p. 191.
37. See for instance A.K. Ramdas and S. Rodriguez, Rep.Prog.Phys. 44:1297 (1981).
38. A. Dalgarno and J.T. Lewis, Proc.R.Soc.London, Ser. A 233:70(1955).
39. E.I. Rashba and V.I. Sheka, Fiz.Tverd.Tela 6:141(1964) [Sov.Phys. - Solid State 3:114(1964)].
40. See for instance J.K. Furdyna, J.Appl.Phys. 53:7637(1982) and references cited therein.
41. R.G. Wheeler and J.D. Dimmock, Phys.Rev. 125:1805(1962).
42. N.B. Brandt and V.V. Moshchalkov, Adv.Phys. 33:193(1984).
43. D. Heiman, P.A. Wolff and J. Warnock, Phys.Rev. B27:4848(1983).
44. J. Defour, "Mémoire de licence en sciences physiques", Université de Liège, 1986 (unpublished).

REVIEW OF NONLINEAR OPTICAL

PROCESSES IN WIDE GAP II – VI COMPOUNDS

C. Klingshirn, M. Kunz, F.A. Majumder, D. Oberhauser,
R. Renner, M. Rinker, H.–E. Swoboda, A. Uhrig and Ch. Weber

Fachbereich Physik der Universität
Erwin–Schrödinger–Straße 1
D–6750 Kaiserslautern
Germany

INTRODUCTION

In this contribution we want to review optically nonlinear properties of wide–gap II–VI compounds. In the first section we outline some rather general ideas of the linear optical properties. The next one is devoted to the description of various optical nonlinearities and of the underlying concepts. In an outlook we finally summarize, how optical nonlinearities can be used to realize pure optical bistability, electro–optic hybrid devices, optical amplifiers or optical oscillators.

We restrict ourselves to wide–gap II–VI materials i.e. the compounds of Cd and Zn with 0,S,Se or Te and mixed crystals of these compounds, which have all a direct gap with band extrema at the Γ–point [1]. The dimensions of the samples will not fall significantly below 1 μm if not stated otherwise. This means, we are mainly dealing with three–dimensional or bulk properties, except for the well known implications at surfaces.

We are aware of the fact that many research groups all over the world contributed to the above outlined field of research. Due to the finite length of this contribution it is not possible to present nor even to cite all. We apologize for this situation and restrict ourselves to some selected examples with some emphasis on recent results from the research group of the author. Some recent reviews, which cover the present or related fields are given e.g. in [2–6].

Optical transitions involve both in absorption and emission always two particles, namely an electron in the energetically higher state and an empty state (i.e. a hole) in the lower one. These two particles interact with each other via the Coulomb–potential, as long as the latter one is not screened in a metallic state reached for high carrier concentrations, e.g. in an electron–hole plasma. In the following, we call interacting individual electron–hole pair–states "excitons". With this rather wide definition of exciton, we can explain most of the linear and nonlinear optical properties of semiconductors in the spectral vicinity of the fundamental absorption edge by excitonic features:

There are free excitons, which are characterized by a plane wave factor of the center of mass motion $\vec{R}$ in eq. (1) where we give the eigenfunction and eigenenergies in the descrete states under the assumptions of nondegenerate, parabolic and isotroptic bands and of a hydrogen–like series of states. We are

167

dealing in all cases with Wannier–excitons, i.e. the Bohr–radius of the exciton is large compared to the lattice constant, and the binding energy Ry* is small compared to the gap.

$$\phi = \Omega^{-1/2}\, e^{i\vec{K}\vec{R}}\, \phi_{n_B,\ell,m}(\vec{r}_e - \vec{r}_h)\, \varphi(\vec{r}_e)\, \varphi(\vec{r}_h)$$

$$E_{ex} = E_g - Ry^*\, n_B^{-2} + \hbar^2 \vec{K}^2 / 2M \qquad (1)$$

with:

E_g	width of the forbidden gap at $\vec{k}_e = \vec{k}_h = 0$
$Ry^* = 13{,}6\ eV\ \mu/(m_0 \cdot \epsilon_0^2)$	effective exciton Rydberg or binding energy
ϵ	effective dielectric constant
$\phi_{n_B,\ell,m}$	hydrogen–like envelope function
$\varphi(r_e), \varphi(r_h)$	Wannier functions of electrons and holes
$\mu = (m_e m_h)/(m_e + m_h)$	reduced mass
$M = m_e + m_h$	translational mass
$\vec{K} = \vec{k}_e + \vec{k}_h$	

Dipole–allowed free exciton states show usually up at low temperature by strong signals in the absorption and reflection spectra. In luminescence, the LO–phonon replica are usually much stronger than the zero–phonon–lines [8,14] The complications brought about by the complex band structure of real crystals are discussed e.g. in [2, 3, 7, 8]

Two excitons may be bound together in a similar way as an H_2 or a positronium molecule to form a socalled excitonic molecule or biexciton. The dispersion relation of the biexciton is given in simplest approximation by

$$E_{biex}(\vec{K}_{biex}) = 2\, E_{ex}(\vec{K} = 0) - E_{biex}^b + \frac{\hbar^2 K^2}{4M} \qquad (2)$$

where E^b_{biex} is the binding energy of the biexciton. Transitions involving biexcitons give rise to a rather large variety of optical nonlinearities. See below or e.g. [2, 3]. Excitons can be bound to lattice defects. The simpliest bound exciton complexes (BEC) are excitons bound to point defects like neutral donors, acceptors or isoelectronic traps. The wave function of shallow bound excitons can be described by a superposition of free exciton wave–functions.

$$\phi_{\text{bound exciton}} = \sum_{\vec{K}} a_{\vec{K}}\, \phi(\vec{K}) \qquad (3)$$

BEC show up as narrow absorption and emission lines at low temperatures. More details are found e.g. in [9]. Excited states of BEC have recently been investigated in some detail in CdS and ZnO [10]. There are first indications that multiple bound exciton complexes, which are well known in the indirect element semiconductors Si and Ge [11] may exist also in CdS [12].

There are also polycentric bound excitons, i.e. excitons bound to a system of interacting defects. Donor–acceptor pairs can be included in this group. Free–to–bound transitions e.g. of a free hole in the valence band and an electron bound to a donor may be considered as ionized states of a BEC at a charged donor D^+.

In mixed crystals excitons may be localized due to various mechanisms: Composition fluctuations on a length scale comparable to the exciton Bohr–radius which are connected with fluctuation of the width of the forbidden gap may lead to a localization of excitons. Excitons may be localized as a whole in these fluctuations or e.g. a hole localized in a fluctuation in the valenced band may bind an electron, leading to a localized exciton. Pioneering work in this field has been published for the system $CdS_{1-x}Se_x$ [15]. Some recent reviews are found e.g. in [16]. Another concept of localization assumes self–localization of excitons preferentially at single Te atoms or small Te clusters in Te substituted II–VI compounds like $ZnSe_{1-x}Te_x$ ($x \ll 1$) [17] or $CdS_{1-x}Te_x$. This fact is remarkable since selflocalization is usually absent for Wannier excitons in semiconductors.

In reasonably pure II–VI compounds (and of course in many other semiconductors) most of the above mentioned electron–hole pair complexes give rise to distinct spectral features in the optical properties (transmission, reflexion or luminescence) at low temperatures [2, 3, 5, 7, 8]. With increasing temperature the states broaden and partly disappear. The relevant parameters which determine e.g. the broadening of the free exciton resonances with increasing temperature are Ry^*/k_BT and the strength of the exciton–phonon coupling. But even if free excitons are no longer present as discrete structures in the optical spectra, there is still the Coulomb–interaction between electron and hole with its significant influence on the oscillator strength. This fact explains e.g. why the square–root dependence of the absorption coefficient expected for simple band–to–band transitions is never observed. In addition, the exciton energy appears in the frequency and temperature dependence of the absorption coefficient which follows for $\hbar\omega < E_{ex}$ the Urbach–Martienssen rule [13]

$$\alpha(\hbar\omega) = \alpha_0 \exp\left[\sigma(T)\left(E_{ex}(n_B=1, \vec{k}=0) - \hbar\omega\right)/k_BT\right] \quad (4)$$

where α_0, E_{ex} and $\sigma(T)$ are material parameters.

BEC are either thermally ionized with increasing temperataure or may even disappear if the charge state of the defect changes with T.

The proper description of light propagating in a semiconductor is in terms of a mixed state of electromagnetic wave and some (virtual) excitation of matter whenever the real part of the complex refractive index $\tilde{n} = n + ix$ deviates from one. In the spectral region around the fundamental absorption edge these excitations are the electron–hole pair or exciton states. The quanta of these mixed states of excitons and photons are called exciton polaritons. Details of this rather fundamental concept are given e.g. in [2, 3, 7, 8, 14].

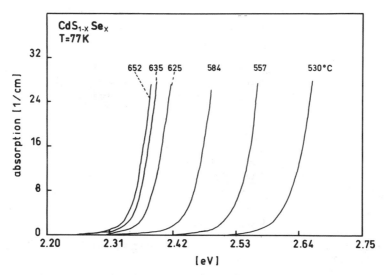

Fig 1. Absorption spectra of Cd, S, Se doped glass samples measured at 77K. Composition equal for all samples. Parameter: annealing temperature T_α. $t_a = 72h$. From [22].

We shall now shortly address the influence of spatial confinement on the linear optical properties of excitons. Following strong activities of many research groups on excitons confined in one dimension in (multiple) quantum wells (MQW) or superlattices SL in III–V materials [18] there are now analogous attempts in II–VI materials which are covered e.g. by various contributions to this volume [19]. There are to our knowledge no results on exciton–confinement in two dimensions (quantum–wires) in II–VI compounds, but a lot of activities concerning excitons in quantum dots, see e.g. [20]. These quantum–dots are usually realized in the form of tiny crystallites of II–VI compounds imbedded in a glass matrix. The size–distribution can be engineered by a heat treatment of the glass [21]. As an example we present in Fig 1 the absorption edges of a Cd, S and Se doped zinc silicate glass. The semiconductor crystallites belong to the quarternary system $Cd_{1-y}Zn_yS_{1-x}Se_x$. The composition of the glass melt is the same for all samples. The parameter is the temperature T_a of the annealing procedure. At low temperature only small crystallites are formed. The localization energy [21] is seen in the blue shift of the absorption edge with decreasing T_a. For $T_a > 920K$ the absorption edge becomes almost indepent of T_a indicating that the crystallite size is so large that the influence of the confinement on the exciton energy is negligible.

OPTICAL NONLINEARITIES

In this section we present selected results on optical nonlinearities in wide gap II–VI compounds. By "optical nonlinearities" (ON) we mean all reversible changes of the optical properties of semiconductors induced by illumination, i.e. in the spectra of luminescence, transmission and reflection or of the complex dielectric function $\epsilon = \epsilon_1 + i\epsilon_2$ or of the complex index of refraction $\tilde{n} = n + i\varkappa$ .We distinguish between photo–electronic and photo–thermal ON. By the first term we mean processes induced by real or virtual optical excitation in the electronic system of a semiconductor; the second one consequently describes ON which result from sample heating through the absorption of light.

There are basically two approaches to describe ON: if the ON follow instantaneously the amplitude(s) of the incident light field(s) $\vec{E}$, it is possible to expand the susceptibility $\chi = \epsilon -1$ in a power series of $\vec{E}$ as shown schematically in (5).

$$\frac{1}{\epsilon_0} P_i = \chi_{ij}^{(1)} E_j + \chi_{ijk}^{(2)} E_j E_k + \chi_{ijkl}^{(3)} E_j E_k E_l + \ldots . \tag{5}$$

The first term on the rhs $\chi^{(1)}$ describes the usual linear optical properties. $\chi^{(2)}$ and $\chi^{(3)}$ give e.g. second harmonic generation and four wave mixing, respectively. The approach of eq (5) is applicable if states are excited only virtually and thus have a very short lifetime determined by the uncertainty relation. These effects are also called "coherent" effects.

The concept breakes down, if states with a finite lifetime are really excited. In this case it is more appropriate to investigate χ, ϵ or $\tilde{n}$ as a function of the density N of some optically excited species i.e.

$$\epsilon(\omega) = \epsilon(\omega,N) \tag{6}$$

The quantity N is given by an integral over the generation rate G weighted by some decay function e.g. an exponential:

$$N(t) = \int_{-\infty}^{t} G(t') e^{-t'/\tau} dt' \tag{7}$$

$$G(t') = \alpha \left(\hbar \omega_{exc} \right) I_{exc}(t') + \beta \left(\hbar \omega_{exc} \right) I_{exc}^2(t') \qquad (8)$$

The parameters in (7) and the one– and two– photon absorption coefficients α and β in (8) may depend on N, leading thus to a rather complex set of differential equations.

To reach the values of $\vec{E}$ or G necessary to observe ON usually pulsed or CW lasers are used. The two most widely used experimental techniques to detect ON are pump–and–probe beam spectroscopy and spectroscopy with laser–induced gratings (LIG).

In the first case one measures the transmission or reflection spectrum of a sample with a weak and spectrally broad probe beam once without excitation and once with excitation. The difference of the two spectra contains the information about changes of the reflectivity R and of the absorption coefficient α (induced absorption, bleaching of absorption and optical amplification i.e. gain). If the samples have plane-parallel surfaces a shift of the Fabry Perot modes gives information about dispersive ON. LIG–spectroscopy or wave mixing work in the following way. Two beams of equal (different) frequencies interfere in the sample. If there are any ON this results in a stationary (moving) grating. In selfdiffraction the pump beams are deflected from the grating which they produce. It is also possible to read the grating by a third independent probe beam. In case of a moving grating there is a spectral Doppler–shift of the diffracted beams. In the case of coherent ON it is possible to describe LIG experiments both in the quasi–particle picture as scattering between identical or different polaritons under energy and momentum conservation or as χ^3 effect.

In this subsection we restrict ourselves to intrinsic excitations and some aspects of localized excitons in mixed crystals. ON connected with BEC are treated e.g. in [10]. We consider mainly quasi–stationary excitation i.e. laser pulses which are longer than the lifetime of the excited species and give only some references for recent time–resolved results. If we excite at low temperature in the exciton resonance or in the band–to–band transition region we increase the density of e–h pairs and the rate of collisions between them. This leads to a decrease of the phase–relaxation time τ_p and a broadening of the exciton resonances. This phenomenon has been observed e.g. by LIG spectroscopy in transmission [24] or reflection [25]. Some of these scattering processes, the so called inelastic ones, give rise to the appearance of new emission bands, of moderate optical gain ($\leq 10^3$ cm^{-1}) and of induced absorption. In these processes which have been observed in many different II–VI compounds an exciton–like polariton is scattered to a photon–like state giving rise to luminescence, while another particle (a second exciton or a free carrier i.e. an exciton in the continuum states) is scattered into a higher energy state under energy and momentum conservation. Details of the experimental results and the theoretical description are found e.g. in [2]. The exciton–exciton scattering bands P_2, P_3, P_∞ which are attributed to a collision where one of the excitons is scattered into states with $n_B = 2, 3, \infty$ have become a subject of discussion in CdS. The energetic distances of the A– and B valence bands and the exciton– and biexciton binding energies allow in this material also an interpretation in terms of a biexciton decay in which one of the excitons is transformed in a B–exciton [26]. If two excitons collide at low temperature ($k_B T < Ry^*$) they may stick together to form a biexciton if they can get rid of the binding energy e.g. by phonon emission [2, 3]. The radiative decay of biexcitons to an exciton and a photon–like polariton contributes to the emission in the spectral region of the M–band. Other processes, which contribute in this range are inelastic scattering processes of free carriers and BEC or the acoustic sidebands of BEC [2]. The inverse process, i.e. the absorption of a photon which transforms an exciton into a biexciton gives rise to induced absorption in the spectral region around

$$\hbar \omega_{abs} = E_{biex} - E_{ex} \qquad (9)$$

The shape of the absorption and gain–spectra around $\hbar\omega_{abs}$ is determined by the population in the exciton and biexciton bands [3]. The corresponding ON has been observed for CdS recently by LIG spectroscopy [24]. The above mentioned ON are typical examples for real excitations described in context of eq (6).

A very direct and precise method to determine the properties of biexcitons is two–photon absorption spectroscopy (TPA): polariton state with energy $\hbar\omega_{exc}$ is populated in the sample by illumination with a laser. Absorption of quanta with a photon energy $\hbar\omega_{abs}$ becomes possible according to (10)

$$\hbar\omega_{abs} = E_{biex} - \hbar\omega_{exc} \qquad (10)$$

The oscillator strength of the resonance at $\hbar\omega_{abs}$ depends on the polariton density at $\hbar\omega_{exc}$. In II–VI materials a variant of TPA has been used, the socalled luminescence assisted two–photon spectroscopy (LATS) in which the sample is excited by a laser at $\hbar\omega_{exc}$ and the TPA signal shows up as a reabsorption dip in the luminescence according to (10) [2, 3, 27]. The dispersion and the diamagnetic shift of the biexciton in CdS has also been determined by LATS experiments [28, 29]. A further nonlinear optical process, in which the biexciton is created only virtually by two incident light quanta, is known as hyper– or two–photon Raman scattering. It is a $\chi^{(3)}$ process. The virtually excited intermediate biexciton state decays spontaneous (or stimulated by an additional test beam) in two new quanta under energy– and momentum conservation between initial and final states [2, 3] according to

$$\hbar\omega_{R} = \hbar\omega_{P1} + \hbar\omega_{P2} - \hbar\omega_{f}$$
$$\vec{K}_{R} = \vec{K}_{P1} + \vec{K}_{P2} - \vec{K}_{f} \qquad (11)$$

The two pump–quanta $\hbar\omega_{P1,2}$ come usually from one laser beam. $\hbar\omega_{R}$ is the observed Raman–type emission and $\hbar\omega_{f}$ is the other quantum in the final state. By measuring $\hbar\omega_{R}$ as a function of $\hbar\omega_{p}$ and the scattering geometry it is possible to determine precisely the dispersion of polaritons around the exciton resonance and the dispersive ON connected e.g. with the TPA to the biexciton via Kramers–Kronig relations [2, 3].

With these examples we want to leave the discussion of the ON in the intermediate density regime where excitons and biexcitons exist still as good quasiparticles. We proceed now to generation rates which are high enough to form an electron–hole plasma (EHP).

The transition from a low or medium density gas of excitons to an EHP is connected with strong (self–) renormalization effects of various energies and consequently also of the optical properties. This is shown in Fig 2 to 6. In Fig 2 we give schematically the variation for various energies as a function of the electron–hole pair density n_p and the corresponding absorption spectra for low and high (room–) temperature [30]. The width of the forbidden gap is a monotoneously decreasing function of n_p due to exchange and correlation energies [2, 4, 31]. The exciton–binding energy and oscillator strength decrease continously with increasing n_p until they go to zero for a certain value of n_p^M which is also called Mott–density. At low temperature (Fig 2a) it is possible to bring the chemical potential μ of the electron–hole pairs above the reduced gap, resulting in population inversion and gain. At higher temperatures (Fig 2b) it is sometimes only possible to bring μ in the vicinity of the reduced gap which results together with the Urbach–Martienssen rule behaviour of the absorption edge of the unexcited crystal in a bleaching of absorption. For quantitative calculation of the renormalization by many particle effects and the resulting optical spectra see e.g. [2, 5, 31].

The experimental verification of the predictions of Fig 2 is shown by pump– and probe beam experiments in Fig 3 for CdS [32]. It should be noted that Fig 3

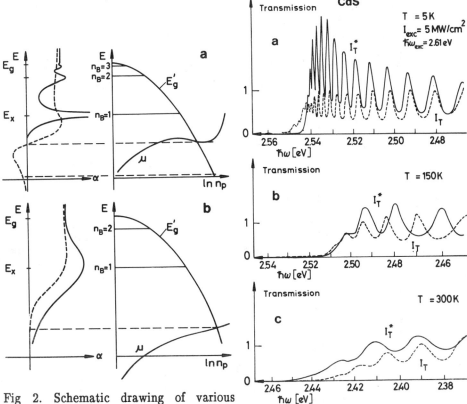

Fig 2. Schematic drawing of various energies as a function of the electron–hole pair density n (right) and the corresponding absorption spectra (left). Liq. He temperature (a), room temperature (b). From [30, 32].

Fig 3. Transmission spectra of CdS without $(- - -)=$ and with $(——)$ additional pump beam for various temperatures. From [32].

gives transmission–spectra in contrast to the spectra of the absorption coefficient in Fig 2. In the transparent spectral region $T(\omega)$ is modulated due to Fabry–Perot interferences since the sample has a thickness of a few μm and as grown, plane-parallel surfaces. We observe at low temperature as a consequence of pumping the disappearence of the exciton absorption, a red shift of the absorption edge and optical gain. At higher temperatures the absorption edge is flatter according to eq. (4), the gain decreases and the red–shift of the edge turns to a blue–shift or a bleaching of absorption. There is a blue–shift of the Fabry–Perot modes, i.e. a decrease of the refractive index at all temperatures. The temperature dependence of the shift of the absorption edge is given for CdS under constant excitation conditions in Fig 4. The absorption edge is defined in this context by T=0.1. The analysis of the gain spectra as a function of excitation intensity and temperature show that the EHP does not reach a liquid–like state, due to its short lifetime, which is in the 100–200 ps region [33]. It has been found by various spectroscopic techniques, that the drift length l_D of excitons and EHP in CdS under inhomogeneous excitation conditions are as follows [33, 34]: l_D is about 1,5 μm for excitons with a trend to decrease with increasing generation rate and lattice temperature. For a degenerate EHP at low temperature l_D is about 7 μm decreasing up to room–temperature where the EHP is no longer degenerate to $l_D \approx 0,5$ μm. The experimental techniques used to obtain these results are: spatially resolved pump and probe beam experiments, the investigation of the EHP gain and of the disappearance of the excitonic reflection structure on the unexcited surface of the sample as a function of the sample thickness and the investigation of the efficiency of LIG as a function of the grating constant 1. Fig 5 gives l_D in CdS under excitation conditions where a EHP is formed

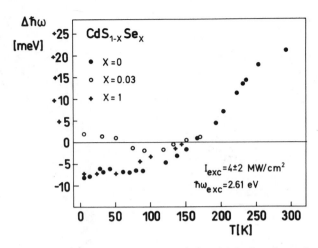

Fig 4. Shift of the absorption edge caused by high band–to–band excitation ($I_{exc} \approx (3 \pm 1)$ MWcm^{-2}) as a function of temperature for CdS, CdSe and CdS.$_{97}$Se.$_{03}$.

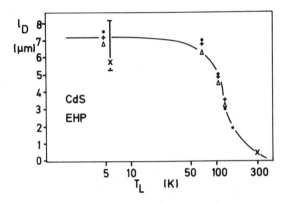

Fig 5. Drift length l_D of the EHP in CdS under inhomogeneous excitation as a function of lattice temperature T . Data from thickness dependence of disappearance of exciton reflection signal on the unexcited surface of the $A\Gamma_5$ exciton ($\bullet$) and the $B\Gamma_5$ exciton (Δ). Data from LIG experiments (x). From [34, 35].

as a function at the lattice temperature [35]. In [36] the expansion of a high density electron–hole pair system has been investigated in CdS by ps resolved LIG experiments. The obtained data for the diffusion constant $D \approx 10$ cm^{2}s^{-1} indicate that no degenerate EHP state has been reached. The above value is compatible with excitons or a nondegenerate EHP. Recent calculations on plasma expansion are reviewed in [37]. Similar results concerning the various properties of an EHP have been obtained in other II–VI compounds like CdSe [33] and Fig 4, ZnO or ZnSe [2]. As an example, we show in Fig 6 the red–shift of the absorption edge and the disappearance of the excitonic reflection structure as a function of I_{exc} at low temperature in ZnSe.

To conclude this section on optical nonlinearities of the EHP we show two results obtained with LIG spectroscopy. In Fig 7 we plot the intensity of the first diffracted order in case of a non–degenerate four–wave mixing experiment as a function of $\Delta\hbar\omega = \hbar\omega_{P1} - \hbar\omega_{P2}$. The diffracted orders show the expected frequency shift. We observe a very narrow peak around zero with a HWHM below 50 μeV. We

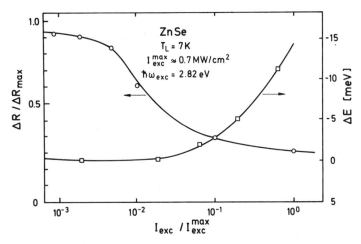

Fig 6. Decrease of the excitonic reflection signal on the excited surface and shift of the absorption edge with I_{exc} in ZnSe. (The edge is defined here by $T = .5$) Circles are averages over results from various samples. From [38].

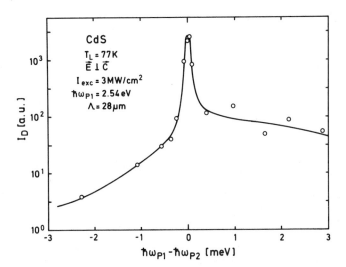

Fig 7. The intensity of the first diffracted order in NDFWM in CdS as a function of the detuning of the two pump beams. From [39].

attribute this peak to a population grating of an EHP. If the grating moves during the EHP lifetime τ more than a grating period, the grating will be smeared out and I_D drops. From Λ and $\Delta\hbar\omega$ at the HWHM point we deduce for the velocity of the grating $v_g < 3.5 \cdot 10^7$ cms^{-1}. Using the relation $v_g \approx \Lambda\tau^{-1}$ we get $\tau \geq 80$ ps, which is a reasonable value [33]. The broad plateau in Fig 7 is attributed to coherent NDFW or a $\chi^{(3)}$ process according to eq. (5). If $\hbar\omega_{P1}$ is shifted to lower values the central peak decreases because the absorption and consequently the population grating decrease. The coherent process can still be observed at lower $\hbar\omega_{P1}$ [39].

In Fig 8 we show schematically the set up for an experiment proofing phase conjugation [40, 41]. The two pump beams P1 and P2 are counter–propagating. The test beam I_T is focussed before the sample. At the focal point there is a diaphragm of variable diameter. If the signal beam I_S is specularly reflected no intensity passes the

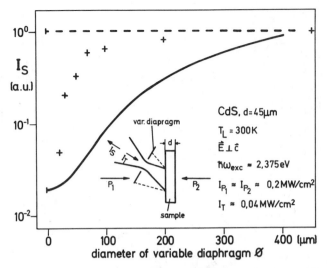

Fig 8. The intensity of the signal beam in a LIG experiment as a function of the diameter ϕ of the variable diaphragm. Schematic of the set–up in the insert. From [40].

diaphragm because the sample is inclined. If the signal beam returns in the direction of the incoming test beam but divergent as indicated by the dashed lines, one would expect a behaviour as shown by the calculated solid curve. If the signal beam is really phase–conjugate to the test beam it will return in itself and I_S should be independent of ϕ (dashed line). The experimental points closely follow this behaviour indicating phase conjugation. The steep drop around $\phi \approx 20$ μm is connected with the finite diameter of the focal point, with diffraction and aberration in the imaging system.

In the mixed crystal system $CdS_{1-x}Se_x$ no EHP was created at low temperatures under quasistationary excitation conditions, which result in an EHP in CdS and CdSe [42]. The reason is that the electron–hole pairs relax into localized states, where many–body effects are considerably reduced. At higher temperatures $(T_L \geq 60 \text{ K})$ thermal reemission into extended states takes place and the behaviour of the mixed crystals approaches the one of the pure constituents $x = 0$ and $x = 1$. An example is given in Fig 4 where the shift of the absorption edge under excitation is measured as a function of temperature for pure and mixed crystals. The optical nonlinearities observed in mixed crystals at low temperature are mainly due to saturation of the transitions in the exponential tail of the density of states, reaching eventually population inversion, gain and stimulated emission [42].

We now want to mention shortly some results on time–resolved measurements of optical nonlinearities in wide gap II–VI compounds. There was a lot of luminescence decay measurements after ps excitation mainly in CdSe due to the available laser pump wavelengths (e.g. [43]). The decay–times turned out to be in the range from 100 ps to 1 ns with a tendency to decrease at increasing excitation. Time–resolved luminescence and reflection measurements in CdS in [44] confirmed the previous results [33] that the EHP does not reach a liquid like state. The investigation of the disappearance of the excitonic reflection with sub ps resolution showed, that the screening of the exciton works in three–dimensional systems faster than the time–resolution, which was 100 fs [45]. Time–resolved LIG experiments in CdS [36] have already been shortly mentioned.

The phase–relaxation time τ_2 has been measured in the exciton–resonance in CdSe in reflection and in the TPA and the induced transition to the biexciton by LIG self diffraction with two delayed ps pulses [46]. Values of τ_2 up to 40 ps have been found which decrease rapidly both with increasing temperature and excitation due to interparticle scattering.

The dynamics of the leading and the trailing edges of the luminescence in $CdS_{1-x}Se_x$ has been investigated by various authors [47, 48]. At low temperatures ($T_L < 60K$) it has been found that after band–to–band excitation with ps pulses both the rise– and decay times increase with increasing localization depth. For higher temperatures, the dynamics are independent of the energy again due to thermal reexcitation to higher levels.

Recent measurements of τ_2 in the region of the localized states of $CdeS_{1-x}Se_x$ gave values up to 80 ps [49]. The values of τ_2 decrease rapidly both with increasing temperature and excitation.

The above mentioned experimental facts can be interpreted consistently with the assumption that the excitons in the "localized states" are not completely immobile, but are rather states of "reduced mobility". The energetic relaxation after band–to–band excitation down to the deepest localized states by phonon assisted tunneling [47, 48] indicates e.g. that there is still some possibility of spatial motion. LIG diffusion measurements of $CdS_{1-x}Se_x$ samples (however showing a modified luminescence behaviour) gave values of the diffusion length, which are approximately a factor three to four smaller than those of pure CdS ($l_D \approx 1.5 \ \mu m$) but not zero [50, 34].

In semiconductor doped glasses of the type addressed in Fig 1, a bleaching of the absorption edge has been found under ps–excitation with a recovery times in the ps and sub–ps range [51, 52]. A new class of transient phenomena in this type of materials has been predicted theoretically and partly verified experimentally in [53]. Calculations and measurements of the dynamic or ac Stark effect are reported e.g. in [54].

To conclude this section, it should be mentioned that photo–thermal optical nonlinearities have been observed e.g. in CdS or ZnSe [55, 56]. In this case cw–illumination with the green or blue lines of an Ar^+ laser leads to an excitation induced increase of the absorption and of the refractive index.

OUTLOOK

We have presented or reviewed now a rather rich variety of ON in wide gap II–VI compounds. Such nonlinearities connected with a suitable feedback are the necessary but not always sufficient ingredients to obtain optical bistability (ob). An optically bistable element has for a certain range of input parameters two (or more) different stable and reproducible states of transmission or reflection. A lot of work on ob has been performed on interference filters, which contain evaporated ZnSe and ZnS layers as active medium [56, 57]. We want to review here shortly ob observed in CdS which shows the richest variety of these phenomena. The photo–electronic ON connected with a broadening of the exciton resonance and the transition to an EHP have been exploited to get ob due to excitation induced increase and bleaching of absorption, and due to changes of the refractive index. For recent reviews which contain also a list of the switching parameters and original references see e.g. [58–60]. Dispersive ob presumably due to bleaching of a BEC in CdS has been reported in [61], while ob in degenerate six–wave mixing was observed in [62]. Photo–thermal effects lead also to ob. There is the shift of a BEC absorption line with temperature, which gave induced absorptive ob at T around 4K [63]. At room temperature, the green (514 nm) line of the Ar^+ laser and CdS form an ideally matched pair, because there is a strong excitation induced increase of absorption and dispersive effects. These nonlinearities have been used to measure induced absorptive, dispersive and mixed ob [55, 64, 65, 66]. The critical slowing down in the ob switching has been measured [64, 65] and the unstable branch [64, 67]. Inserting the bistable device in a hybrid ring resonator gave various types of self–oscillations

[64] and in the case of absence of ob a Feigenbaum route to deterministic chaotic optical oscillation [64]. A hybrid photo–thermal SEED device has been realized with CdS [68] and differential optical amplification [69]. Optical nonlinearities and bistability in evaporated or epitaxial CdS layers have been described in [69, 70]. Optical bistability due to induced absorption has also been reported in semiconductor doped glasses [71, 72]. Due to the reduced heat conductivity, it was possible to observe in this material partial switching connected with the formation of spatial kincks in the absorption coefficient [71, 72]. The latter effect has also been reported for photo–electronic ob in CdS crystals [58]. Ob due to the temperature dependence of the angle of total reflection has been reported in [73].

Acknowledgements: Part of the work reported here has been supported by the Deutsche Forschungsgemeinschaft and by the Commission of the European Communities.

References

[1]   Landolt–Börnstein, New Series, Group III, Vol 17 b, O. Madelung, M. Schulz and H. Weiss eds. Springer (1982)

[2]   C. Klingshirn and H. Haug, Physics Reports 70 315 (1981)

[3]   B. Hönerlage, R. Levy, J. B. Grun, C. Klingshirn and K. Bohnert, Physics Reports 124 161 (1985)

[4]   High Excitation and Short Pulse Phenomena M. H. Pilkuhn ed., J. Luminesc. 30 (1985)

[5]   Optical Nonlinearities and Instabilities in Semiconductors, H. Haug ed., Academic Press (1988)

[6]   C. Klingshirn in Laser Spectroscopy of Solids II W.M. Yen ed., Springer, to be published (1988)

[7]   Excitons, K. Cho ed., Topics in Current Physics, 14 Springer (1979)

[8]   Excitons, E.I. Rashba and M.D. Sturge eds, Modern Problems in Condensed Matter Sciences 2, North–Holland (1982)

[9]   P. J. Dean and D.C. Herbert in [8] p 55

[10]   G. Blattner, C. Klingshirn, R. Helbig and R. Meinl phys. stat. sol. b 107 105 (1981)
       R. Baumert, I. Broser, J. Gutowski and A. Hoffmann, Phys. Rev. B 27 6263 (1983), J. Gutowski, Phys Rev B 31 3611 (1985), J. Gutowski and I. Broser, Phys Rev. B 31 3621 (1985), J. Gutowski, J. Crystal Growth 86 528 (1988)

[11]   M.L.W. Thewalt, Proc. 14th Intern. Conf. Phys. Semicond. Inst. Phys., Conf. Ser. 43, 605 (1979) B.L.H. Wilson ed., and J. Phys. Soc. Japan. 49 Suppl. A 441 (1980)

[12]   J. Gutowski, Proc. 18th Intern. Conf. Phys. Semicond. O. Engström ed., World Scientific (1987) p 1413

[13]   F. Urbach, Phys. Rev. 92 1324 (1953)
       W. Martienssen, J. Phys. Chem. Sol. 2 257 (1957)
       For a recent theoretical treatment see e.g. G. Liebler, S. Schmitt–Rink and H. Haug, J. Luminesc. 34 1 (1985)

[14]   C. Klingshirn in Energy Transfer Processes in Semiconductors, B. Di Bartolo ed., Plenum Press, NATO ASI Series B 114 p 285, (1984)

178

[15]   E. Cohen and M.D. Sturge, Phys. Rev. B 25 3828 (1982)
       S. Permogorov, A. Reznitsky, S. Verbin, G.O. Müller, P. Flögel and M.
       Nikiforova, phys. stat. sol. b 113 589 (1982)

[16]   S. Permogorov and A. Reznitsky, Proc. Intern. Conf. "Excitons 84", p 194,
       Güstrow, GDR (1984)
       E. Cohen, Proc 17th Intern. Conf. Phys. Semicond.
       J.D. Chadi and W.A. Harrison eds., p 1221, Springer (1984)
       C. Klingshirn, Proc. Intern. Summer School on "Disordered Solids: Structures
       and Processes", Erice, Sicily (1987), B. Di Bartolo ed. Plenum Press, to be
       published

[17]   A. Reznitsky, S. Permogorov, S. Verbin, A., Naumov, Yu. Korostelin, V.
       Novozhilov and S. Prokovev, Solid State Commun. 52 13 (1984)
       D. Lee, A. Mysyrowicz, A.V. Nurmikko, B.J. Fitzpatrick, Phys. Rev. Lett. 58
       1475 (1987)
       T. Yao, M. Kato, J.J. Davies, H. Tanino
       J. Crystal Growth 86 552 (1988)

[18]   For recent reviews see e.g. IEEEJ.QE 22 (1986) D.S. Chemla and A. Pinczuk
       eds.

[19]   See the contributions of e.g. Konagai, McGill or Cornelissen to this volume or
       D. J. Olego, J. P. Faurie and P.M. Raccah, Phys. Rev. Lett. 55 328 (1985),
       H. Yang, H. Fujiyasi, A. Ishida and H. Kuwaraba, J. Lumenesc 40/41 717
       1988)

[20]   See e.g. the contribution of Ironside to this volume or A.I. Ekimov, Al.L.
       Efros and A.A. Onushchenko, Solid State Commun. 56 921 (1985)
       S. Schmitt–Rink, DAB Miller and D.S. Chemla Phys. Rev. B. 35 8113 (1987)

[21]   I.M. Lifshitz and V.V. Slezov, Sov. Phys. JETP 35 331 (1959)

[22]   A. Uhrig, Diploma–Thesis, Kaiserslautern (1989) to be published

[23]   L. Brus in Ref [18] p 1909
       E. Hanamura, Phys. Rev. B. 37 1273 (1988)

[24]   H. Kalt, V.G. Lyssenko, R. Renner and C. Klingshirn JOSA B2 1188 (1985),
       H. Kalt, R. Renner and C. Klingshirn, IEEEJ.QE 22 1312 (1986)

[25]   R. Renner, Ch. Weber, U. Becker and C. Klingshirn, J. Crystal Growth 86
       581 (1988)

[26]   J. Gutowski and I. Broser, J. Phys. C. 20 3771 (1987)

[27]   H. Schrey, V.G. Lyssenko, C. Klingshirn, Solid State Commun. 32 897 (1979)
       J.M. Hvam, G. Blattner, M. Reuscher and C. Klingshirn, phys. stat. sol. b
       118, 179 (1983)

[28]   G. Kurtze, V.G. Lyssenko and C. Klingshirn phys. stat. sol. b 110, K 103,
       (1982)

[29]   V.G. Lyssenko, K. Kempf, K. Bohnert, G. Schmieder, C. Klingshirn and S.
       Schmitt–Rink. Solid State Commun. 42 401 (1982)

[30]   C. Klingshirn in Ref [4] p 13

[31]   H. Haug and S. Schmitt–Rink, Prog. Quant. Electr. 9, 3, (1984)

[32]   H.E. Swoboda, F.A. Majumder, V.G. Lyssenko, C. Klingshirn and L. Banyai,
       Z.Physik B 70 341 (1988)

[33] K. Bohnert, M. Anselment, G. Kobbe, C. Klingshirn, H. Haug, S.W. Koch, S. Schmitt–Rink and F.F. Abraham, Z. Physik, B 42, 1, (1981)
F.A. Majumder, H.E. Swoboda, K. Kempf and C. Klingshirn, Phys. Rev. B 32, 2407 (1985)

[34] Ch. Weber, U. Becker, R. Renner and C. Klingshirn, Appl. Phys. B 45 113 (1988) and Z. Physik B (1988) in press

[35] M. Rinker, Diploma Thesis, Frankfurt (1988)

[36] H. Saito, J. Luminesc. 30 303 (1985), H. Saito and E.O. Göbel, Phys. Rev. B 31 2360 (1985)

[37] G. Mahler, T. Kuhn, A. Forchel and H. Hillmer in Ref [4] p 159

[38] M. Kunz, Ph. D. Thesis, Frankfurt/M (1989)

[39] R. Renner, Ph. D. Thesis, Kaiserslautern (1989)

[40] D. Oberhauser, Diploma–Thesis, Frankfurt (1988)

[41] The Principles of Nonlinear Optics, Y.R. Shen, J. Wiley and Sons, (1984)
D.J. Eichler and P. Gunter, Laser Induced Gratings, Springer Series in Optical Sciences 50 (1986)

[42] F.A. Majumder, S. Shevel, V.G. Lysenko, H.E. Swoboda and C. Klingshirn, Z. Physik B 66 409 (1987)

[43] H. Yoshida, H. Saito and S. Shionoya, J. Phys. Soc. Japan 50 881 (1981)

[44] H. Yoshida and S. Shionoya, phys. stat. sol. b 115 203 (1983)

[45] J.G. Fujimoto, S.G. Shevel and E.P. Ippen, Solid State Commun. 49 605 (1984)

[46] J. M. Hvam, I. Balslev and C. Dörnfeld, J. Luminesc. 38 76 (1987) and 40/41 529 (1988)

[47] J.A. Kash, A. Ron and E. Cohen, Phys. Rev. B 28 6147 (1983)

[48] S. Shevel, E.O. Göbel, G. Noll, P. Thomas, R. Fischer and C. Klingshirn, J. Luminesc. 37 45, (1987), H.E. Swoboda, F.A. Majumder, C. Klingshirn, S. Shevel, R. Fischer, E.O. Göbel, G. Noll, P. Thomas, A. Reznitsky and S. Permogorov, ibid. 38 178 (1987)

[49] J.M. Hvam, C. Dörnfeld and H. Schwab, to be published Proc. Intgern. Conf. on "Optical Nonlinearity and Bistability of Semiconductors", Berlin, Aug (1988)
H.–E. Swoboda, F.A. Majumder, C. Weber, R. Renner, C. Klingshirn, G. Noll, E.O. Göbel, S. Permogorov and A. Reznitsky, to be published Intern. Conf. Phys. Semicond. Warsaw (Aug 1988)

[50] H. Schwab, Diploma Thesis, Frankfurt (1988)

[51] M.C. Nuss, W. Zinth and W. Kaiser, Appl. Phys. Lett. 49 1717 (1986)

[52] M. Wegener, C. Dörnfeld and J.M. Hvam, private communication (1987)

[53] N. Peyghambarian and S.W. Koch, Revue Phys. Appl. 22 1711 (1987)

[54] M. Lindberg and S.W. Koch, JOSA B5 139 (1988) and M. Joffre, D. Hulin,

A. Migus, A. Antonetti, C. Benoit a la Guillaume, N. Peyghambarian, M. Lindberg and S.W. Koch, Optics Lett. 13 276 (1988), R. Zimmermann, phys. stat. sol. b 146 545 (1988)

[55] M. Lambsdorff, C. Dörnfeld and C. Klingshirn, Z. Physik B 64 409 (1986)

[56] H.M. Gibbs, Optical Bistability, Controlling Light with Light, Academic Press, New York (1985)

[57] "From Optical Bistability Towards Optical Computing the EJOB Project", P.Mandel, S.D. Smith and B. Wherett eds, North Holland (1987)

[58] M. Wegener, C. Klingshirn, S.W. Koch and L. Banyai, Semiconductor Science and Technology 1, 366 (1986)

[59] M. Wegener, C. Dörnfeld, M. Lambsdorff, F. Fidorra and C. Klingshirn, Proc. SPIE 1986 Quebeck Intern. Symp. on Optical Chaos, 667 102 (1986)

[60] F. Henneberger, phys. stat. sol. b 137 371 (1986)

[61] M. Dagenais and W.F. Sharfin, Appl. Phys. Lett. 46 230 (1985)

[62] A. Borshch, M. Brodin, V. Volkov, and N. Kukhtarev, Optics Comm. 41 213 (1982)

[63] M. Dagenais, Appl. Phys. Lett. 45 1267 (1984)

[64] M. Wegener and C. Klingshirn, Phys. Rev. A. 35 1740 and 4247 (1987) and M.Wegener, C. Klingshirn and G. Miller–Vogt, Z. Physik B 68 519 (1987)

[65] I. Haddat, M. Kretzschmar, H. Rossmann and F. Henneberger, phys. stat. sol. b 138, 235, (1986)
F. Henneberger, J. Puls, H. Rossmann, Ch. Spiegelberg, M. Kretzschmar and I. Haddad, Physica Scripta T 13 195 (1986), V.S. Dneprovskii, A.I. Furtichev, V.I. Klimov, E.V. Nazvanova, D.K. Okorokov and U.V. Vandishev, phys. stat. sol. b 146 341 (1988)

[66] I. Broser and J. Gutowski, Appl. Phys. B 46 1 (1988)
J. Gutowski, to be published in Proc. Intern. Conf. Phys. Semicond., Warshaw (1988)

[67] M. Wegener, C. Klingshirn, A. Daunois, J.–Y. Bigot, N. Cherkaoui and J.B. Grun, J. Apppl. Phys 52 685 (1988)

[68] C.Klingshirn, M. Wegener, A. Witt, D. Gnass and D. Jäger, Appl. Phys. Lett. 52 342 (1988)

[69] A. Witt, M. Wegener, V.G. Lyssenko, C. Klingshirn, G. Wingen, Y. Iyechika, D. Jäger, G. Müller–Vogt, H. Sitter, H. Heinrich, H.A. MacKenzie, IEEEJ.QE, to be published

[70] Cz. Koepke, phys. stat. sol. b 141 K 139 (1987)

[71] G.R. Olbright, H.M. Gibbs, H. Peyghambarian, H.E. Schmidt, S.W. Koch and H. Haug, Optical Bistability III, Springer Proceedings in Physics 8 186 (1985)

[72] C. Klingshirn, M. Wegener, C. Dörnfeld and M. Lambsorff in Ref [57] p 2252

[73] H. Rossmann, Intern. Workshop on "Nonlinear Optics and Excitation Kinetics in Semiconductors" Bad Stuer, GDR Nov (1987)

OPTICAL NONLINEARITIES, COHERENCE AND DEPHASING IN WIDE GAP

II-VI SEMICONDUCTORS [*]

Jørn M. Hvam and Claus Dörnfeld[a]

Fysisk Institut, Odense Universitet
DK-5230 Odense M, Denmark

INTRODUCTION

The direct bandgap near the visible part of the optical spectrum, makes the wide gap II-VI semiconductors potentially very interesting from the applications point of view. One can foresee widespread use in future all-optical or electro-optical devices, if some of the current problems in the preparation of these materials (purity, doping, etc.) find their solution with the new crystal growth techniques that are developing these years.[1,2] With these perspectives in mind, it is of particular importance to investigate the nonlinear optics of the II-VI semiconductors.

The optical properties of these material around the bandgap, near the onset of strong linear absorption, are dominated by excitonic features at low temperatures and at low and medium excitation intensities, and by e-h plasma effects at higher temperatures and/or intensities.[3] The strong optical nonlinearities associated with these transition processes are exciton interactions (state-filling, exciton-exciton collisions, biexciton formation) in the excitonic range, and many-body exchange and correlation effects (screening, bandgap renormalization) in the plasma range.[4] In the mixed crystals, like $CdSe_x S_{1-x}$, special localization phenomena occur,[5] due to the potential fluctuations associated with the random variation in the local atomic configurations.

In the present work, we will discuss the optical nonlinearities of the wide gap II-VI semiconductors at low temperatures and in the low and medium intensity range, where the excitonic features dominate. We shall use CdSe as an example, but the results are quite similar to those obtained in other materials like CdS and ZnO. We will show results on nonlinear luminescence, optical gain, and degenerate four-wave mixing. The latter with picosecond time resolution, which allows us to investigate also the coherence and dephasing of the excitonic transitions.

[*] Work supported by the Danish Natural Science Research Council

[a] On leave from Universität Kaiserslautern, Fachbereich Physik
Erwin Schrödinger Strasse, D-6750 Kaiserslautern, FR-Germany.

EXPERIMENTAL TECHNIQUES

The CdSe samples used in the experiments were all thin (10 - 50μm) platelets, grown from the vapour-phase. They were of good optical quality with the c-axis in the plane of the sample, revealing the optical anisotropy for light propagating perpendicular to the slab. The experiments reported here are all for light polarized perpendicular to the c-axis (E⊥c), allowed for optical transitions involving the uppermost valence band (A excitons). The samples were placed either in the helium exchange gas (4.2K) or in the superfluid helium (1.8K) of a variable temperature He-cryostat.

The optical gain measurements were performed under quasi steady-state conditions with the nanosecond excitation from a Nitrogen-gas laser pumped dye-laser. The optical gain spectra were recorded by the excitation length modulation technique, previously described.[6]

The nonlinear luminescence and four-wave mixing experiments were performed with the picosecond time resolution of a dye-laser, synchronously pumped by a mode-locked Ar-ion laser. The nonlinear contribution to the luminenscence was separated from the (larger) linear contribution by a special modulation and heterodyne detection technique.[7] In four-wave mixing the nonlinear signal is emitted in directions with no linear background and is therefore easily isolated by spatial filtering.[8]

For luminescence and gain measurements, the dye laser was tuned to the absorbing region of the spectrum and the emission was spectrally analyzed and detected. In the four-wave mixing experiments, the spectral information was obtained by slowly scanning the dye-laser wavelength while accumulating the nonlinear signal in an optical multichannel analyzer. The temporal information was in the form of correlation traces, obtained by scanning the delay between the exciting (and/or interfering) beams and recording the nonlinear signal as a function of the delay.

OPTICAL NONLINEARITIES

The predominant optical nonlinearities in the wide gap II-VI semiconductors at low temperatures and at low and medium excitaion intensities are due to exciton interactions. At exciton densities $N_x \geq 10^{-4} \cdot r_x^{-3}$, where r_x is the exciton Bohr radius, excitons no longer behave as noninteracting ideal Bosons. This implies that the optical response from the light-exciton interaction develops a nonlinear contribution.[9] The fact that excitons are composed of electrons and holes becomes important,[10] and state filling effects appear with increasing excitation density.[10] These effects can be observed as broadenings and shifts of the exciton resonance in excite-and-probe transmission and reflection experiments,[11,12] but maybe even more clearly they are revealed in purely nonlinear four-wave mixing experiments,[8] as we shall see.

Direct exciton interactions may also give rise to new and inherently nonlinear optical transitions, for example via inelastic exciton-exciton (x-x) collisions[13,14] or formation of biexcitons (excitonic molecules).[15,16] Both of these processes result in nonlinear emission, or induced absorption, in the P- and M-band, respectively, and they are easily inverted to give optical gain.[17,18] Biexciton formation also results in very strong, resonantly enhanced (giant), two-photon absorption (TPA) and emission processes.[19,20] The nonlinear resonances just below the free exciton energy E_x are thus: the P-band an exciton binding energy E_x^b below, the M-band a biexciton binding energy E_M^b below, and the TPA resonance $0.5 \cdot E_M^b$ below.

The above resonances are all due to intrinsic, extended (free) electronic excitations. Localized (bound) excitations also play a role in the optical bandgap transitions at low temperatures. In the pure compound semiconductors they are predominantly impurity bound excitons (e.g. excitons bound to neutral donors and acceptors, I_2 and I_1 lines.[7,15] In the mixed crystals they are excitons localized by potential fluctuations.[5] The bound excitons also contribute to the optical nonlinearities, since they tend to saturate at even lower intensities than the free excitons.[7,21]

In the following, we shall show results of the nonlinearities discussed above, using CdSe as an example. In CdSe, the relevant excitation energies are: $E_x(\Gamma_{ST})$= 1.825eV, E_x^b= 15meV, E_M^b= 5meV, $E_{I_1}^b$ = 9meV, and $E_{I_2}^b$ = 4meV.

Nonlinear Luminescence

In Fig.1 is shown the nonlinear contribution to the luminescence spectrum in CdSe at 1.8K excited by a picosecond pulse train of 0.05nJ per pulse (focused to $\cong 2\mu J/cm^2$). The full curve is the total luminescence intensity and the dashed curve is the quadratic term in the expansion of the luminescence intensity in orders of the excitation intensity:

$$I_{lum}(\hbar\omega) = a(\hbar\omega)\cdot I_{exc} + b(\hbar\omega)\cdot I_{exc}^2 + \cdots \qquad (1)$$

The latter is obtained by the excitation correlation technique, previously described.[7] Under these conditions, the luminescence is dominated by bound exciton recombination (I_2), P-band emission, and recombination of free excitons with simultaneous emission of one LO-phonon (E_x-LO), and it is predominantly linear in the excitation intensity. The strongest nonlinear contributions are due to saturation of the bound exciton emission (negative signal) and superlinear emission (positive signal) in the M- and P-bands. With increasing excitaion intensity, the latter processes are easily stimulated, giving rise to optical gain and/or induced absorption depending on the populations in the initial and final excitonic levels.[17,18]

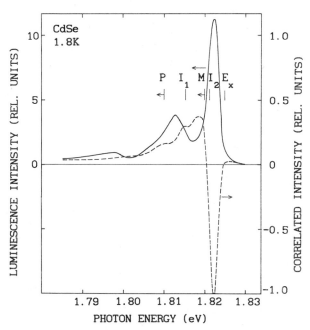

Fig.1   Nonlinear luminescence in CdSe at 1.8K with picosecond excitation. The full curve is the total luminescence, and the dashed curve is the leading nonlinear contribution.

Optical Gain

We have investigated the optical gain in the II-VI semiconductors under quasi steady-state (≅10ns) excitation conditions. In Fig.2 are shown optical gain spectra, for different excitation densities, in CdSe at 1.8K. The spectra were automatically recorded by the excitation length modulation technique, where the signal intensity $I(gL)$ is a simple function of the gain coefficient g and the excitaiton length L.[17]

$$I(gL) = \log(e^{gL} + 1)/\log 2 \tag{2}$$

Under the excitation conditions of our experiment ($I_{exc} = 10\text{-}500 \text{kW/cm}^2$, L = 300μm), it is clear from Fig.2 that the M-band exhibits no gain. The population is all the time larger in the exciton level than in the biexciton level, resulting in induced absorption. This is observed also in the four-wave mixing experiments, discussed below. In the P-band, on the other hand, we observe optical gain at low intensities changing into induced absorption at higher intensities shifting the optical gain to lower photon energies. This behaviour is rather similar to the performance in CdS.[17] In ZnO, on the other hand, the biexciton binding energy is sufficiently large to obtain population inversion and gain also in the M-band.[18] The magnitude of the gain coefficients observed are typical for exciton transitions in the II-VI compounds.

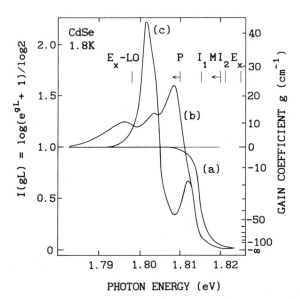

Fig.2   Optical gain in CdSe at 1.8K with nanosecond excitation. I_{exc} = 10kW/cm^2 (a), 100kW/cm^2 (b) and 500kW/cm^2 (c). The excitation length L = 300μm. Note the nonlinear scale of gain coefficient to the right.

Four-Wave Mixing

In order to investigate the optical nonlinearities in, or near, the exciton resonance itself, where there is strong linear absorption, we have performed degenerate four-wave mixing (DFWM) experiments in a back scattering geometry.[8] We have applied the two-beam configuration, where two incoming beams with wave vectors k_1 and k_2 interact via the nonlinear medium and create a self-diffracted signal, e.g. in the direction $2k_2 - k_1$. The result of such an experiment is shown in Fig.3, where also the linear reflection curve is shown (dashed curve). Three resonances are clearly seen, associated with the exciton, the TPA, and the M-band transitions. As already discussed, we explain these nonlinearities by state-filling, two-photon absorption to the biexciton states, and induced exciton-biexciton transitions, respectively. The origin of the back scattered signals in Fig.3 is different for the three different resonances. In the exciton resonance, it is real back-diffraction from a two-dimensional grating, because the absorption length of the incident beams is comparable to the wavelength of the light. In the other two resonances, on the other hand, the absorption length is comparable to the sample thickness (5-10µm). Accordingly, the signal is the forward diffraction, reflected in the back surface of the sample.

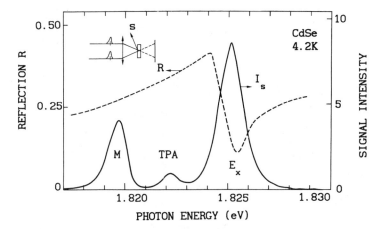

Fig.3   Excitation spectrum of the signal intensity (full curve) in back scattering of DFWM in CdSe at 4.2K with picosecond excitation (I_1 = 10kW/cm^2 and I_2 = 40kW/cm^2). The dashed curve shows the linear reflection coefficient R (scale to the left).

From a forward DFWM experiment in the TPA and M-band resonances, we can actually estimate the magnitude of the nonlinearity, which is governed by the third order nonlinear susceptibility $\chi^{(3)}$. Assuming quasi steady-state conditions, and that the nonlinear absorption is small compared to the linear absorption, we find the following expression:[22]

$$|\chi^{(3)}|^2 = \frac{4\varepsilon_0^2 n^4 c^4 \alpha^2}{\omega^2 I_1(0) \cdot I_2(0)} \cdot \frac{I_s(d)}{I_2(d)} \qquad (3)$$

where ε_0 is the vacuum dielectric constant, n is the refractive index, c is the velocity of light, α is the absorption coefficient, and ω is the optical frequency. $I_{1,2}(0)$ are the incident intenities and $I_{2,s}(d)$ are the pump and signal intensities transmitted through the sample, the signal in the $2\mathbf{k}_2 - \mathbf{k}_1$ direction. From the experimental data, we find $\chi^{(3)} \cong 10^{-9} \mathrm{cm}^2/\mathrm{V}^2 \cong 10^{-5}\mathrm{esu}$, which is a large value for a wide gap semiconductor.[23]

We have also performed DFWM experiments in mixed crystals of $CdSe_x S_{1-x}$. In these materials we find only one nonlinear resonance near the absorption edge. Spectraly it coincides with the observed luminescence and the width is about 5meV.[24] We associate it with excitons localized in the randomly fluctuating potential of the alloy and we suggest the nonlinearity to be a saturation effect.

COHERENCE AND DEPHASING

In the two-beam DFWM experiment described above the diffracted signal beam is generated by the coherent interaction of the incident laser beams, consisting of picosecond pulse trains. The signal beam therefore probes the coherence between the polarizations set up by light pulses from the two beams. By varying the delay between the two pulse trains one can get information about the coherence and dephasing of the resonant polarizations with respect to the generating electromagnetic fields.[8,11,22] In Fig.4 are shown correlation traces, i.e. signal intensity as a function of the delay τ_{12} between the incident pulses, for the exciton resonance in back scattering geometry. We observe strongly asymmetric correlation traces, indicating that the coherence, or dephasing, time T_2 is larger than the laser pulse width $\tau_L = 1.4\mathrm{ps}$ (FWHM).

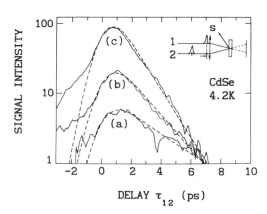

Fig.4   Experimental (full curves) and calculated (dashed curves) correlation traces. The parameters are Γ = 9meV and (a): $N_x = 4 \cdot 10^{15} \mathrm{cm}^{-3}$, $T_2 = 12\mathrm{ps}$; (b): $N_x = 8 \cdot 10^{15} \mathrm{cm}^{-3}$, $T_2 = 7\mathrm{ps}$; (c): $N_x = 1.4 \cdot 10^{16} \mathrm{cm}^{-3}$, $T_2 = 5\mathrm{ps}$.

The dashed curves in Fig.4 are calculated correlation traces, based on a two-level model with inhomogeneous broadening.[8] The fitting parameters in the calculations are the dephasing time T_2 (corresponding to a homogeneous line width T_2^{-1}), the inhomogeneous line with Γ, and τ_L. Thus, from the fits we can extract T_2 and Γ. In the inhomogeneously broadened two-level system, the decay of the correlation traces for $\tau_{12} \gg \tau_L$ is simply given by $\exp\{-4\tau_{12}/T_2\}$.

The experimental curves in Fig.4 are for different excitation intensities, i.e. for different densities N_x of excitons, created by linear absorption of the incident beams. In the M-band resonance, we have determined the dephasing time over a larger dynamic range in exciton density by applying an advanced pump pulse of variable intensity before the DFWM experiment. The result is shown in Fig.5, where the dephasing rate $1/T_2$ is plotted as a function of N_x. The linear behaviour suggests that the dephasing is governed by exciton-exciton collisions in the corresponding intensity range. Hence, we may express the relationship by

$$1/T_2 = 1/T_2^0 + v_{th}\sigma_x N_x \tag{4}$$

where $v_{th} = 10^6$ cm/s is the thermal velocity of excitons, σ_x is their scattering cross section, and T_2^0 is the low density value for the dephasing time, governed by impurity (and phonon scattering). From the fit in Fig.5, we determine $T_2^0 = 50$ps and $\sigma_x = 6.25 \cdot 10^{-12}$cm^2. The latter value is in good agreement with a recent theoretical calculation.[25]

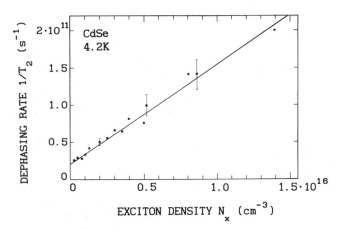

Fig.5   Dephasing rate $1/T_2$ in CdSe at 4.2K for the exciton and M-band resonances as a function of exciton density.

In the mixed crystals, the correlation traces show a two-component behaviour: a fast component from the freely propagating polaritons and a slow component ($T_2 \cong 80$ps) from the localized excitons with an apparently reduced scattering rate.[24]

CONCLUSIONS

The optical nonlinearities in the wide gap II-VI semiconductors, of which a variety has been presented and discussed in the present paper, are readily attainable with present day light signal levels. One can therefore easily think of applications in devices like laser, amplifiers, modulators, mixers, switches, etc.. One advantage is the fast response time of devices based on II-VI's.[23] The recombination lifetime of excitations in these materials is already in the nanosecond or subnanosecond range. Devices based on the coherent nonlinearity, investigated by degenerate four-wave mixing in this work, can have response times in the picosecond range, being limited only by the dephasing time.

The main obstacle for large scale applications is the material and device preparation technology. One can also argue that the nonlinearities based on excitonic resonances are impractical, because they need cooling to low temperatures. Again, materials technology may help us if high quality two-dimensional layers can be grown by for example an MBE technique.[2] In two (or lower) dimensions, excitons may bind sufficiently strong to be present even at room temperature, as has been demonstrated in the III-V semiconductors.[9,10,26]

ACKNOWLEDGEMENTS

We want to thank M. Jørgensen and H. Schwab for taking part in some of the experiments, I. Balslev for contributing to the theoretical calculation, and D.C. Reynolds for supplying the high-quality CdSe crystals. One of us (C.D.) wishes to thank the Danish ministry of Education (Scholarship section) and the DAAD (Deutscher Akademischer Austausch Dienst) for financial support.

REFERENCES

1. J.B. Mullin, S.J.C. Irvine, J. Giess and A. Royle, Recent Developments in the MOVPE of II-VI Compounds, J. Crystal Growth 72:1 (1985).
2. T. Yao, Characterization of ZnSe Grown by Molecular-Beam Epitaxy, J. Crystal Growth 72:31 (1985).
3. C. Klingshirn and H. Haug, Optical Properties of highly Excited direct gap semiconductors, Phys. Reports 70:315 (1981).
4. E. Hanamura and H. Haug, Condensation Effects of Excitons, Phys. Reports 33:209 (1977).
5. S. Permogorov, A. Reznitskii, S. Verbin, G.O. Müller, P. Flögel and M. Nikiforova, Localized Excitons in $CdS_{1-x}Se_x$ Solid Solutions, phys.stat.sol. (b) 113:589 (1982).
6. J.M. Hvam, Direct Recording of Optical Gain Spectra from ZnO, J.Appl.Phys. 49:3124 (1978).
7. M. Jørgensen and J.M. Hvam, Time-resolved Nonlinear Luminescence Spectroscpy by Picosecond Excitation Correlation, Appl.Phys.Lett. 43:460 (1983).
8. C. Dörnfeld and J.M. Hvam, Optical Nonlinearities and Phase Coherence in CdSe Studied by Transient Four-Wave Mixing, to be published.
9. W.H. Knox, R.L. Fork, M.C. Downer, D.A.B. Miller, D.S. Chemla, C.V. Shank, A.C. Gossard and W. Wiegmann, Femtosecond Dynamics of Resonantly Excited Excitons in Room-Temperature GaAs Quantum Wells, Phys.Rev.Lett. 54:1306 (1985).

10. S. Schmitt-Rink, D.S. Chemla and D.A.B. Miller, Theory of Transient Excitonic Optical Nonlinearities in Semiconductor Quantum-Well Structures, Phys.Rev. B 32:6601 (1985).

11. L. Schultheis, J. Kuhl, A. Honold and C.W. Tu, Ultrafast Phase Relaxation of Excitons via Exciton-Exciton and Exciton-Electron Collisions, Phys.Rev.Lett. 57:1635 (1986).

12. R. Leonelli, J.C. Mathae, J.M. Hvam, F. Tomasini and J.B. Grun, Transient Phase-Space Filling by Resonantly Excited Exciton Interactions in CuCl, Phys.Rev.Lett. 58:1363 (1987).

13. D. Magde and H. Mahr, Exciton-Exciton Interaction in CdS, CdSe, and ZnO, Phys.Rev.Lett. 24:890 (1970).

14. J.M. Hvam, Exciton-Exciton Interaction and Laser Emission in High-Purity ZnO, Solid State Commun. 12:95 (1973).

15. S. Shionoya, H. Saito, E. Hanamura and O. Akimoto, Anisotropic Excitonic Molecules in CdS and CdSe, Solid State Commun. 12:223 (1973).

16. J.M. Hvam, Exciton Interaction in Photoluminescence from ZnO, phys.stat.sol. (b) 63:511 (1974).

17. J.M. Hvam, Induced Absorption and Gain from High Density Excitons in CdS, Solid State Commun. 26:373 (1978).

18. J.M. Hvam, Optical Gain and Induced Absorption from Excitonic Molecules in ZnO, Solid State Commun. 26:987 (1978).

19. T. Itoh, Y. Nozue and M. Ueta, Giant Two-Photon Absorption and Two-Poton Resonance Raman Scattering in CdS, J.Phys.Soc.Japan 40:1791 (1976).

20. J.M. Hvam, G. Blattner, M. Reuscher, and C. Klingshirn, The Biexciton Levels and Nonlinear Optical Transitions in ZnO, phys.stat.sol. (b) 118:179 (1983).

21. M. Dagenais, Low Power Optical Saturation of Bound Excitons with Giant Oscillator Strength, Appl.Phys.Lett. 43:742 (1983).

22. J.M. Hvam, I. Balslev and B. Hönerlage, Optical Nonlinearity and Phase Coherence of Exciton-Biexciton Transition in CdSe, Europhys. Lett. 4:839 (1987).

23. C. Dörnfeld and J.M. Hvam, Strong Nonlinear Resonances in CdSe with Picosecond Response Time, J.de Physique (in press).

24. J.M. Hvam, C. Dörnfeld and H. Schwab, Optical Nonlinearities and Phase Coherence in CdSe and CdSe$_x$S$_{1-x}$, phys.stat.sol. (b), to be published.

25. S.G. Elkomoss and G. Munschy, Exciton-Exciton Elastic Scattering Cross-Section for Different Semiconductors, J.Phys.Chem.Solids 42:1 (1981).

26. D. Chemla and D.A.B. Miller, Room-Temperature Excitonic Nonlinear-Optical Effect in Semiconductors, J.Opt.Soc.Am. B2:1155 (1985).

FREQUENCY DEPENDENCE OF INTERBAND TWO-PHOTON ABSORPTION MECHANISMS IN ZnO AND ZnS

I.M.Catalano, A.Cingolani and M.Lepore

Dipartimento di Fisica – Universita'di Bari
Via Amendola 173 – 70126 Bari (Italy)

ABSTRACT

The transition mechanism changes involved in direct interband two-photon absorption (TPA) processes have been studied in two II-VI compounds:ZnO and ZnS. The experimental results have indicated that the transition mechanism nature changes when the excitation energy is varied. The transitions are of an allowed-allowed type near the energy gap and become of an allowed-forbidden type far from it. In particular, for ZnO when $2\hbar\omega - E_g > \sim 400$ meV the first experimental evidence of a dominant mechanism of forbidden-forbidden type has been obtained. Furthermore, the spectral behaviour of the TPA coefficient ($\alpha^{(2)}$) has been well described by a parametric formula containing terms (with different energy dependences) related to the above-mentioned different transition mechanisms.

INTRODUCTION

The different theoretical models proposed to calculate the TPA coefficient ($\alpha^{(2)}$) predict different transition mechanisms depending upon the symmetries of the valence, conduction and virtual bands and, therefore, upon the appropriate matrix elements involved. In particular, the following transition types have been considered [1]:
(a)"allowed-allowed" (a-a), when transitions both from the valence band to the intermediate state (v-i) and from intermediate state to the conduction band (i-c) are considered of the allowed type. In this case, both the dipole matrix elements ($|P_{vc}|^2$ and $|P_{ic}|^2$) involved in $\alpha^{(2)}$ theoretical evaluation can be assumed as $\vec{K}$-independent at least in the region of the $\vec{K}$-space near a critical point;
(b)"allowed-forbidden" (a-f), when (v-i) and (i-c) transitions

are assumed to be allowed and forbidden, respectively. In this case, the dipole matrix element for the forbidden transitions ($|\vec{P}_{ic}|^2$) has to be taken as linearly dependent on $\vec{K}$;

(c) "forbidden-forbidden" (f-f), when both the (v-i) and (i-c) transitions are of the forbidden type. In this case both matrix elements are to be considered $\vec{K}$-dependent.

It is worth noting that the above-mentioned two-photon transition types are characterized by different spectral dependences. In particular, for the (a-a) transitions an $\alpha^{(2)}$ dependence nearly equal to $(2\hbar\omega - E_g)^{1/2}$, for the (a-f) ones an $\alpha^{(2)}$ frequency dependence roughly equal to $(2\hbar\omega - E_g)^{3/2}$ and for the (f-f) ones an $\alpha^{(2)}$ frequency dependence nearly equal to $(2\hbar\omega - E_g)^{5/2}$ have been predicted, (where E_g is the energy-gap). Which of the aforementioned transition types is the dominant TPA mechanism, has until now, been believed to be dependent on material paramenters [2]. On the contrary, for a direct gap semiconductor, GaAs, it has been shown that the dominant TPA transition mechanism changes in the same material, when the excitation energy becomes greater than E_g [3]. In particular, transitions seem to be of the (a-a) type near the energy gap and become of the (a-f) one far from it, while no experimental evidence has yet been given of (f-f) type transitions.

The aim of this paper is to obtain $\alpha^{(2)}$ dispersion curves in order to point out the different frequency dependence. In this way, it will be shown that every theoretical model proposed to evaluate the $\alpha^{(2)}$ frequency dependence is valid only in a limitated energy range when large-gap semiconductors are considered and that all intermediate states have to be considered to explain the spectral dependence of the absorption coefficient.

EXPERIMENTAL PROCEDURE

The TPA spectral dependence has been investigated in two different II-VI compounds: ZnO and ZnS in the energy range experimentally available to our dye laser 2.32 eV < $2\hbar\omega$ <4.25 eV. Both samples are direct gap semiconductors, the former has a wurtzite structure (E_g=3.43 eV at 80 K, m_c=.19m, $m_{v\perp}$= .21m, $m_{v\perp}$=.2m, m_{v_3}=.71m, n=2) while the latter has a zinc blende one with a mainly ionic bonding (E_g=3.82 at 80 K, m_c=.3m, m_v=.49m, n=2.37). Measurements have been carried out in high quality crystals to prevent damage from high-power density excitation and to avoid dependence on sample purity degree. TPA line shapes have been measured in all samples at 80 K by means of nonlinear luminescence technique [4]. Detailed information on this experimental technique and set-up have been reported in ref.1.

RESULTS AND DISCUSSION

The TPA line shapes (dots) of ZnO and ZnS as a function of $(2\hbar\omega - E_g)$ have been reported in figs.1 and 2. These experimental results clearly shows that in these large gap semiconductors $\alpha^{(2)}$ raises by increasing the difference $(2\hbar\omega - E_g)$. In particular, the $\alpha^{(2)}$ energy dependence is sublinear near the energy gap, while it becomes superlinear far from it. An attempt to give a general interpretation of these line shapes can be made by fitting the experimental data with a parametric formula which takes into account all the three possible transition mechanisms predicted by the theories, i.e.

$$\alpha^{(2)} = \sum_{i=1}^{3} c_i \left(2\hbar\omega - E_g \right)^{i-1/2} \qquad (1)$$

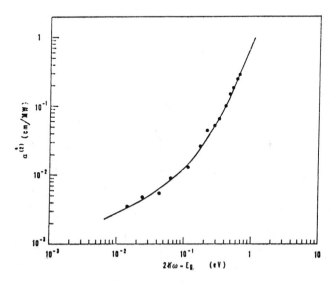

Fig. 1 - Two-photon absorption coeff. vs $(2\hbar\omega - E_g)$ for a ZnO single crystal. The full line is the best fit curve obtained by means of eq. 1. The dots indicate experimental results (The χ^2 value for degree of freedom is .48 and the fit probability is .92)

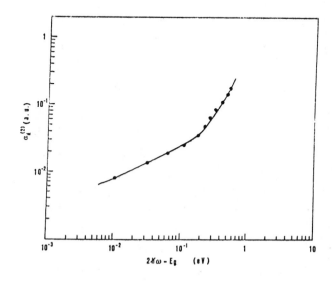

Fig. 2 - The same of fig. 1 for ZnS with $\alpha^{(2)}$ in arbitrary units. (The χ^2-value for degree of freedom is .86 and the fit probality is .63)

195

The results of the fits are shown by the continuous curves in figs.1 and 2. The agreement is very good for both materials as shown by the χ^2 and by the fit probability reported in the figure captions.

For ZnO these results shows that in the $\alpha^{(2)}$ dependence on $(2\hbar\omega-E_g)$ all terms of eq.(1) contribute, suggesting that all the three TPA mechanisms corresponding to a different choice of matrix elements contribute simultaneously to the transition amplitude in the energy range considered here though with different weights at different $2\hbar\omega$. In particular, the $(2\hbar\omega-E_g)^{5/2}$ trend is predominant for $2\hbar\omega \gg E_g$. This last dependence has not been detected for ZnS because its larger gap does not allow us to excitate this sample very far from the gap. As a consequence, the third term of eq.(1) has not been considered in the ZnS fitting procedure. It is worth noting that the analysis of the ZnO and ZnS band structures [1] allow us to exclude that in these two semiconductors the $\alpha^{(2)}$ dependences could be ascribed to other singularities in the joint density of states.

The interpretation of our experimental $\alpha^{(2)}$ line shapes, which has been qualitatively carried out by means of the functional form (eq.(1)), can be further improved by the comparison between the theoretical and the experimental $\alpha^{(2)}$.

In figure 3 and 4, the theoretical $\alpha^{(2)}$ line shapes and the experimental points are reported for ZnO and ZnS, respectively. In figure 3 the $\alpha^{(2)}$ experimental values have been obtained by normalizing our data with the absolute values $\alpha^{(2)}$ =.021 cm/MW at $(2\hbar\omega-E_g)$ =.13 eV measured in ref.[5]. Conversely for ZnS an analogous normalization procedure cannot be carried out because no absolute values is available. The theoretical curves (a-b) have been calculated for both materials by means of formula 10 of ref.[6], formula 3 of ref. [7] and curve c for ZnO by means of formula 9 of ref. [8]. The first equation (curve a) is derived by a three-band model with parabolic and nondegenerate bands, and both the v-n and the n-c transition of the allowed type ($\alpha^{(2)} \propto (2\hbar\omega-E_g)^{1/2}$). The second equation (curve b) is obtained from a nonparabolic two-band model and the transitions are of the allowed and forbidden type ($\alpha^{(2)} \propto (2\hbar\omega-E_g)^{3/2}$). Finally, for ZnO the third equation (curve c) considers a three-band model with both the transition forbidden ($\alpha^{(2)} \propto (2\hbar\omega-E_g)^{5/2}$). It must be noted that this last equation does not consider band nonparabolicity and degeneration effects. Thus, to take these effects into account in numerical comparison of fig.3 , a rough estimate has been made as indicated in ref.[9]. Some experimental points with very close to the gap have been reported in this figure. These points stand away from the curve which is obtained above the edge and confirm the importance of excitonic contribution in TPA processes, as pointed out in ref.[1].

From figs. 3 and 4 the following feature can be deduced for ZnO and ZnS:

a) In the energy region just above the direct gap ($10 < (2\hbar\omega-E_g)$ 120 meV for ZnO and $9 < (2\hbar\omega-E_g) < 150$ meV) for ZnS where the choice of constant matrix elements is a reasonable assumption, the experimental data well agree with the theoretical model giving a ($2\hbar\omega-E_g)^{1/2}$ dependence [6]. For ZnO, this model gives good quantitative agreement.

b) In the intermediate energy region, for $2\hbar\omega-E_g$ ranging between 120 and 400 meV for ZnO and 150 and 420 meV for ZnS,

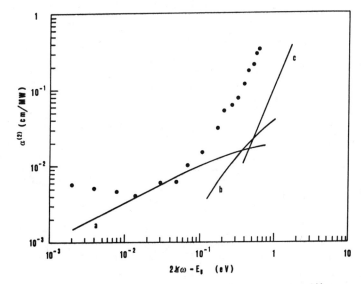

Fig.3. - Quantitative comparison between the $\alpha^{(2)}$ experimental (dots) and theoretical (full line) line shapes for ZnO.

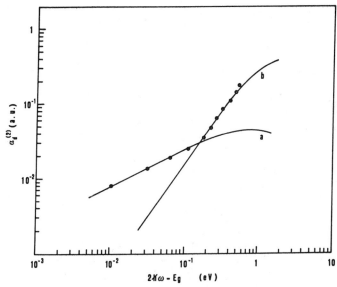

Fig.4 - Qualitative comparison between the $\alpha^{(2)}$ experimental (dots) and theoretical (full line) line shape for ZnS.

an accurate proportionality is found between our experimental data and the model, assuming one of the matrix elements constant and the other linearly $\vec{k}$-dependent (spectral dependence $(2\hbar\omega - E_g)^{5/2}$)) [7]). However the theoretical values underestimate the ZnO experimental data by a factor of 5. c) In the range beyond 400 meV for ZnO a good qualitative agreement is found only with the theoretical model assuming both matrix elements linearly $\vec{k}$-dependent, with a spectral dependence approximately equal to $(2\hbar\omega - E_g)^{5/2}$. Also in this case the theoretical values underestimates the experinmental ones by a factor of 10.

This underestimate can be ascribed to the fact that only some intermediate states have been considered in all the model used for TPA virtual transitions, while the complete set of states of the crystal must be considered. Furthermore at high excitation energy the choice of constant matrix elements is no more a reasonable assumption and its $\vec{k}$-dependence must be considered to give the correct $\alpha^{(2)}$ energy dependence.

CONCLUSIONS

All the results reported here clearly show that the dominant transition mechanism changes when the the excitation energy is varied. In particular, this mechanism is of the a-a type for energies just above the gap, it becomes of the a-f type far from the gap and for well above E_g the f-f transition type are prevailing. Furthermore, it has been proved that in these large gap semiconductors $\alpha^{(2)}$ dispersion curve cannot be described by anyone of the available theoretical models. Conversely, the complete $\alpha^{(2)}$ line shape can be fully described by a parametric formula taking all the three different TPA mechanism into account.

REFERENCE

[1] I.M.Catalano, A.Cingolani and M.Lepore, Phys.Rev.B, 33, 7270, (1986);
I.M.Catalano, R.Cingolani and M.Lepore, Solid State Comm. 60, 385, (1986)
I.M.Catalano, R.Cingolani and M.Lepore, Il Nuovo Cimento, 9D, 1313, (1987)
I.M.Catalano, A.Cingolani, R.Cingolani and M.Lepore, Physica Scripta, 37, 579, (1988)
[2] G.D.Mahan, Phys.Rev. 170, 825, (1968)
[3] J.P.van der Ziel, Phys.Rev.B, 16, 2775, (1976)
[4] I.M.Catalano and A.Cingolani, J.Appl.Phys. 50, 5638, (1968)
[5] G.Kobbe and C.Klingshirn, Z. Phys. B, 37, 9, (1980)
[6] A.R.Hassan, Il Nuovo Cimento, 70B, 21, (1970)
[7] A.Vaidyanathan, T.Walker, A.H.Guenther, S.S.Mitra and R.R.Narducci, Phys.Rev.24, 2259, (1981)
[8] R.Braunstein and N.Ockman, Phys.Rev. 134, A449, (1964)
[9] A.Vaidyanathan, A.H.Guenther and S.S. Mitra, Phys.Rev.B, 22 6480, (1980)

II-VI SEMICONDUCTOR-DOPED GLASS: NONLINEAR OPTICAL PROPERTIES AND DEVICES

C. N. Ironside

Department of Electronics and Electrical Engineering
University of Glasgow
Glasgow G12 8QQ UK

INTRODUCTION

II-VI semiconductor doped glass consists of nanocrystallites of semi-conductors, such as CdS_xSe_{1-x} and CdTe, in a glass matrix. Depending upon the manufacturing process the crystallites can be between 2 and 10 nm in size and up to 0.1% of the total volume. They appear to be randomly orientated in the glass and of a fairly uniform size distribution. Due to the optical absorption characteristics of the semiconductor crystallites these glasses have a sharp cut off in their transmission for photon energies larger than the band-gap. The wavelength of the cut off can be altered primarily according to value of x in CdS_xSe_{1-x} and the range 500 to 695 nm. This feature makes the glasses particularly useful as wavelength filters and they are commercially available in this form. Recent investigations (Yao et al., Olbright et al. and Roussignol et al.) of the nonlinear optical properties of these semiconductor doped glass (SDG) have revealed a significant nonlinear refractive index which recovers on a picosecond time-scale. It has also been possible to manufacture both optical waveguides and optical fibres using SDG. The advantage of guided wave configurations for nonlinear optics is that high intensities can be maintained over longer distances than bulk optics and interaction distances can be absorption rather than diffraction limited. In summary SDG is attractive as nonlinear optical material primarily because it offers a rapid recovery time and room temperature operation; it is also commercially available, inexpensive and has an available guided wave technology. The combination of a material with a fast nonlinear refractive index and waveguide technology opens up the possibility of high speed all-optical logic devices utilising guided wave geometries.

This paper is a review of work in this field and concentrates on work carried out by the author and his collaborators. It reports on the preparation and nonlinear optical properties of CdS_xSe_{1-x} doped glasses waveguides and optical fibres and on the operation of an all-optical switch fabricated in this material.

COMPOSITION AND CRYSTALLITE DEVELOPMENT IN THE GLASS

The semiconductor doped glass is manufactured by adding a few percent of cadmium, selenium and sulphur to what would otherwise be a transparent base glass. The base glass can be either soda-lime or silicate glass; the usual commercial SDG is a silicate glass. The batch of materials is typically melted within the range 1300-1400 C and the glass formed with conventional casting techniques. Part of the process is an annealing step at about 500 C to produce a stress free glass. A subsequent heat treatment in the range 575-750 C for 30 to 240 minutes produces the nano-crystalline phase within the glass. It is by altering the conditions at this stage that various crystallite sizes can be produced. Generally, the higher temperature and the longer the temperature is maintained then the larger the crystallites and the larger the density of crystallites.

QUANTUM CONFINEMENT

Evidence for quantum confinement phenomena in these glasses was produced by a group at Corning (Borrelli et al.); by limiting the heat treatment stage to 30 minutes and adjusting the temperature in the range 550 to 700 C or alternatively fixing the temperature at 593 C and varying the duration between 30-240 minutes it was shown that spectral features could be produced which corresponded to quantum confinement effects. The crystallite sizes were around 3 nm with size distributions of standard deviation, typically 1.2nm. The effect of quantum confinement on the non-linear optical properties of CuBr doped glasses was investigated by Henneberger et al. and enhancement of the optical nonlinearity due to quantum confinement was observed. Although, at the moment, there is no evidence that quantum confinement influences the nonlinear optical properties of CdS_xSe_{1-x} doped glass.

WAVEGUIDES AND FIBRES

Both optical fibres and ion-exchange waveguides have been produced with SDG. The optical fibres (Cotter et al 1987) are produced by making a rod of the multi-component glass. This glass rod is used as the core material for the optical fibre which is made by the rod and tube method. The resulting fibre is clear and low-loss subsequent heat treatment is used to effect the nucleation and growth of the CdS_xSe_{1-x} crystallites which are confined to the core region. By localising the heat treatment it is possible to confine crystal growth to short (~1 mm) lengths in otherwise clear fibre. Single mode optical fibres at 600 nm with SDG cores have been produced by this method.

The ion-exchange optical waveguides (Cullen et al.) have been produced in specially prepared soda-lime semiconductor doped glass and in commercially available SDG. The waveguides are fabricated by Potassium/Sodium ion exchange. The process is as follows: first of all the substrate has a metal diffusion mask made on the surface by photolithographic techniques which allow features down to 2µm to be defined. The substrate is then placed in a bath of molten KNO_3 at 375 C where the sodium ions in the glass diffuse out to be replaced by potassium ions and thus regions of higher refractive index are created. By this method single mode channel waveguides can be fabricated. The commercially available glasses are made from silicate glass and have a diffusion constant for potassium/sodium ion exchange which is 20 times smaller than the soda-lime glasses. The diffusion constant, D, for the silicate glass is 2.68×10^{-17} m^2/s; this compares with soda-lime glass with D= 4.5×10^{-16} m^2/s which means that the silicate glasses take 24 hours in molten KNO_3 to produce a single mode waveguide compared to 1 hour for the soda-lime glass.

The nonlinear optical properties of SDG in bulk form have been invest-
gated by several groups (Yao et al., Olbright et al. and Roussignol et al.)
and the material has an intensity/fluence dependent refractive index and
absorption coefficient for which the mechanism appears to be bandfilling
in the semiconductor crystallites. In bandfilling the states at the
bottom of the conduction band in the semiconductor are filled by photoexcited carriers and according to the Pauli exclusion principle, further
transitions into these occupied states are blocked, thereby causing the
effective band-gap of the semiconductor to increase. For longer wavelengths close to the band-gap resonance the absorption coefficient is
decreased and also through the Kramers-Kronig relationship the refractive
index is decreased. The time taken for the optical nonlinearities to
recover is determined by the recombination rate for the photoexcited
carriers.

In good quality direct-gap semiconductors recombination is of the
order of a few nanoseconds. In semiconductor doped glasses the recombination time is few tens of picosecond. The difference is usually
ascribed to the larger density of defect sites in the SDG associated with
the much larger surface area. It should also be noted that the size of
the effect in the SDG is about three orders of magnitude smaller than in
the bulk crystalline material.

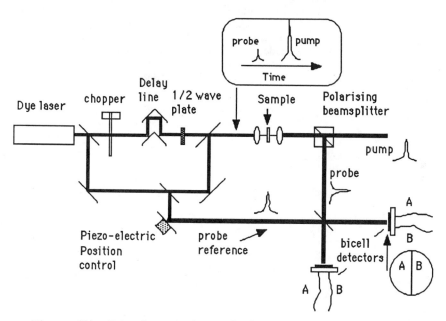

Figure (1)   Interferometric optical arrangement of measuring
              optical nonlinearities

The optical layout for the interferometric system for measuring
the absorption and refractive index nonlinearities with ultrashort
time resolution and absolute calibration of magnitude and phase.

It is important to realise that for this resonant optical nonlinearity the effect is proportional to the density of photoexcited
carriers and not directly to the intensity of light. The optical nonlinearity is only directly proportional to the intensity of the light if
the recovery time of the photoexcited carriers is short compared to the
timescale of the measurement. Specifically in this paper, we are considering experiments where the pulsewidth of the light is the same order as
the recovery time of the photoexcited carriers in which case the optical
nonlinearity is proportional to the fluence (energy/area) of the optical
pulse and therefore throughout this paper optical nonlinearities will be
measured in terms of fluence.

The techniques used to measure the optical nonlinearities (and in
particular the nonlinear refractive index) in bulk glass such as optical
phase conjugation become difficult to apply in guided wave configurations.
Therefore a new method was devised (Cotter et al. 1988) which can measure
the optical nonlinearities in guided wave geometries with picosecond time
resolution and with absolute calibration in magnitude and sign. The
technique is illustrated in Figure (1); it is basically pump/probe
experiment with the added refinement that the probe beam is further split
to provide a probe phase reference. The probe and probe reference beams
form the arms of a Mach-Zehnder interferometer with the guided wave sample
placed in the probe arm of the interferometer. It is necessary to
separate the pump and probe at the output of the waveguide and for this
purpose the pump and probe are polarised orthogonally and separated at
the output by a polarising beam splitter. Refractive index changes
induced by the pump cause the path length in one arm of the interferometer to be altered and this in turn leads to a movement of the fringes
at the output of the interferometer. This fringe movement is detected by
a bicell position sensitive detector which also provides an error signal
that is used in a feedback loop which stabilises the interferometer
against long term drift. An alternative method of measuring fringe movement is to use a computer controlled camera.

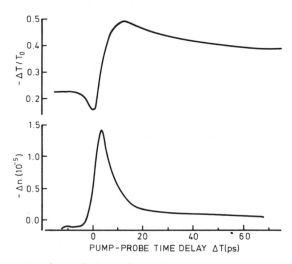

Figure (2)   Results of interferometric measurement of SDG optical
              fibre nonlinearities

Measured transmission change ΔT and induced refractive index Δn in a
10 mm long semiconductor-doped glass fibre as a function of time delay
after excitation with 4ps, 40pJ pulse at 610 nm. The absorption
coefficient is $2cm^{-1}$.

The experiments were all performed using a mode-locked cavity dumped dye laser as the source with Rodamine 6G dye at wavelengths between 590-630 nm. This experiment gives information on both the nonlinear absorption coefficient through the intensity of the probe and the nonlinear refractive index through the induced fringe shift. The time response of the optical nonlinearities can be obtained by varying the delay between pump and probe. The time resolved results for both the nonlinear absorption and the nonlinear refractive index of a SDG optical fibre are shown in Figure (2); the results for both types nonlinearity are directly comparable because they are measured under identical conditions. Similar results were obtained for ion exchanged waveguides, the only difference being that the induced absorption had a very similar time behaviour to the induced refractive index which, as can be seen in Figure (2), is not the case for the fibres. The reason for the difference between the time behaviour of the absorption and refractive index nonlinearities in the optical fibre is not yet clear. Although it is known that heating the material causes a shift in the band-edge to longer wavelengths and the increase in absorption which becomes apparent in Figure (2) after about 10 ps may be associated with this thermal effect. The subsequent long decay time would also be consistent with a thermal effect.

A series of measurements on SDG waveguides has revealed that the optical nonlinearity saturates at a fluence of around $2mJ/cm^2$ and the maximum induced refractive index, Δn_{sat}, is approximately -3×10^{-5}.

ALL-OPTICAL LOGIC DEVICES

The all-optical switching device can be designed by analogy with already existing electro-optic devices because they also employ small refractive index changes to obtain intensity switching. An electro-optic directional coupler (sometimes called the $\Delta\beta$ switch) consists of two waveguides close together that are coupled through their evanescent fields; after a distance L_c, the coupling length, all the light in the input waveguide crosses over the other waveguide. This only happens if the waveguides have the same propagation constant $\beta = 2\pi n/\lambda$ (n, refractive index and λ, wavelength). If a small difference in the propagation constants can be introduced, in the electro-optic device through an applied electric field and in the nonlinear optical device by a change in optical fluence, then coupling between the waveguides will be destroyed because there will not be phase synchrony between the two waveguides. The nonlinear optical device is called the nonlinear coherent coupler and was first discussed by Jensen. Figure (3) shows a diagram of the device which was fabricated in SDG.

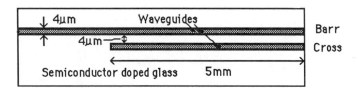

Figure (3)   Dimensions of the nonlinear coherent coupler in SDG

The nonlinear coherent coupler fabricated in SDG.

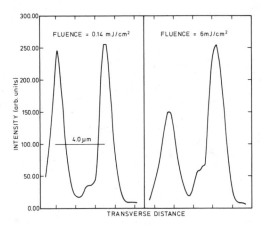

Figure (4)　A Scan of the output from the nonlinear coherent
　　　　　　 coupler

The output beams of each waveguide in the nonlinear directional
coupler for high and low fluence.

SWITCHING CHARACTERISTICS

Figure (4) shows a scan of the output of the device for low and high input
fluence. In the low fluence case the two output beams have almost equal
energy. This is because the device has approximately half a coupling
length overlap between the guides and at high fluence the ratio of energy
between the two guides is 65.35. More energy has been retained in the
input guide because the waveguides have been to some extent decoupled due
to the difference induced in the propagation constants by the nonliner
refractive index, although this is not the only effect which is present.
As can be seen more clearly in Figure (5) the transmission in both waveguides is increasing as a function of input energy due to the saturation
of the absorption coefficient.

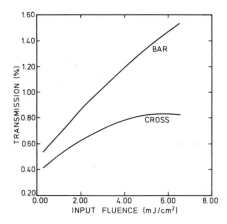

Figure (5)　Switching characteristics of the SDG nonlinear directional
　　　　　　 coupler

Transmission of each waveguide versus fluence

Since the induced refractive index in this experiment is small($\sim 10^{-5}$) compared to the refractive index associated with the waveguiding ($\sim 10^{-3}$) then we are in the weak nonlinearity regime, i.e. the optical nonlinearity does not significantly perturb the waveguiding characteristics and coupled wave theory is appropriate. The coupled wave equations, which describe the evolution of the field amplitudes as a function of distance, z, in the nonlinear coherent coupler, are as follows:-

$$\frac{da_1}{dz} = \frac{i\pi L}{2L_c} a_2 - \frac{\propto(\rho_1) a_1}{2} + iK_o \Delta n(\rho_1) a_1 \qquad (1)$$

and

$$\frac{da_2}{dz} = \frac{i\pi L}{2L_c} a_1 - \frac{\propto(\rho_2) a_2}{2} + iK_o \Delta n(\rho_2) \cdot a_2 \qquad (2)$$

where the subscripts 1 and 2 refer to the input (barr) guide and the other guide (cross). a_1 and a_2 are the field amplitudes in the respective guides, $\propto(\rho)$ is the absorption coefficient which a function of photocarrier density, ρ, similarly $\Delta n(\rho)$ is the photocarrier dependent refractive index, L is the length of the device and K_o is the free space propagation constant.

If the Banyai-Koch theory is used to model $\propto(\rho)$ and $\Delta n(\rho)$ then the results shown in Figure (6) are obtained for the numerical solution to equations (1) and (2). The full solution to equations (1) and (2) is shown in Figure (6) as a solid line and results of this model show qualitative agreement with the experimental data. The dotted line, which shows the solutions to equations (1) and (2) without the inclusion of the $\propto(\rho)$ terms, does not fit the observed data. This clearly indicates the importance of absorption saturation in the operation of this device.

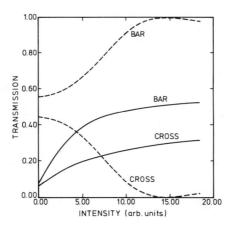

Figure (6)   Numerical solution to Coupled Wave Equations

The solid line includes absorption nonlinearity and the dotted line assumes only a refractive index nonlinearity.

CONCLUSIONS

The nonlinear optical properties of semiconductor doped glass wave-guides and fibres have been explored and it has been shown that these materials have a high-speed response of a few tens of picosecond, Δn_{sat} of -3×10^{-5} at a fluence of around 2 mJ/cm^2 and a fluence dependent absorption coefficient; these are important parameters for guided-wave all-optical logic device design.

A nonlinear coherent coupler has been fabricated in semiconductor doped glass and its switching characteristics measured and qualitatively modelled.

Future work in this area will include the investigation of the effects of quantum confinement of the nonlinear optical properties of CdS_xSe_{1-x} doped glass. It may also be possible to increase the maximum induced refractive index, Δn_{sat}, by increasing the density of crystallites in the material.

ACKNOWLEDGEMENTS

This paper is a review of work which has been carried out with the help of several collaborators; these are: at Glasgow University T.J. Cullen, M. O'Neill, J. Bell and B. Bhumbra; at the Optical Sciences Centre, University of Arizona G.I. Stegeman, C.T. Seaton, N. Finlayson and W. Banyia; at British Telecom Research Laboratories D. Cotter, B.J. Ainslie and H.P. Girdlestone.

The work was supported at Glasgow University by the SERC and the University of Arizona/University of Glasgow collaboration was supported by NATO. The author also acknowledges British Telecom for the receipt of a short-term fellowship.

REFERENCES

Banyia L. and Koch S.W., 1986, "A simple theory for the effects of plasma screening" Z.Phys.B. - Condensed Matter 63:283.

Borrelli, N.F., Hall, D.W., Holland, H.J. and Smith, D.W.,1987, "Quantum confinement effects of semiconductor microcrystallites in glass", J.Appl.Phys.,61:5399.

Cotter, D., Ainslie, B.J. and Girdlestone, H.P.,1987, "Semiconductor-doped fibre waveguides exhibiting picosecond optical nonlinear-ity", Electronic Letts., 23:405.

Cotter, D., Ainslie, B.J., Girdlestone, H.P. and Ironside, C.N., submitted for publication 1988, "Pump-probe interferometric measurements of nonlinear-refractive materials with ultrafast time resolution".

Cullen, T.J., Ironside, C.N., Seaton, C.T. and Stegeman, G.I.,1986, "Semiconductor-doped glass ion-exchanged waveguides", Appl.Phys. Letts., 49:1403,

Henneberger, F., Woggon, U., Puls, J. and Spiegelberg, Ch., 1988, "Exciton-related optical nonlinearities in semiconductors and semiconductor microcrystallites", App.Phys.B., 46:19

Jensen, S.M., 1982, "The nonlinear coherent coupler", IEEE J.Quant. Electron, 18:1580.

Olbright, G., Peyghambarian, N., Koch, S.W. and Banyai, L.,1987, "Optical nonlinearities of glass doped with semiconductor microcrystallites", Opt.Lett., 12:413

Roussignol, P., Ricard, D., Lukasik, J. and Flytzanis, C., 1987, "New results on optical phase conjugation in semiconductor-doped glasses, J.Opt.Soc.Am.B., 4:5.

Yao, S.S., Karaguleff, C., Gabel, A., Fortenberry, R., Seaton, C.T. and Stegeman, G.I., 1985, "Ultrafast carrier and grating lifetimes in semiconductor-doped glasses", Appl.Phys.Lett., 46:801.

MBE AND ALE OF II-VI COMPOUNDS:
GROWTH PROCESSES AND LATTICE STRAIN
IN HETEROEPITAXY

T. Yao, M. Fujimoto*, K. Uesugi*, and S. Kamiyama*

Electrotechnical Laboratory
1-1-4 Umezono, Tsukuba, Ibaraki 305
Japan

INTRODUCTION

Recently, dynamical behaviour of reflection high-energy electron diffraction (RHEED) patterns, i.e., RHEED intensity oscillations during molecular beam epitaxy (MBE) growth[1,2] and RHEED intensity variations during atomic layer epitaxy (ALE) growth[3,4] of II-VI compounds have been reported. The period of the RHEED intensity oscillation exactly corresponds to a monolayer growth rate of the crystal by MBE. This technique has been utilized to calibrate beam flux, to control growth rate, and to measure quantum well thickness[5]. The temporal behaviour of specular beam intensity of RHEED during ALE[3,4] reflects adsorption processes of impinging molecular beams. Adsorption time for monatomic layer is estimated from the temporal dependence of specular beam intensity. The specular beam intensity varies with time during suspension of ALE at high substrate temperature. This is due to sublimation processes of surface atoms. The sublimation time for the surface layer can be estimated[3,4,6,7]. Thus the investigation of dynamical behaviour of RHEED intensity gives important informations relevant to surface processes during MBE and ALE.

This paper reports detailed investigations on dynamical behaviour of RHEED pattern during MBE and ALE growth of II-VI compounds. We have compared the growth processes at the initial stages of heteroepitaxy of II-VI compounds on As-stabilized GaAs surface with those on Ga-stabilized GaAs surface using a RHEED observation system. We have investigated effects of RHEED observation conditions on dynamical behaviors of RHEED intensity during MBE and ALE growth. We suggest that the dynamical behaviour of RHEED intensity comes from the dynamical effect[10], although some of the phenomena can be understood in terms of a simple kinematical picture. We have investigated effects of growth conditions (substrate temperature and molecular beam flux) on the RHEED intensity variations. It is found that the behaviour of RHEED intensity variation are strongly correlated with surface stoichiometry of growing layer. The layer-by-layer sublimation processes of surface atomic layer and of mono-molecular layer occur at high substrate temperature. The relaxation of surface lattice strain as the growth proceeds is investigated 'in situ', and its correlation with the growth processes is discussed.

EXPERIMENT

The experiments were carried out in a three-chamber system which has two growth chambers where GaAs buffer layers and II-VI epilayers were grown in each growth chamber. We grew GaAs buffer layers to obtain an As-stabilized GaAs surface, while a Ga-stabilized surface was obtained by thermally cleaning of GaAs substrate without impinging As flux. The growth conditions for ZnSe and ZnTe were as follows: substrate temperature varying in the range 120 - 500 °C and the molecular beam flux ratio of group VI to Zn ranging from 0.3 to 20.

We have constructed a RHEED observation system to analyze RHEED images quantitatively. The RHEED patterns were observed on a phosphor screen using an electron gun at 20 kV. RHEED patterns during growth are recorded using a CCD camera and a video-tape recorder. The RHEED image is digitized into 256x256 pixels having 6-bit resolution. The digitized signals are processed by a personal computer and can be displayed in various modes: digitized picture, three-dimensional image, horizontal line scan, and so on. The intensity and half width of diffraction spots/streaks were measured. The surface lattice constant was determined from the separation of bulk reflections.

RESULTS AND DISCUSSION

MBE surface phase diagram.

The surface stoichiometry is characterized by surface reconstruction. Figure 1 shows a surface phase diagram for the growth of ZnSe on predeposited (001)ZnSe layers on (001)GaAs substrates[9]. The (2x1) reconstructed RHEED pattern was observed up to the Se-to-Zn beam flux ratio of 20. At around the boundary, which is indicated by the dashed curve, a mixture of (2x1) and c(2x2) reconstructed pattern was observed. The Se-stabilized (2x1) reconstructed surface is stable at lower substrate temperatures, while the Zn-stabilized c(2x2) surface is stable at higher substrate

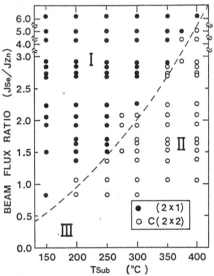

Fig.1. Surface phase diagram for MBE growth of ZnSe on predeposited ZnSe on (001)GaAs. The RHEED pattern in the region I shows a streaky (2x1) reconstructed pattern, while the region II shows c(2x2)[9].

temperatures. This is because, as will be shown below, the sublimation process of Se monatomic layer occurs at higher substrate temperatures, while the sublimation of Zn monatomic layer scarcely occurs[4].

Growth processes at the initial stages of heteroepitaxy on (001)GaAs

We have examined heteroepitaxy of ZnSe both on Ga-stabilized GaAs surface and on As-stabilized GaAs surface. The former was formed by thermal cleaning of GaAs substrate without impinging As flux. It has been reported that the initial stages of heteroepitaxy of ZnSe on thermally cleaned surface are dominated by the Stranski-Krastanov growth mode[8]. General features of the evolution of RHEED pattern during growth of ZnSe on thermally cleaned GaAs substrates are as follows. The first few layers show streaky RHEED patterns. With the deposition of more than 2 layers, RHEED pattern becomes spotty and diffused, but eventually becomes streaky again. Figure 2 shows the variation of RHEED intensity and half width of the (10) rod during growth of ZnSe on thermally cleaned GaAs substrate[9]. The substrate temperature is 250 °C and the molecular beam flux ratio of Se to Zn is 4. Both the diffraction intensity and half width remains at constant values from the very beginning of heteroepitaxy until the deposition of 2 MLs. As the deposition exceeds more than 2MLs, the intensity and half width increases. These features can be attributed to the commencement of the three dimensional island growth. The intensity and half width decreases after the deposition of about 50 MLs, due to the coalescence of islands. After the deposition of 100 MLs, they become almost the same values as the initial values at the very beginning of growth, which indicates that the two-dimensional layer-by-layer growth occurs. The observed variations of the diffraction intensity and half width confirm the results which are obtained by a simple observation of the evolution of RHEED pattern during growth.

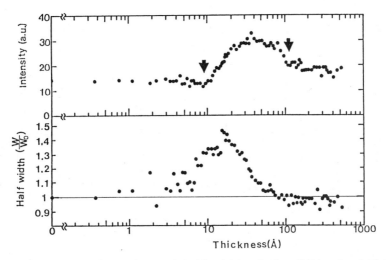

Fig.2. Variation of the intensity and half width of the (10) rod of ZnSe layer grown on Ga-stabilized GaAs. The substrate temperature during growth was 250 °C, the Se-to-Zn beam pressure ratio 2, and the growth rate was 0.65 Å/s[9].

The growth process of ZnSe on As-stabilized GaAs surface has close correlation with the surface stoichiometry of grown ZnSe layer. Figure 3 shows the variation of RHEED intensity and half width of the (00) rod during the growth of ZnSe epilayer on an As-stabilized GaAs buffer layer[9]. The substrate temperature was 320 °C and the Se-to-Zn flux ratio was 4. These growth conditions yielded a faint c(2x2) reconstruction pattern, which indicates that the growth conditions are near stoichiometry and slightly on the Zn-rich side. General features are almost the same as the previous figure with less variation in half width and diffraction intensity. These features would indicate that although the initial growth stage is dominated by the Stranski-Krastanov mode, two-dimensional growth mode

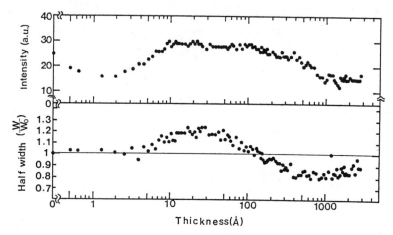

Fig.3. Variation of the intensity and half width of the (00) rod of ZnSe layer grown on As-stabilized (001)GaAs. The substrate temperature during growth was 350 °C, the Se-to-Zn beam pressure ratio 4, and the growth rate was 0.65 Å/s. RHEED pattern shows a weak c(2x2) reconstruction[9].

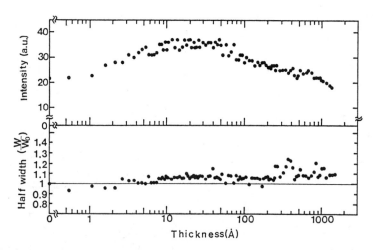

Fig.4. Variation of the intensity and half width of the (10) rod of ZnSe layer grown on As-stabilized (001)GaAs. The substrate temperature during growth was 350 °C, the Se-to-Zn beam pressure ratio 20, and the growth rate was 0.53 Å/s. RHEED pattern shows a (2x1) reconstruction[9].

is more predominant in the growth on As-stabilized GaAs surface than in the growth on Ga-stabilized surface. As the growth proceeds, the island eventually coalesces into a continuous layer and the two-dimensional growth dominates.

Figure 4 shows the variation of RHEED intensity and half width of the diffraction spot from ZnSe grown at 350 °C with the Se-to-Zn beam flux ratio at 20[9]. These growth conditions yielded a Se-stabilized surface. The diffraction intensity and half width shows almost constant values from the very beginning of heteroepitaxy. The corresponding RHEED pattern showed a streaky pattern throughout the growth of ZnSe from the very beginning of heteroepitaxy. The growth is obviously dominated by the two-dimensional layer-by-layer growth mode. As is shown in Figs. 2-4, the growth process is strongly dependent on growth conditions which yield different surface stoichiometry. The growth conditions for the Se-stabilized surface is preferable for two-dimensional layer-by-layer growth.

RHEED intensity oscillations during MBE

Figure 5 shows RHEED intensity oscillations during MBE growth of ZnTe on ZnTe predeposited layer in which the growth of ZnTe initiates on impinging Zn flux on the Te-covered surface. The substrate temperature is 350 °C and the Te-to-Zn beam pressure ratio is almost 1, which yielded Zn-stabilized surface. The azimuth of electron beam is along the <110> direction with the incidence angle at 0.6°, which is off-Bragg diffraction condition. Very stable RHEED oscillations more than 150 are observed. The appearance of the RHEED oscillation indicates that the growth is controlled by the two-dimensional nucleation and growth mechanism[10].

Incidence angle dependence. As is well known in the case of GaAs, the dynamical behaviors of RHEED intensity oscillation is strongly dependent on the incidence angle of electron beam[10]. In particular, the phase of the oscillation changes with the incidence angle. This is an indication of the presence of dynamical effect (multiple scattering). Figure 6 shows how the oscillating behavior varies with the incidence angle on ZnTe surface. The phase and the amplitude of the oscillation differs with the incidence angle. The oscillation is considered to

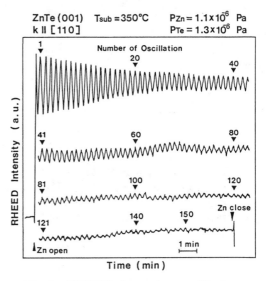

Fig.5. RHEED intensity oscillations during MBE growth of ZnTe on ZnTe predeposited layer on (001)GaAs with k//[110]. The growth is initiated by impinging Zn flux on the Te-covered surface. The substrate temperature is 350 °C and the Te-to-Zn flux ratio is 1.

consist of the diffused scattering and contribution from the specular beam. The contribution of these two components changes with the incidence angle. Such a feature reflects the dynamical effect.

Growth rate. The oscillation period precisely corresponds to the growth rate of mono-molecular layer, which has been confirmed by comparing the measured growth rate of the epilayer and the oscillation period. Figure 7 shows the growth rate of ZnTe as a function of either Zn (▲) or Te (●) K-cell temperatures, where the other K-cell temperature were kept constant. The substrate temperature was 350 °C. The results agree well with those reported previously[11], where the growth rate was measured from the layer thickness. When the Te K-cell temperature is low, the growth rate increases monotonically with an increase in Te K-cell temperature, while it tends to saturate, when the Te-cell temperature is high. The low Te-cell temperature region corresponds to the Zn rich condition as is characterized by c(2x2) reconstructed RHEED pattern, while the high Te K-cell temperature condition corresponds to the Te rich region, where a (2x1) reconstruction pattern was observed. The dotted line shows the boundary between the Zn-stabilized and Se-stabilized surfaces. This corresponds to the condition where a high quality epilayer grows[11], since it gives the stoichiometric growth conditions.

Effects of stoichiometry. The oscillation behavior is influenced by the group VI-to-Zn beam flux ratio, i.e. stoichiometry. Figure 8 shows RHEED intensity oscillation

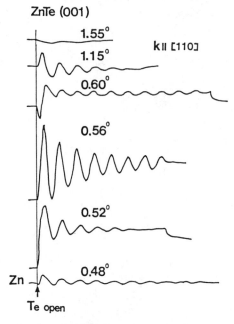

Fig.6. RHEED intensity oscillations during MBE of ZnTe with various incidence angles with k//[110]. The growth is initiated by impinging Zn beam on the Te-covered surface.

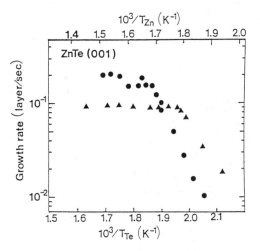

Fig.7. Growth rate of ZnTe measured from periodicity of RHEED intensity oscillations against either Zn (Δ) or Te (o) K-cell temperatures. The substrate temperature is fixed at 350 °C.

behaviors from the Zn- and Te-covered surfaces with k//[110]. The Zn-covered surface shows stronger intensity than the Se-covered one. On impinging the Te beam on the Zn-covered surface, the RHEED intensity decreases and saturates at a value, while the intensity from the Te-covered surface increases on impinging Zn beam and saturates at the same intensity level. The saturation intensity level is correlated with the surface stoichiometry. When the growth conditions are near the stoichiometric conditions, the intensity lies at around the middle point between the two levels. As the growth conditions go to the Zn-rich side, the intensity level approaches that of the Zn-covered surface, while the intensity approaches that of the Te-covered surface as the growth conditions go to the Te-rich side. The phase of the oscillation is also dependent on the stoichiometry. The variations of the intensity with surface stoichiometry can be interpreted in terms of the formation of two-dimensional growth steps extending to the direction of the dangling bond of the outermost surface.

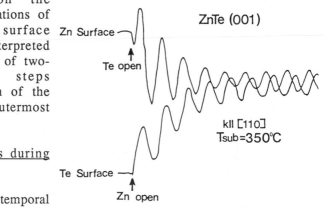

Fig.8. RHEED intensity oscillations during MBE of ZnTe with k//[110]. The growth is initiated by impinging Te and Zn fluxes on the Zn- and Te-covered surfaces, respectively. The Zn-covered surface shows stronger intensity than the Se-covered one.

RHEED intensity variations during ALE

Figure 9 shows the temporal response of the specular beam intensity during ALE of ZnSe with k//[100][4]. When the surface is covered with Zn and the Zn beam is turned off, the next impinging Zn beam does not change the specular beam intensity, but the next impinging Se beam reduces the specular beam intensity, which eventually saturates at a certain value during the impingement of the Se beam. When the Se beam is stopped and the Zn beam is turned on again, the specular beam intensity increases and saturates at the same intensity level as that of the previous Zn-covered surface. When the Zn beam is turned off after the formation of the Zn-covered surface and the Se-beam is turned on, the specular beam intensity decreases again and saturates at the same intensity level as the previous Se-covered surface.

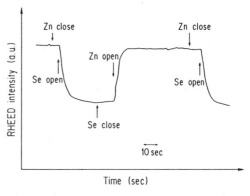

Fig.9. Temporal response of the specular beam intensity during ALE of ZnSe with k//[100][4].

Simultaneously with the change in intensity the surface reconstruction pattern changes completely from a (2x1) to c(2x2) as the specular intensity changes from the level of the Se-covered surface to the Zn-covered surface and vice versa. The time constant for the change of these two intensity levels should be correlated with the adsorption time for impinging atoms. Using the intensity variations of the specular beam during the deposition of a constituent element, accurate "in situ" measurements of the adsorption time or the coverage time of the impinging molecular beam become possible. The variations of the intensity with the azimuthal direction can be interpreted again in terms of the formation of two-dimensional growth steps extending to the direction of the dangling bond of the outermost surface.

Layer-by-layer sublimation

When the Se shutter is turned off after the Se-covered surface has been formed, the specular beam intensity gradually increased near the level of the Zn-covered surface as shown in Fig.10[4]. The RHEED reconstruction pattern changes from (2x1) to c(2x2) together with the change in intensity. This fact indicates the change of the surface from the Se-covered to the Zn-covered one. The time constant for this variation should correspond to the desorption time for the Se layer. The specular beam intensity shows only a slight change on closing the Zn shutter after the Zn-covered surface is formed. This fact suggests that neither sublimation of Zn layer nor the sublimation of ZnSe molecular layer occurs at high substrate temperature. At high substrate temperature, the growth rate decreases. This is due to the enhancement of Se layer sublimation.

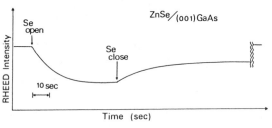

Fig.10. Temporal response of the specular beam intensity during sublimation of Se layer[4]. The RHEED reconstruction pattern changes from (2x1) to c(2x2) during sublimation.

When the substrate temperature of ZnTe is high, RHEED intensity oscillation is observed after suspension of growth. This is due to the molecular layer-by-layer sublimation as was observed in GaAs[12]. Figure 11 shows RHEED oscillation behaviors during sublimation of ZnTe layer at the substrate temperature of 450 °C. The period corresponds to the sublimation time for outermost layers. The sublimation time decreases with increase in

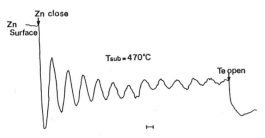

Fig.11. RHEED oscillation behaviors during sublimation of ZnTe layer at the substrate temperature of 450 °C.

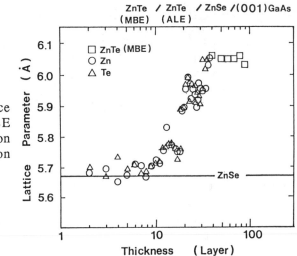

Fig.12.
Variations of surface lattice parameter during ALE growth of ZnSe epilayer on ZnTe predeposited layer on (001)GaAs substrate.

substrate temperature. The estimated activation energy is 28.6 kcal/mol, which is almost the same value as the bulk sublimation energy (29.8 kcal/mol)[13]. This is consistent with the idea that Zn and Te layers sublimate simultaneously. Therefore, the decrease in growth rate at high substrate temperature is due to the simultaneous sublimation of Zn and Te layers.

Lattice strain relaxation

The relaxation of lattice strain at the initial stages of heteroepitaxy is strongly correlated with growth mechanisms. The transition from the strained to dislocated states causes a change from two-dimensional layer-by-layer growth to three-dimensional island growth. Figure 12 shows variations of surface lattice parameter during ALE growth of ZnSe epilayer on ZnTe predeposited layer on GaAs substrate. The lattice parameter of the ZnSe epilayer coincides with that of the ZnTe layer until the thickness of ZnSe layer exceeds 6 - 7 atomic layers. Simultaneously with the lattice strain relaxation, the RHEED intensity and half width of diffraction spot increases, indicating the onset of three-dimensional growth. The implication of this result to the growth of superlattice structure is obvious: The control of each layer thickness within the critical thickness is crucial in order to realize smooth interface and to grow a good quality layer without dislocations.

CONCLUSIONS

We have made detailed investigations on dynamical behaviors of RHEED pattern during MBE and ALE growth of II-VI compounds. Growth processes at the initial stages of heteroepitaxy of II-VI compounds on As-stabilized GaAs surface are dominated by two dimensional layer-by-layer growth, while those on Ga-stabilized GaAs surface is dominated by the Stranski-Krastanov mode. Dynamical behaviors of RHEED intensity during

MBE and ALE growth are strongly dependent on the incidence angles of electron beam and surface stoichiometry. Dynamical scattering effect should be taken into account to fully understand the phenomena. The layer-by-layer sublimation processes of surface atomic layer and mono-molecular layer occurs at higher substrate temperatures. The relaxation of surface lattice strain during the ALE growth of ZnTe layer on ZnSe layer is investigated 'in situ' as the growth proceeds. The misfit strain relaxation induces three-dimensional island growth.

REFERENCES

*) on leave from Department of Electronic Engineering, Tokai University, 1117 Kitakaname, Hiratsuka, Kanagawa 259-12.

1. T. Yao, H. Taneda, and M. Funaki, Jpn. J. Appl. Phys. 25, L952 (1986).
2. R.L. Gunshor, L.A. Kolodziejski, M.R. Melloch, M. Vaziri, C. Choi, and N. Otsuka, Appl. Phys. Lett. 50, 200 (1987).
3. T. Yao, Jpn. J. Appl. 25, L942 (1986).
4. T. Yao and T. Takeda, J. Cryst. Growth 81, 43 (1987).
5. L.A. Kolodziejski, R.L. Gunshor, N. Otsuka, B.P. Gu, Y. Hefetz, and A.V. Nurmikko, J. Cryst. Growth 81, 491 (1987).
6. K. Menda, I. Takayasu, T. Minato, and M. Kawashima, Jpn. J. Appl. Phys. 26, L1326 (1987).
7. H.J. Cornelissen, D.A. Cammack, and R.J. Dalby, J. Vac. Sci. Technol. B6, 769 (1988).
8. T. Yao and T. Takeda, Appl. Phys. Lett. 48, 160 (1986).
9. T. Yao, M. Fujimoto, and H. Nakao, to appear in Proc. SPIE 944 (1988).
10. P.J. Dobson, B.A. Joyce, J.H. Neave, and J. Zhang, J. Cryst. Growth 81, 1 (1987).
11. T. Yao, in *The Technology and Physics of Molecular Beam Epitaxy*, edited by E.H.C.Parker (Plenum, New York, 1985), Chap. 10.
12. T. Kojima, N.J. Kawai, T. Nakagawa, K. Ohta, T. Sakamoto, and M. Kawashima, Appl. Phys. Lett. 47, 286 (1985).
13. K.C. Mills, in *Thermodynamic Data for Inorganic Sulphides, Selenides and Tellurides* (Butterworths, London, 1974).

WIDE BANDGAP II-VI COMPOUND SUPERLATTICES PREPARED BY MBE AND MOMBE

Makoto Konagai, Shiro Dosho, Yasushi Takemura,
Nobuaki Teraguchi, Ryuhei Kimura and Kiyoshi Takahashi

Department of Electrical and Electronic Engineering
Tokyo Istitute of Technology
2-12-1, Ohokayama, Meguro-ku,Tokyo 152, Japan

INTRODUCTION

II-VI compound semiconductors, especially ZnS, ZnSe and ZnTe, are considered to be the promising materials for LEDs and laser diodes emitting blue light. Recently, ZnSe pn diodes have been successfully grown on GaAs by MOCVD using lithium nitride as a p-type dopant[1]. To make a pn junction in the wide bandgap II-VI compound semiconductors, we selected a ZnSe-ZnTe superlattice structure[2]. We have prepared ZnSe-ZnTe strained-layer superlattices(SLS) by MBE with a modulation doping technique[3]. The concept of producing both p and n-type conduction using a superlattice is illustrated in Fig.1. Modulation doping of the ZnSe layers with Ga is thought to result in a n-type SLS. P-type SLS can similarly be obtained by modulation doping with Sb in the ZnTe layers. This opens up the possibility of fabricating a pn junction by using ZnSe-ZnTe SLSs. The modulation doping technique can be also applied to the ZnS(ZnSSe)- ZnTe system.

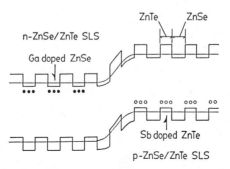

Fig.1 The concept of making a pn junction using ZnSe-ZnTe strained-layer superlattices.

In this paper, we have grown ZnSe and ZnTe by atomic layer epitaxy(ALE) using MBE to control precisely the layer thickness. The original concept of ALE ZnSe and ZnTe was reported by Yao et al[4]. In ALE, constituent elements are alternately deposited on the substrate. It was found that the layer thickness per one cycle of opening and closing the shutters of the constituent elements corresponds precisely to one monolayer growth. We achieved a drastic improvement in the interface abruptness and optical properties compared with SLSs grown by the conventional MBE.

Furthermore, ZnS-ZnSe and ZnS-ZnTe SLSs have been grown by metalorganic molecular beam epitaxy(MOMBE). In the ZnS-ZnSe SLS, intense and sharp photoluminescence related to the quantized levels was observed at 71K. We calculated the band discontinuities between ZnS and ZnSe for each strain. In the ZnS-ZnTe SLS, broad but very strong blue emission at around 460nm attributed to the deep levels was observed even at 400K.

CURRENT STATE OF ZnSe-ZnTe SLSs GROWN BY MBE

We have been investigating the properties of ZnSe-ZnTe SLSs prepared by MBE for several years and obtained the following results to date:-
1)The PL peak position was shifted by tailoring the structure of the superlattice. The luminesence color changed from blue-green to red[2].
2)The lattice mismatch in SLSs was accommodated by the tetragonal distortion of successive layers in the superlattice system[5].
3)SLSs with modulation doping of ZnTe with Sb and ZnSe with Ga exhibited p and n-type conduction, respectively, with carrier concentrations of $(0.5-1.0) \times 10^{14} cm^{-3}$ [3].

It is encouraging that we have successfully made both p and n-type SLSs. However, the carrier concentration we obtained is low for fabricating pn junctions. In this work, at first, we tried to clarify the limiting factors of carrier concentration. In general, there are several factors limiting the optical and electrical properties of superlattices as shown by the following:-
1.Structural degradation of the SLS at the growing temperature(interdiffusion).
2.Thermal diffusion of impurities in the SLS.
3.Interface abruptness, interface uniformity and irregularity of superlattice period.

We found that the first and the second are not the main factors limiting the carrier concentration as described in the following section, and to improve the third factor, ALE technology was adapted for preparing SLSs.

Interdiffusion in ZnSe-ZnTe SLSs

A detailed study has been made of interdiffusion in ZnSe-ZnTe SLSs grown by MBE at a growth temperature of 320°C. In X-ray diffraction measurements, the satellite peak intensities relative to the 0-order peak intensity decreased with annealing time. By using a linear diffusion model, the interdiffusion coefficient of ZnSe-ZnTe SLSs was determined to be $D = 2 \times 10^{-19} cm^2/s - 3 \times 10^{-21} cm^2/s$ at 500°C. It was found that the interdiffusion was negligible at a growth temperature of 320°C.

Impurity diffusion in ZnSe-ZnTe SLSs

Figures 2(a) and 2(b) show the Ga and Sb profile of the as grown

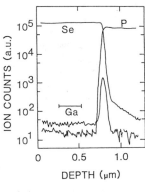 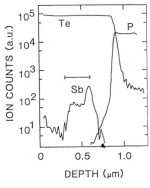

(a) ZnSe:Ga on InP (b) ZnTe:Sb on InP

Fig.2 The SIMS analysis of the as-grown ZnSe(a) and ZnTe(b) films grown on InP substrate. Only the middle epilayers were selectively doped with Ga(a) and Sb(b).

ZnSe and ZnTe epitaxial layers, respectively. In these samples only the middle epilayers(3000Å) are doped with Ga and Sb. Ga atoms spread into the whole ZnSe layer and pile up at the ZnSe/GaAs interface. Sb atoms are rather stable in the ZnTe layer. We found also that the extent of the thermal diffusion of dopants is strongly dependent on the interface stress.

Figures 3(a) and 3(b) show dopant profiles of the ZnSe-ZnTe SLSs grown on InP substrates. In this experiment, undoped ZnSe-ZnTe SLSs were introduced between the InP substrate and doped ZnSe-ZnTe SLSs. We

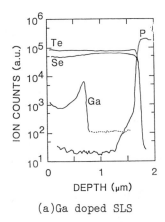 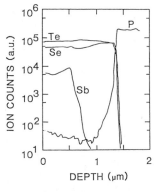

(a)Ga doped SLS (b) Sb doped SLS

Fig.3 The SIMS analysis of doped-SLS/undoped SLS/InP substrate structure. The superlattice period is 20Å, which is smaller than the depth resolution of SIMS.

221

found that the SLS structure blocks the thermal diffusion of Ga atoms, even if Ga has a large thermal diffusion coefficient in the ZnSe layer.

Besides the thermal diffusion of dopants in SLSs, the reduction of impurities from the substrate might be one of the key issues to enhance the SLS film quality. The diffusion of phosphorus from the InP substrate is clearly observed in Fig.3. Thus we need to develop a low temperature epitaxy technique for SLSs. We found that ALE is a very useful technique to reduce the epitaxial temperature as described in the next section.

ATOMIC LAYER EPITAXY OF ZnSe-ZnTe SLSs

Experiment

We used a conventional MBE system with solid sources for ALE growth. Figure 4 shows a schematic illustration of the MBE system used for ALE growth. Four K-cells containing Zn, Se, Te and $ZnCl_2$ were used as beam sources, and four shutters above each cell were controlled by a personal computer with stepping motors. In this experiment, $ZnCl_2$ was used as a n-type doping source instead of Ga. The shutters of Zn and chalcogens were alternatively opened and closed with intervals of 1 sec. The substrate temperature was 150-400°C.

Figure 5 shows the growth rate per cycle as a function of the substrate temperature. The number of cycles used was 1000 and the thickness profile was measured using an SEM. The one monolayer/cycle condition was achieved over a range of substrate temperatures Tsub=250-350°C. At a Tsub higher than 350°C, the growth rate decreased gradually due to the desorption effect. At a Tsub lower than 200°C, the growth rate increased rapidly.

Figure 6 shows the film thickness grown at a Tsub of 280°C against the total number of opening and closing cycles of the shutters. It shows clearly that the thickness of deposit grown per cycle by opening and closing the shutters of the constituent elements corresponds to one monolayer thickness for both ZnSe and ZnTe.

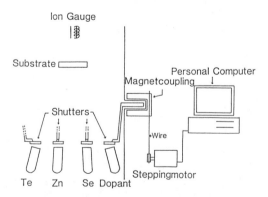

Fig.4 Schematic diagram of MBE system used for ALE growth.

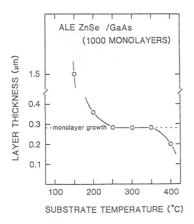

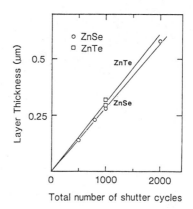

Fig.5 Growth rate per cycle of ZnSe as a function of the substrate temperature.

Fig.6 Film thickness of ZnSe and and ZnTe grown at Tsub of 280°C vs. the total number of opening and closing cycles of the shutters.

Figure 7 shows surface morphology of Cl-doped ZnSe films prepared by ALE on GaAs substrates. A sharp influence of ALE growth on the surface morphology could be obtained. The samples prepared by ALE conditions(Tsub=250-350°C) showed a mirrorlike surface. For samples prepared at substrate temperatures above 400°C or below 200°C, a rough surface morphology was observed.

Properties of ALE-ZnSe

Photoluminescence measurements were used to evaluate the optical properties of ALE-ZnSe. The PL spectrum of undoped ZnSe obtained at 4.2K was dominated by a donor bound-exciton emission I_2 at 2.796 eV with a FWHM of 6.303 meV. Figure 8 shows the PL spectra of Cl-doped ZnSe as a function of substrate temperature. The donor bound-exciton emission significantly increased at a substrate temperature of 250°C suppressing deep level emission.

The electrical properties of these samples were measured using the Van der Pauw method. All the samples doped with $ZnCl_2$ showed n-type conduction. The carrier concentration of $1.5 \times 10^{18} cm^{-3}$ could be obtained with the electron mobility of $480 cm^2 V^{-1} s^{-1}$ at a $ZnCl_2$-cell temperature of 115°C. It indicates that $ZnCl_2$ is an effective dopant of ALE-ZnSe.

ZnSe-ZnTe SLSs prepared by ALE

We prepared ZnSe(1-4 monolayers)-ZnTe(1-4 monolayers) SLSs with periods of 200-1000 on InP substrates by ALE at a growth temperature of 260°C which is about 60°C lower than that in conventional MBE. Direct observation of $(ZnSe)_4$-$(ZnTe)_2$ ALE-SLS was performed by TEM. Figure 9(b) shows the bright field image of the ALE-SLS. This figure exhibits

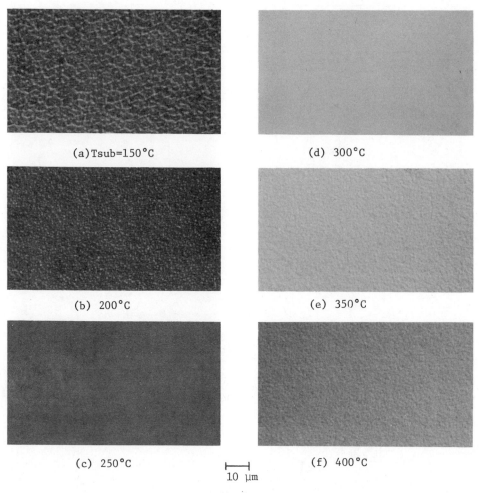

(a) Tsub=150°C   (d) 300°C

(b) 200°C   (e) 350°C

(c) 250°C   (f) 400°C

├──┤
10 μm

Fig.7 Surface morphology of Cl-doped ZnSe films prepared by ALE on GaAs
substrates.

the presence of more fine superlattice structures and well-defined
interface compared with a TEM photograph(Fig.9(a)) of ZnSe-ZnTe SLSs
prepared by the conventional MBE which we have reported previously[5].
Thus a drastic improvement in the abruptness and flatness at the
interface could be achieved by using the ALE technique.

Figure 10 shows the low temperature(71K) PL spectra of the ZnSe-
ZnTe SLSs. It is interesting to note that the PL intensity of the SLS
grown by ALE is more than ten times greater than that of the SLS grown
by the conventional MBE method. The FWHM value of a PL peak for ALE
SLSs(70 meV) is smaller than that of MBE SLS with an equivalent
structure(95 meV).

Modulation doped n-type SLSs have been prepared by ALE using Cl as
a n-type dopant. The maximum carrier concentration we obtained is
$3.8 \times 10^{16} cm^{-3}$ which is 3 orders of magnitude higher than that of MBE
grown n-type SLSs.

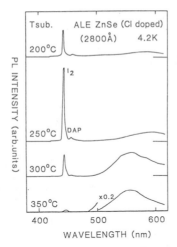

Fig.8 PL spectra of Cl-doped ZnSe as a
function of substrate temperature.

(a) Conventional MBE
ZnSe(30A)-ZnTe(20A)

(b) ALE
$(ZnSe)_4-(ZnTe)_2$

Fig.9 Bright field TEM images of ZnSe-ZnTe SLSs.

In this paper we report, for the first time, the synthesis and properties of $(ZnSe)_1-(ZnTe)_1$ superlattice grown by ALE. The PL intensity from $(ZnSe)_1-(ZnTe)_1$ superlattice is much stronger than that from mixed crystals with the same composition($ZnSe_{0.5}Te_{0.5}$). Figure 11 shows the variation of PL spectra with temperature. We could observe the strong green-yellow emission even at the room temperature. These results indicate that the monolayer superlattice is superior in optical quality and could be used in optical device applications. However, there are no reports about the theoretical analysis of electronic states in the ultrathin superlattices for ZnSe-ZnTe SLS system. It is important,therefore, to investigate theoretically the electronic properties, especially bandgap, of this new tailored material.

SLSs GROWN BY MOMBE

MOMBE which combines MOCVD and MBE is considered to be a powerful epitaxial growth method. We have previously reported on the growth of ZnS and ZnSe epitaxial films[7,8] and ZnS-ZnSe SLSs. In this paper, the growth and characterization of ZnS-ZnSe and ZnS-ZnTe SLSs by MOMBE are reported.

Experiments

Figure 12 shows the MOMBE system used in this work. Diethylzinc(DEZn), diethylsulfer(DES), diethylselenium(DESe) and diethyltellium(DETe) were used as source gases for Zn, S, Se and Te, respectively. The pyrolysis of DES, DESe and DETe was carried out in cracking cells at the outlet of the tube. The DEZn beam was kept on during the entire growth, while the DES and DESe(DETe) beams were alternated to produce the ZnS and ZnSe(ZnTe) layers, respectively. The switching of DES and DESe(DETe) flows was operated by the on/of operation of pneumatic valves by a sequence controller. The superlattices were grown on (100) GaAs substrates directly, without a buffer layer. The substrate temperature was 400°C.

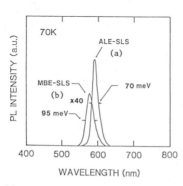

Fig.10 Photoluminescence spectra of ALE-SLS compared with MBE-SLS. (a)ZnSe(4 monolayers)-ZnTe(2 mono-layers, (b)ZnSe(10A)-ZnTe(5A).

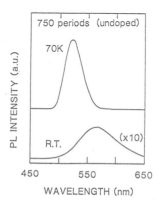

Fig.11 Variation of PL spectra of $(ZnSe)_1-(ZnTe)_1$ SLS with the temperature.

ZnS-ZnSe SLSs

Photoluminescence measurements of the ZnS-ZnSe SLSs were carried out at 4.2K and 70K.   PL was excited by a 2mW He-Cd laser(325nm).   Only excitonic emission peak was observed with no luminescence from deep levels in the low energy region at 4.2K.

Figure 13 shows the relationship between the thickness of the ZnSe well layer and the PL peak energy shift measured at 70K.   The thickness of the ZnS layer was fixed at 30A for all samples. The PL peak position shifted towards higher energy as the thickness of the ZnSe well layer was decreased.   The solid curve represents the calculated PL peak. Here, the valence band discontinuity(ΔEv) between ZnS and ZnSe was assumed to be 800meV.   This line-up provided the best agreement between the experimental values and calculated ones.   The values of conduction band discontinuity(ΔEc) for each layer thickness are also indicated in Fig.13.   From these results, we concluded that ΔEv between ZnS and ZnSe doesn't depend on the strain and is equal to 800meV and that ΔEc increases with an increase in strain.

ZnS-ZnTe SLSs

Figure 14 shows the temperature dependence of the PL spectra of a ZnS-ZnTe(20Å-20Å) SLS sample. The PL spectra of ZnS and ZnTe films were dominated by broad luminescence bands between 400-700nm and 600-900nm, respectively.   In the ZnS-ZnTe SLSs, however, a broad but very strong blue emission at around 460nm was observed in the PL measurements. This visible luminescence was observed even at 400K.   However, the peak wavelength of the spectra didn't vary with layer thickness.   Hence, we concluded that this luminescence located at 475nm at 70K is not attributed to the quantized levels but due to the deep levels formed in SLS layers due to a large lattice mismatch(13%) between ZnS and ZnTe.

CONCLUSION

We have successfully grown ZnSe-ZnTe SLSs by atomic layer epitaxy. A drastic improvement in the interface abruptness was obtained compared with SLSs grown by the conventional MBE.   The PL intensity was much greater than that of MBE grown SLSs. $(ZnSe)_1$-$(ZnTe)_1$ monolayer SLSs which showed the strong emission at room temperature have been prepared for the first time.

Furthermore, we have grown ZnS-ZnSe and ZnS-ZnTe SLSs on GaAs substrates by MOMBE.   In PL measurements of ZnS-ZnSe SLSs, we observed a strong quantum effect shift.   In ZnS-ZnTe SLSs, a strong blue emission at 460nm was observed even at 300K.

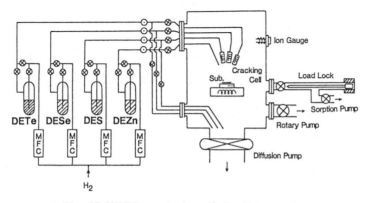

Fig.12 MOMBE system used in this work.

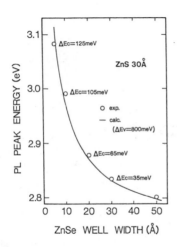

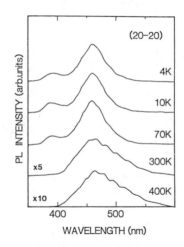

Fig.13 Relationship between ZnSe well layer thickness and PL peak energy shift. The solid curve was calculated by the Kronig-Penny model

Fig.14 Temperature dependence of the PL spectra of ZnS(20A)-ZnTe (20A) SLS with 200 periods on GaAs substrate.

ACKNOWLEDGMENTS

This work is supported by Grand-in-Aid for Scientific Research on Priority Areas, New Functionality Materials-Design, Preparation and Control, The Ministry of Education, Science and Culture, 491940111653.

REFERENCES

1. T.Yasuda, I.Mitsuishi, and H.Kukimoto,Metalorganic vapor phase epitaxy of low-resistivity p-type ZnSe, Appl. Phys. Lett.52:57(1988)
2. M.Kobayashi, N.Mino, H.Katagiri, R.Kimura, M.Konagai, and K.Takahashi, Photoluminescence study of ZnSe-ZnTe strained-layer superlattices grown on InP substrates, J. Appl. Phys.60:773(1986)
3. M.Kobayashi, S.Dosho, A.Imai, R.Kimura, M.Konagai, and K.Takahashi, Realization of both p- and n-type conduction for ZnSe-ZnTe strained-layer superlattices, Appl. Phys. Lett. 51:1602(1987)
4. T.Yao and T.Takeda, Growth process in atomic layer epitaxy of Zn chalcogenide single crystalline films on (100)GaAs, Appl. Phys. Lett. 48:160(1986)
5. M.Kobayashi, M.Konagai, and K.Takahashi,Lattice strain and lattice dynamics of ZnSe-ZnTe strained-layer superlattices, J. Appl. Phys. 61:1015(1987)
6. A.Imai, M.Kobayashi, S.Dosho, M.Konagai, and K.Takahashi,Inter-diffusion in ZnSe-ZnTe strained-layer superlattices, J. Appl. Phys.63: (1988) in press
7. H.Ando, A.Taike, M.Konagai, and K.Takahashi,Metalorganic molecular-beam epitaxy of ZnSe and ZnS, J. Appl. Phys. 62:1251(1987)
8. A.Taike, N.Teraguchi, M.Konagai, and K.Takahashi,Growth of ZnSe-ZnS Strained-Layer Superlattices by Metalorganic Molecular Beam Epitaxy, Jpn. J. Appl. Phys. 26:L989(1987)

II-VI / III-V HETEROINTERFACES: EPILAYER-ON-EPILAYER STRUCTURES

R. L. Gunshor, L. A. Kolodziejski,* and M. R. Melloch

School of Electrical Engineering
Purdue University
West Lafayette, Indiana 47907 USA

N. Otsuka

Materials Engineering
Purdue University
West Lafayette, Indiana 47907 USA

A. V. Nurmikko

Division of Engineering and Department of Physics
Brown University
Providence, Rhode Island 02912 USA

INTRODUCTION

The integration of several optoelectronic device functions onto a common substrate material is an area which is currently being actively pursued. In an effort to achieve this objective, experiments are under way to examine the epitaxial growth and material properties of a variety of both II-VI and III-V compounds grown on a substrate where the II-VI/III-V heterostructure can be utilized. This paper describes some recent developments involving the molecular beam epitaxial (MBE) growth and characterization of two important II-VI/III-V heterostructures: ZnSe/GaAs and InSb/CdTe; a comparison is made between epitaxial layer/substrate interfaces and epilayer/epilayer interfaces for both heterostructures. The ZnSe/GaAs heterointerface, having a 0.25% lattice constant mismatch, has potential for use in passivation of GaAs devices. The InSb/CdTe heterointerface possesses an even closer lattice match, ~0.05% (comparable to the (Al,Ga)As/GaAs material system), and is motivated by possible device applications provided by InSb/CdTe quantum wells.

PROPERTIES of the ZnSe/GaAs HETEROJUNCTION

Recently we have reported the successful interrupted MBE growth of ZnSe on MBE-grown GaAs epilayers[1]. Problems of autodoping are avoided by providing for the growth of each material in a separate growth chamber. Two methods have been used to fabricate the II-VI/III-V heterojunctions. In one case, the as-grown GaAs surface was preserved by means of arsenic passivation[2] allowing transfer in air from one MBE system to the other. Arsenic passivation has been shown to minimize surface contamination problems[3] which occurred during interrupted growth; such contaminants can result in the introduction of undesireable interface states. A second fabrication approach involved the use of a modular MBE system where the ZnSe and GaAs growth chambers are connected by ultrahigh vacuum transfer modules. In order to provide an interfacial region that is free of dislocations, the "insulating" ZnSe is grown as a pseudomorphic layer providing a coherent interface. Reflection high energy electron diffraction (RHEED) intensity oscillations are used to monitor the character of the ZnSe nucleation, while transmission electron microscopy is used to confirm the microstructural quality of the interface. The device potential of the II-VI/III-V heterointerface is demonstrated by capacitance-voltage measurements on a metal-insulator-semiconductor (MIS) structure and by operation of a GaAs depletion-mode field effect transistor (FET). The C-V measurements indicate that the Fermi level at the ZnSe/GaAs interface can be varied from accumulation to inversion and that the interface state density can be made to be less than 10^{11} cm^{-2}. In these II-VI/III-V device configurations, the high resistivity ZnSe serves as a passivating 'insulator.'

The nucleation of ZnSe on MBE-grown GaAs epilayers was monitored by RHEED. Viewing in the [210], it was possible to observe (with the unaided eye) strong periodic variations in the intensity of the specular spot upon opening the Zn and Se source shutters[4]; the oscillations lasted for approximately 120 periods, corresponding to one-third of the growth duration for the 1000 Å pseudomorphic ZnSe layer. The intensity oscillations are characteristic of two-dimensional nucleation and layer-by-layer growth where each oscillation represents the completion of the growth of one monolayer of ZnSe[1,5]. A (2x4) reconstructed surface is observed on the GaAs epilayer prior to arsenic passivation. Depending on the temperature and duration of the amorphous arsenic desorption step, two surface reconstructions, (4x6) and (2x4), have been observed on the MBE-grown GaAs epilayer. Two-dimensional nucleation occurs on both (4x6) and (2x4) reconstructed surfaces of the GaAs epilayer. When nucleating ZnSe on an As-deficient bulk GaAs substrate, having a (4x6) surface reconstruction (resulting from the common practice of desorbing the oxide in the absence of impinging As for II-VI MBE), the intensity oscillations are not observed as three-dimensional nucleation occurs[1,5]. The ZnSe is nucleated at 320 °C with a Se-to-Zn flux ratio close to unity as measured with a quartz crystal monitor; the growth rate was ~0.4 μm/hr. One possible factor, contributing to the nucleation, is the density of surface nucleation sites as suggested by the different types of nucleation observed under conditions of similar surface stoichiometry for either GaAs substrates or epilayers. In a related study by Tamargo et al.[6], ZnSe was nucleated on two different surface conditions, As-rich and Ga-rich, for both GaAs substrates and epilayers. A Se-to-Zn beam equivalent pressure ratio of 4 or 5-to-1 and a substrate temperature of 270 °C was employed. In this case two-dimensional nucleation was reported to occur for both As-rich GaAs epilayers, and As-rich GaAs substrates (for which oxide desorption is performed in the presence of an arsenic flux). Under certain growth conditions it seems clear that the surface stoichiometry plays an important role in determining the nucleation behavior.

The microstructure of the pseudomorphic ZnSe/GaAs heterointerface was examined with cross-sectional and plan-view transmission electron microscopy[4,5]. As viewed in a cross-sectional bright field image[7], the interface between the 1000 Å ZnSe epilayer and the MBE-grown GaAs layer appears as a sharp straight line with no evidence of threading dislocations in the ZnSe.

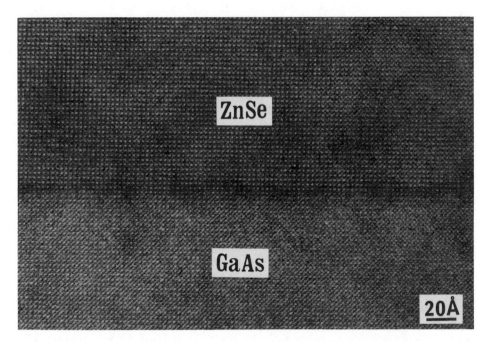

Fig. 1. High resolution TEM micrograph of the 1000 Å pseudomorphic
ZnSe/epitaxial GaAs heterointerface observed in the [100] projection.

Figure 1 shows a high resolution image of the ZnSe/GaAs interface in the [100]
projection. A defect-free, coherent contact between the two crystals is observed,
suggesting a structural perfection similar to that seen between (Ga,Al)As and
GaAs. Photoluminescence studies[1] of the 1000 Å ZnSe layer, the TEM
investigations, and x-ray diffraction measurements of the lattice parameter[8] all
confirm the pseudomorphic nature of the ZnSe film.

Recently, we have fabricated a prototype device structure wherein
pseudomorphic ZnSe formed the "insulator" in a GaAs depletion-mode field
effect transistor[7,9]. In the past (Al,Ga)As, having a direct bandgap of 2.0 eV for
an Al mole fraction of 0.5, had been widely used as an insulator for GaAs field
effect transistors. The (Al,Ga)As/GaAs heterojunction however, presented very
low interfacial barriers to carrier flow from GaAs into the (Al,Ga)As. ZnSe,
having a direct bandgap of 2.7 eV, was closely lattice matched (0.25%) to GaAs;
in a variety of device applications the ZnSe/GaAs heterointerface could provide
an alternative to (Al,Ga)As. The FET curves showed good depletion-mode
characteristics with complete pinch-off and current saturation[7,9]. The
modulation of the channel carrier concentration indicated that the Fermi level
positioning at the ZnSe/n-GaAs (epi) interface could be varied by at least 0.6 eV.

The opportunity afforded by the MBE growth technique is the ability to
systematically control the interface and film properties to reduce the number of
states that contribute to charge trapping. A range of interface state densities
have been measured in different Al/ZnSe/GaAs MIS capacitor samples; the high
frequency Terman method to analyze the capacitance versus voltage data was
employed to determine the interface state density[10]. At the present time, we
have measured for our best samples interface state densities which average

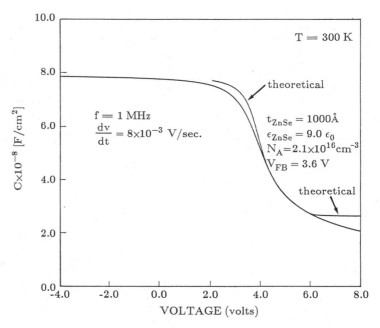

Fig. 2. Capacitance versus voltage for Al/ZnSe/GaAs MIS capacitors for p-type GaAs exhibiting accumulation. (These data were taken at 1 MHz in the dark at room temperature.)

below 10^{11} cm^{-2} eV^{-1} over the bandgap; these densities are comparable to densities reported for the (Ga,Al)As/GaAs interface[11-13]. Figure 2 shows the high frequency (1MHz) C-V characteristics for p-type GaAs obtained in the dark. The experimental C-V characteristics for a ZnSe/p-GaAs MIS structure is compared to a theoretical curve for the evaluation of the interfacial state density. In addition, the C-V characteristics for the case of an n-type sample exhibits behavior representative of the presence of an inversion layer[10]. It is important to note that such low interface state densities can be achieved at an interface between two compound semiconductors composed of elements having different valences. These measurements clearly support the use of ZnSe as an appropriate passivating material for a variety of GaAs-based devices.

MBE GROWTH of InSb/CdTe HETEROSTRUCTURES

One of the more interesting quantum well structures which have been proposed involves InSb as the well material with CdTe as the barrier layer. Theoretical predictions[14] of band offsets agree fairly well with experimental measurements[15] and suggest that these quantum wells will be of Type I with substantial conduction and valence band confinement. Minimal strain effects are expected as these two materials are very closely lattice matched (~0.05%), while a perfect lattice match can be achieved by incorporating a few percent of either Zn or Mn into the CdTe barrier layer. Large quantum shifts in the bandgap energy are predicted for relatively wide quantum wells as a result of the small effective mass of electrons and light holes. For example, a 75 Å quantum well has a ground state transition energy twice that of bulk InSb[16]. Structures involving reasonable well dimensions allow a wavelength range of 2-5.5 μm to be accessed. The high carrier mobilities and the large de Broglie wavelength provide the possibility for a wide variety of interesting devices. The realization of proposed device structures however, has been hampered by the significant materials problems associated with this II-VI/III-V material system.

Several research groups have reported the MBE growth of InSb on CdTe substrates[17], and CdTe on both InSb substrates[15,18-21] and InSb epilayers[19]. The majority of studies have primarily focussed on interfaces between epitaxial layers and substrates; the work reported here, as well as the work of Golding et al.[22], involves epilayer/epilayer interfaces. Previous studies employed InSb and CdTe bulk substrates which were ion etched and thermally annealed prior to epitaxy. The resultant epitaxial layers were of very high quality; however close examination of the interfacial region revealed a variety of problems. These difficulties must be eliminated, as in quantum well stuctures the interfaces can completely dominate the electronic and optical behavior. Many of the difficulties arise due to the widely differing optimum growth temperatures for the two materials; high quality CdTe on InSb is grown at temperatures as low as 160 °C[23], whereas InSb with superior electrical properties is grown at temperatures at or above 400 °C. Interfacial problems occurring at the CdTe/InSb interface include interdiffusion[15,17,20], precipiate formation (metallic indium or Sb segregation)[19-21], and intermediate layer formation (In_2Te_3)[15,21]. In the case of epilayer/epilayer interfaces, it has been shown that In_2Te_3 does form at the interfaces under certain growth conditions[24]. In addition to the interface problem, a fundamental difficulty associated with the CdTe/InSb system is the tendency for autodoping. A study performed by Williams et al.[25] indicated that when both compounds are grown in the same MBE chamber, Te seriously contaminated the Sb source such that it was difficult to control the carrier concentration of the InSb material.

As a means to circumvent some of the aforementioned problems, our approach has been: i) to employ two separate Perkin-Elmer model 430 growth chambers, connected by an ultrahigh transfer module, to eliminate problems of autodoping, and ii) to use an antimony cracker[26] for a source of Sb_2 for low temperature growth of InSb (a single growth chamber was employed for both the CdTe and InSb growth). In order to avoid interdiffusion problems, the growth temperature of InSb must be greatly reduced to 300 °C or below. Successful efforts have been reported[27,28] for the low temperature growth of GaAs where the use of As_2 resulted in GaAs which had comparable electrical properties to GaAs grown at higher temperatures (550-600 °C) using As_4. The growth of GaAs has also benefitted by a much reduced growth rate at lower growth temperatures[29]. As a consequence growth rates of 0.18-1.2 Å/sec have been employed during the MBE growth and we have used Sb_2 in an attempt to reduce generation of point defects during the low temperature growth of InSb. To avoid use of ion bombarded and thermally annealed substrates, epitaxial InSb grown on chemically etched InSb (100) substrates has been used as a base for our structures. The chemical etching and thermal treatment of the InSb substrates prior to growth can be found in detail in an earlier report[30].

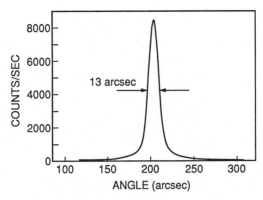

Fig. 3. Double crystal x-ray rocking curve measurement obtained on a 2.5 μm thick epilayer of InSb. The FWHM is 13 arc seconds.

The InSb epilayer has been characterized by both transmission electron microscopy (TEM)[30], photoluminescence[30], and double crystal x-ray rocking curve measurements. Figure 3 shows a rocking curve for a 2.5 μm epitaxial layer of InSb grown on an InSb substrate at 300 °C. The full width at half maximum (FWHM) is measured to be 13 arc seconds indicating the very high crystalline quality of the epitaxial layer.

When two growth chambers where used for the growth of CdTe/InSb single quantum well structures, the InSb epilayer was transferred from the InSb growth chamber via an ultrahigh vacuum transfer module (4x10^{-10} Torr) to a separate growth chamber for the CdTe epitaxy. The barrier layer of CdTe is grown from a compound source with a growth temperature of 240 °C (although nucleation of the CdTe on the InSb epi occurs at 200 °C) at a rate of ~0.9 Å/sec. Following the CdTe growth, the structure is transferred back to the InSb chamber for formation of the quantum well, after which the second CdTe barrier or cap layer is formed. Multiple quantum wells (20-25 periods) of InSb/CdTe have also been grown in a single growth chamber; in this case the Sb cracker is employed for the growth of the InSb layers with substrate temperatures of 280 °C. (The substrate temperature is calibrated for each run by observing the eutectic phase change of a 500 Å Au film on Ge.) The entire structure, including buffer layers, can be grown in the single chamber eliminating the necessity of transfer between the growth of each layer.

To study the various epilayer/epilayer interfaces present in the single quantum well and multiple quantum well structures, cross-sectional TEM specimens have been prepared. Significant problems have been encountered in the argon ion milling step due to defect generation in the two compound semiconductors and due to the tendency of InSb to be sputtered at a faster rate than CdTe. In agreement with Chew et al.[31], the defect generation problem is greatly reduced when iodine is used in the milling step. Figure 4 shows a dark field TEM micrograph of a 20 period CdTe/InSb multiple quantum well. The light contrast is the 167 Å CdTe barrier layer, while the dark contrast is the 163 Å InSb

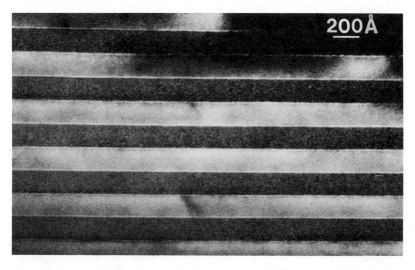

Fig. 4. Dark field TEM micrograph of a 163 Å InSb/ 167 Å CdTe
multiple quantum well consisting of 20 periods.

well layer.   Although this study is in an initial stage, the TEM examination does
not reveal the presence of any intermediate layers forming at the CdTe/InSb
interface.   The presence of an In_2Te_3 interfacial layer (having a 5% lattice
constant mismatch), which has been reported to be tens of angstroms thick[15], is
expected to be visible in these TEM images;   future work will continue to address
this problem.

Figure 5 shows a low temperature infrared photoluminescence spectrum
obtained from a 5600 Å single quantum well structure fabricated using the
modular MBE approach.   The double heterostructure consists of a 0.42 μm InSb
buffer layer grown on an InSb substrate, followed by a 1.63 μm CdTe buffer
layer, the active InSb layer, and a 2200 Å CdTe cap layer.   When compared
against luminescence from a thick InSb epitaxial layer or substrate, the
quantum efficiency from the double heterostructure is lower but still
reasonable.   (It is important to note that the structure was fabricated by the
interrupted growth approach using two separate growth chambers.)   The
spectrum contains two features of which the higher energy peak is assigned to
band-to-band recombination.   In comparison with bulk InSb, the spectrum is
broader, however, measuring approximately 20 meV at T=10K.   The inset shows
the Raman resonant excitation at the E_1 transition of InSb.   In this case
(backscattering from (100) face) the LO-phonon peak of InSb (191 cm^{-1})
dominates over the CdTe LO-phonon scattering (both indicated by the arrows in
the inset).   The relatively low and featureless background in the Raman
spectrum suggests that large scale formation of chemically intermixed species at
the heterointerfaces has not occurred here.   Clearly, however, more detailed
studies are needed to systematically investigate the dependence of both the
luminescence and Raman spectra on growth conditions.

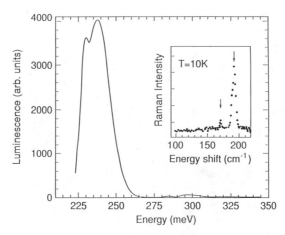

Fig. 5. Infrared photoluminescence spectrum obtained from a 5600 Å single quantum well of InSb compared to the luminescence emitted from an InSb substrate.

SUMMARY

Two technologically important compound semiconductor heterojunctions, ZnSe/GaAs and InSb/CdTe, have been studied with emphasis on epilayer/epilayer interfaces and the differences observed from comparisons with epilayer/substrate interfaces. The nature of epilayer/epilayer interfaces has important consequences for future potential applications involving II-VI/III-V heterojunctions and their implementation into integrated optoelectronic device structures. Both heterojunction systems are closely lattice matched and are predicted to provide adequate conduction and valence band offsets necessary for many device applications. Using pseudomorphic ZnSe as an insulator, prototype GaAs depletion-mode field effect transistors have been fabricated. Capacitance-voltage measurements of Al/ZnSe/GaAs capacitors provide interface state densities comparable to interface densities reported for interfaces of (Al,Ga)As/GaAs. Preliminary studies of the InSb/CdTe heterojunction have concentrated on investigations of the microstructure of the heterointerface. For the growth of these heterostructures an Sb cracker has been used as the source of Sb_2, and the InSb growth temperatures have been kept relatively low at 300 °C. Future work will continue to characterize the InSb/CdTe single quantum well both optically and electrically.

ACKNOWLEDGEMENT

The work briefly summarized above has resulted from the effort of a large number of people. The authors would like to acknowledge several students who have made substantial contributions: J. Qiu, G. D. Studtmann, J. L. Glenn, S. O., J. M. Gonsalves, M. Haggerott, and T. Heyen. We would like to acknowledge Q. Qian for the development of the MIS capacitors with low surface state density. We would also like to thank J. A. Cooper, R. F. Pierret for collaborating in the understanding of the electrical characteristics of the ZnSe/GaAs interface, and D. Lubelski for his dedication to the operation of the MBE facility. We would like to acknowledge M. Kobayashi for participation in all of the above II-VI/III-V heterojunction work. Research support at Purdue was provided by Air Force Office of Scientific Research (AFOSR), U.S. Office of Naval Research (ONR), SDIO/IST-Naval Research Laboratory, Defense Advanced Research Projects Agency (DARPA)/ONR University Research Initiative Program, a National Science Foundation-MRG Grant, and a joint AFOSR/ONR Research Instrumentation Grant. The research at Brown was supported by DARPA.

REFERENCES

*current address: Department of Electrical Engineering and Computer Science, Massachusetts Institute of Technology, Cambridge, Massachusetts, 02139

1. R. L. Gunshor, L. A. Kolodziejski, M. R. Melloch, M. Vaziri, C. Choi, and N. Otsuka, Appl. Phys. Lett. 50:200 (1987).
2. S. P. Kowalczyk, D. L. Miller, J. R. Waldrop, P. G. Newman, and R. W. Grant, J. Vac. Sci. Technol. 19:255 (1981).
3. D. L. Miller, R. T. Chen, K. Elliott, S. P. Kowalczyk, J. Appl. Phys. 57:1922 (1985).
4. L. A. Kolodziejski, R. L. Gunshor, A. V. Nurmikko, and N. Otsuka, RHEED Intensity Oscillations and the Epitaxial Growth of Quasi-2D Magnetic Semiconductors, in: "Thin-Film Growth Techniques for Low Dimensional Structures," R. F. C. Farrow, S. S. P. Parkin, P. J. Dobson, J. H. Neave, and A. S. Arrott, eds., Plenum Publishing, New York (1987).
5. R. L. Gunshor and L. A. Kolodziejski, IEEE J. Quantum Electronics, Special Issue on Quantum Well Heterostructures and Superlattices, in press, (1988).
6. M. C. Tamargo, J.L. de Miguel, D. M. Hwang, B. J. Skromme, M. H. Meynadier, R. E. Nahory, Mat. Res. Soc. Symp. Proc. 102:125 (1988).
7. G. D. Studtmann, R. L. Gunshor, L. A. Kolodziejski, M. R. Melloch, J. A. Cooper, R. F. Pierret, and D. P. Munich, Appl. Phys. Lett. 52, 1249 (1988).
8. T. Yao, Y. Okada, S. Matsuri, K. Ishida, and I. Fujimoto, J. Crystal Growth 81:518 (1987).
9. G. D. Studtmann, R. L. Gunshor, L. A. Kolodziejski, M. R. Melloch, N. Otsuka, D. P. Munich, J. A. Cooper, Jr., and R. F. Pierret, paper presented at the 45th Device Research Conference, Santa Barbara, CA, June 22-24, 1987.
10. Q. Qian, J. Qiu, M. R. Melloch, J. A. Cooper, R. L. Gunshor, and L. A. Kolodziejski, Appl. Phys. Lett., to be published.
11. P. M. Solomon, T. W. Hickmott, H. Morkoç, and R. Fischer, Appl. Phys. Lett. 42: 821 (1983).
12. T. J. Drummond, R. J. Fischer, W. F. Kopp, D. J. Arnold, J. F. Klem, H. Morkoç, and M. S. Shur, IEEE Trans. on Electron Devices ED-31:1164 (1984).
13. Sang-Koo Chung, Y. Wu, K. L. Wang, N. H. Sheng, C. P. Lee, and D. L. Miller, IEEE Trans. on Electron Devices ED-34:149 (1987).
14. J. Tersoff, Phys. Rev. Lett. 56:2755 (1986).
15. K. J. Mackey, P. M. G. Allen, W. G. Herrenden-Harker, R. H. Williams, C. R. Whitehouse, and G. M. Williams, Appl. Phys. Lett. 49:354 (1986).
16. R. G. van Welzenis and B. K. Ridley, Solid-State Electron. 27:113 (1984).
17. Koichi Sugiyama, J. Crystal Growth 60:450 (1982)
18. R. F. C. Farrow, G. R. Jones, G. M. Williams, and I. M. Young, Appl. Phys. Lett. 39:954 (1981).
19. S. Wood, J. Greggi, Jr., R. F. C. Farrow, W. J. Takei, F. A. Shirland, and A. J. Noreika, J. Appl. Phys. 55:4225 (1984).
20. G. M. Williams, C. R. Whitehouse, N. G. Chew, G. W. Blackmore, and A. G. Cullis, J. Vac. Sci. Technol. B3:704 (1985).
21. D. R. T. Zahn, K. J. Mackey, R. H. Williams, H. Munder, J. Geurts, and W. Richter, Appl. Phys. Lett. 50:742 (1987).

22. T. D. Golding, M. Martinka, and J. H. Dinan, J. Appl. Phys. 15 August issue (1988).

23. R. F. C. Farrow, A. J. Noreika, F. A. Shirland, W. J. Takei, S. Wood, J. Greggi, Jr., and M. H. Francombe, J. Vac. Sci. Technol. A2:527 (1984).

24. D. R. T. Zahn, T. D. Golding, K. J. Mackey, J. Geurts, J. H. Dinan, W. Richter, and R. H. Williams, J. Appl. Phys., in press.

25. G. M. Williams, C. R. Whitehouse, T. Martin, N. G. Chew, A. G. Cullis, T. Ashley, D. E. Sykes, and K. Mackey, paper presented at the 4th International Conference on Molecular Beam Epitaxy, York, England, September 7-10, 1986.

26. The Sb cracker was obtained from EPI, Effusion Products, Incorporated.

27. J. H. Neave, P. Blood, and B. A. Joyce, Appl. Phys. Lett. 36:311 (1980).

28. M. Missous and K. E. Singer, Appl. Phys. Lett. 50:694 (1987).

29. G. M. Metze, A. R. Calawa, and J. G. Mavroides, J. Vac. Sci. Technol. B1:166 (1983).

30. L. A. Kolodziejski, R. L. Gunshor, N. Otsuka, and A. V. Nurmikko, Mat. Res. Soc. Symp. Proc. 102:113 (1988).

31. A. G. Cullis, N. G. Chew, and J. L. Hutchison, Ultramicroscopy 17:203 (1985).

GROWTH OF II-VI/III-V MIXED HETEROSTRUCTURES

M. C. Tamargo, J. L. de Miguel,[*] F. S. Turco,
B. J. Skromme, D. M. Hwang, R. E. Nahory and H. H. Farrell

Bellcore
331 Newman Springs Road
Red Bank, New Jersey 07701

ABSTRACT

In this paper we describe results on the growth of ZnSe/GaAs heteroepitaxial structures using a dual chamber molecular beam epitaxy (MBE) system. We have observed a preference for ZnSe to grow on the GaAs (2x4) As rich surface. In addition, we observe one dimensional disorder in the initial nucleation of GaAs when grown on ZnSe layers. We attribute these two behaviors to the existance of an electronic imbalance at the II-VI/III-V interface which can be minimized by tailoring of the surface stoichiometry. We have also obtained extremely narrow photoluminescence linewidths for layers of ZnSe grown on thin buffer layers of GaAs, AlAs and $In_xGa_{1-x}As$. We attribute this to the elimination of inhomogeneous strain associated with partial lattice relaxation.

INTRODUCTION

Two-six semiconductor materials are presently receiving renewed attention due to dramatic progress in epitaxial crystal growth techniques. Among these materials, ZnSe is particularly interesting because it is a wide, direct bandgap semiconductor with potential applications as a blue light emitter. It also has the advantage that its lattice constant closely matches that of the well-known semiconductor GaAs. As a result, GaAs has been used successfully as a high quality substrate for epitaxial ZnSe growth.

The basis for recent solid state device research has involved the epitaxial growth of different materials in layered structures. Depending on the materials involved, this heteroepitaxy can exhibit different characteristics which will determine the approaches to be followed for their growth. Most frequently, only isoelectronic materials, with closely matching lattice parameters, have made up these heteroepitaxial structures. More recently, structures of materials with significantly different lattice constants have been successfully grown. A different type of problem arises when the materials have different electronic structures. Such is the case of GaAs on Si or Ge and of II-VI/III-V heteroepitaxy.

In this paper we will review our recent advances in the fabrication of ZnSe/GaAs heteroepitaxial structures. We have used a multiple chamber molecular beam epitaxial (MBE) system which allows the transfer of the sample, in ultra-high vacuum, between growth chambers. One growth chamber is dedicated to the growth of III-V materials and the other to ZnSe growth. Our system also contains surface analysis tools which allow us to investigate the nature of the heterointerface in situ and to evaluate improvements in the quality of the interfaces. First, we

---

*Present address: Centro Nacional de Microelectronica, C.S.I.C., 28006 Madrid, Spain

Table I. Transformation of the RHEED pattern during ZnSe growth on GaAs

Substrate Type	Initial Pattern	Transition Time
Bulk, Ga rich	spotty	1-3 min
Bulk, As rich	streaky	10 sec
Epi, As rich	streaky	10 sec
Epi, Ga rich	spotty	5 min

have found that two dimensional nucleation of ZnSe on GaAs occurs only on the As rich (2x4) GaAs surface. In addition, we observed a marked anisotropy in the nucleation of GaAs on ZnSe layers. We attribute this behavior to the different electronic structure of the elements making up the two semiconductors, which results in an electronic imbalance at the interface. The electronic imbalance can, in turn, affect the quality of the growth. This imbalance can be reduced by careful tailoring of the surface stoichiometry. Secondly, we have investigated the effects of the small but significant lattice mismatch which exists between ZnSe and GaAs. We have found that the incorporation of buffer layers of various compositions can reduce the presence of inhomogeneous broadening, attributed to the lattice mismatch, which is seen in the photoluminescence spectra of thin ZnSe layers.

GROWTH OF ZnSe ON GaAs AND GaAs ON ZnSe: Surface stoichiometry considerations

We have recently shown[1] that the growth of ZnSe on GaAs is strongly dependent on the stoichiometry of the GaAs surface on which nucleation takes place. Using a dual chamber MBE system, we have investigated the initial stages of ZnSe growth on GaAs bulk substrates and epitaxial layers which were terminated by an As rich surface or a Ga rich surface. Reflection high energy electron diffraction (RHEED) observations, summarized in Table I, indicate a preference for ZnSe to grow on the As terminated surface.

On this surface, the presence of a streaky RHEED pattern indicates that two dimensional growth occurs within seconds of the initiation of growth; while on the Ga rich surface, an initial, long lasting three dimensional growth, characterized by a spotty RHEED pattern, always takes place. The results were confirmed by studies using X-ray photoelectron spectroscopy (XPS),[2] in which ZnSe growth was interrupted after only a few monolayers and analyzed by XPS. The sample could be reintroduced in the MBE chamber for additional growth and then analyzed again. The ratios of the peak intensities of the 3d core level for As to Se and Ga to Zn after growth of approximately 40 A are 1.0 and 1.9, respectively, indicating that island formation occurs when the ZnSe is grown on Ga rich GaAs surfaces. Alternatively, after only 30 A of ZnSe growth on the As rich GaAs surface the ratios are 0.17 and 0.24, respectively, indicating that uniform layer coverage occurs on these surfaces. Cross sections of these heterostructures studied by transmission electron microscopy (TEM) also confirmed the improved interface quality obtained on the As rich interfaces.[1]

The epitaxy of GaAs layers on ZnSe gave a somewhat more complex picture.[2] Upon initiation of GaAs growth on ZnSe surfaces, we observed an anisotropy in the surface reconstruction using RHEED. This anisotropy indicates a degree of disorder along the $[0\bar{1}1]$ direction of the (100) surface. XPS was used to study the initial stages of growth of these layers and the results are summarized in Table II.

Table II. XPS analysis for GaAs grown on ZnSe

GaAs layer thickness	I_{Ga}/I_{Zn}		I_{As}/I_{Se}	
	exp	calc	exp	calc
22 Å	1.09	2.4	0.74	1.9
44 Å	3.17	10.5	2.67	8.4

Again, the ratios of the 3d core level peaks for Ga to Zn and As to Se were determined from the experimental data. A comparison with calculated values confirms the disordered nature of the nucleation process. Details of the calculations will be published elsewhere. These results, along with observations using low energy electron diffraction (LEED), suggest the formation of long ridges or facets along the $[0\bar{1}\bar{1}]$ direction, possibly terminated by (111) surfaces. Cross sections of these interfaces studied by TEM revealed the presence of microtwins, possibly resulting from this initial growth instability.

The observations described above can be interpreted by invoking the existence of an electronic imbalance at the interface of ZnSe and GaAs.[3] Such an imbalance arises when the number of valence electrons at the interface either exceeds, or is insufficient to fill the available number of bonding states. It occurs when the substrate surface is terminated with essentially only one species (e.g. As or Ga), and should be absent when the surface composition is evenly mixed. In practice this mixed composition occurs for the first case (ZnSe on GaAs). The As rich (2x4) surface is expected to be terminated by a partial monolayer of As, possibly near one half of a monolayer. When ZnSe growth begins, Se atoms will complete this layer, forming a mixed layer of As and Se. This will then be followed by alternating full layers of Zn and of Se. Because of the formation of this mixed intermediate layer, the number of electron deficient Zn-As bonds will be roughly equal to the number of electron rich Ga-Se bonds, and an overall charge balance at the interface will be achieved.

On the other hand, in the case of GaAs growth on ZnSe, the ZnSe surface is most likely terminated with essentially a complete monolayer of Zn atoms after it is heated to 300 C for GaAs deposition.[4] As the GaAs growth begins, the next layer will almost certainly be As and thus the interface will be comprised of electron deficient Zn-As bonds. A net flow of charge into the interface region is expected to ocurr in order to fill these incomplete states. This charge is not balanced locally by nuclear charge, thus it can generate extremely large electrostatic fields.[5] Such fields are expected to enhance a variety of phenomena, such as three dimensional growth and enhanced surface diffusion, which can disrupt the growth process. We feel that the surface roughening observed in these interfaces is due at least in part to this effect. Our model indicates that in order to solve the instability at the ZnSe surface for GaAs growth one must carefully tailor the stoichiometry of the ZnSe surface prior to deposition of GaAs.

GROWTH OF ZnSe ON AlAs AND ON $In_xGa_{1-x}As$: Interface strain considerations

Although the mismatch between the lattice constants of ZnSe and GaAs can be considered to be small (0.27%), it is sufficient to affect the quality of the grown ZnSe layer, especially near the interface. It is known that layers of ZnSe grown on GaAs remain coherently strained until a certain critical thickness is surpassed which is experimentally measured to be about 0.17 μm.[6] Beyond this thickness, ZnSe relaxes to its bulk lattice constant and a certain density of misfit dislocations and defects are formed at the interface in order to accomodate the lattice mismatch. As a result of these defects, the photoluminescence linewidths exhibit broadening due to inhomogeneous strain relief. Linewidths of 1.2 meV or greater, which are much broader than those obtained from bulk ZnSe crystals, are typical for ZnSe epitaxial layers.

Table III. Dependence of ZnSe PL linewidth on the composition of $In_xGa_{1-x}As$ buffer layers

x	PL linewidth
0.0 %	0.55 meV
0.6 %	0.50 meV
4.4 %	0.30 meV
8.7 %	2.0 meV

We have recently reported[7] that very high quality pseudomorphic layers of ZnSe can be grown, using a dual chamber MBE system, on buffer layers of GaAs. These thin layers exhibit extremely narrow PL linewidths as can be seen in Figure 1(a). The narrow lines indicate that the layers have very high structural quality, with negligible inhomogenous broadening due to the onset of the relaxation process. In addition, when a thin layer of AlAs was used as the buffer layer, a further improvement in the linewidths of the photoluminscence peaks is observed,[8] as shown in Figure 1(b). Both of these structures are entirely pseudomorphic, conforming to the GaAs lattice constant. When 0.15 μm thick $In_xGa_{1-x}As$ layers with various values of x were used as buffer layers separating the GaAs substrate and the thin ZnSe layers, the linewidth appears to depend on the value of x, with the narrowest lines observed for x of approximately 4%.[7] The results are summarized in Table III.

These extremely sharp, well resolved luminescence spectra are the best ever reported for heteroepitaxial ZnSe layers. The exact mechanism for the improved structural quality of these layers is still not well understood. Differences in the nucleation chemistry of the ZnSe on the different buffer layers or the incorporation of multiple interfaces between the GaAs substrate and the ZnSe layer are some possible effects. The PL spectra, nevertheless, establish the high degree of structural quality achieved with thin layers of ZnSe.

CONCLUSIONS

We have shown that two dimensional nucleation of ZnSe on GaAs occurs on the (2x4) As rich GaAs surface whereas initial island formation is evident when growth takes place on the Ga rich surfaces. We have also grown epitaxial layers of GaAs on ZnSe. In this case, a degree of interface disorder, manifested as microtwins, were present near the interface. A model for interface formation which invokes charge imbalance at the III-V/II-VI heterointerface was proposed, which requires careful control of the surface stoichiometry for good interfaces to form. We have further improved the structural quality of thin ZnSe layers by incorporating an $In_xGa_{1-x}As$ buffer layer.

In view of these results, the possibility of utilizing ZnSe/GaAs heterostructures as active elements of solid state devices appears very promising.[9] Yet, several factors remain to be settled before these materials could become useful. One of them is the question of the alignment of the bands. Although conflicting reports appear in the literature,[10,11] it is believed that a small but positive potential barrier exists in the conduction band leaving most of the bandgap discontinuity to the valence band.[12] With careful tailoring of the interface stoichiometry and accounting for the interface strain effects it should be possible to grow the high quality structures needed to investigate this issue. We are presently pursuing this work in our laboratory. A second problem to be addressed is that of doping of the II-VI material. Here too, the potential of advanced growth techniques, such as MBE, is pointing to new solutions which are likely to succeed in the near future. The resolution of these issues along with the possibility of high quality heteroepitaxy opens new directions in semiconductor materials research.

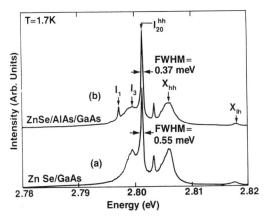

Figure 1. Low temperature exciton photoluminescence for 0.15 μm ZnSe layers grown on (a) GaAs and (b) 0.7 μm AlAs buffer layers. The experimental resolution is 0.09 meV or better.

REFERENCES

1. M. C. Tamargo, J. L. de Miguel, D. M. Hwang and H. H. Farrell J. Vac. Sci. Technol. **B 6**, (1988) 784.

2. M. C. Tamargo, J. L. de Miguel, D. M. Hwang, B. J. Skromme, M. H. Meynadier, R. E. Nahory and H. H. Farrell, Mat. Res. Soc. Symp. Proc. **102**, (1988)125.

3. H. H. Farrell, M. C. Tamargo and J. L. de Miguel, presented at the Physics and Chemistry of Surfaces and Interfaces Meeting, January, 1988.

4. H. H. Farrell, M. C. Tamargo and J. L. de Miguel, J. Vac. Sci. Technol. **B 6**, (1988) 767.

5. W. A. Harrison, E. A. Kraut, J. R. Waldrop and R. W. Grant, Phys. Rev. **B 6**, (1978) 4402.

6. T. Yao, in *The Technology and Physics of Molecular Beam Epitaxy,* edited by E. H. C. Parker (Plenum, New York, 1986) p. 313.

7 B. J. Skromme, M. C. Tamargo, J. L. de Miguel and R. E. Nahory, presented at the Electronic
   Materials Conference, June, 1988.

8. B. J. Skromme, M. C. Tamargo, J. L. de Miguel and R. E. Nahory, Mat. Res. Soc. Symp. Proc. **102** (1988) 577.

9. E. T. Yu and T. C. McGill, Appl. Phys Letters **53**, (1988) 60.

10. G. D. Studtmann, R. L. Gunshor, L. A. Kolodziejski, M. R. Melloch, J. A. Cooper, Jr., R. F. Pierret, D. P. Munich, C. Choi and N. Otsuka, Appl. Phys. Letters **52**, (1988) 1249.

11. K. Mazuruk, M. Benzaquen and D. Walsh, J. Physique **48** (1987) C5-357.

12. S. P. Kowalczyk, E. A. Kraut, J. R. Waldrop and R. W. Grant, J. Vac. Sci. Technol. **21**, (1982) 481.

ZnSe GROWTH ON NON-POLAR SUBSTRATES BY MOLECULAR BEAM EPITAXY

Robert M. Park

Dept. of Materials Science and Engineering
University of Florida
Gainesville, FL. 32611

ABSTRACT

The following paper is essentially a review of an investigation carried out by the author and his former colleagues at 3M Canada Inc. into the growth of ZnSe by molecular beam epitaxy on both Ge and Si substrate materials. In-situ substrate preparation procedures developed particularly for Ge and Si are described together with techniques developed for ZnSe epitaxial growth on these materials. The primary characteristic differences observed experimentally between ZnSe layers grown on non-polar substrates as opposed to polar substrates (in particular, GaAs) will be reported and discussed. Characterization techniques employed in this study included 4.2K photoluminescence measurements, double-crystal rocking-curve x-ray analysis and, transmission electron microscopy analysis, in both planar and cross-sectional modes.

INTRODUCTION

Although the primary focus of the research on ZnSe has been to develop the material in order to fabricate blue light-emitting devices, it is the author's belief that a broader view should be taken of the material which could lead to other interesting device applications. Significant steps have already been taken in this direction. For example, ZnSe has been used as a lattice-matched insulator in a ZnSe/GaAs MISFET device[1] which exhibited promising transistor characteristics. ZnSe may, in fact, replace AlGaAs in a number of applications due to the fact that ZnSe is somewhat easier to grow than AlGaAs and also since a larger band-edge discontinuity is provided by the ZnSe/GaAs system compared to the AlGaAs/GaAs system. In addition, ZnSe exhibits interesting non-linear optical effects and, consequently, the material could find application in the areas of integrated optics and all-optical computing systems[2].

Another potential application for ZnSe is in the area of ZnSe/Ge hetero-junction IR detector devices. The ZnSe/Ge heterostructure system is interesting in this regard on account of the following: (a) ZnSe is transparent over a wide range of wavelengths, namely, $20\mu m$ to $0.5\mu m$, (b) large band-edge discontinuities should be available, and (c) ZnSe and Ge are closely lattice-matched, the lattice mismatch being approx. 0.17%.

Finally, in keeping with current thinking regarding the usefulness of semiconductor materials integration, e.g. GaAs/Si integration, the author believes that the areas of application of ZnSe-based devices could be expanded by virtue of them being integrated in a monolithic fashion with Si electronic devices. However, as in the case of GaAs/Si integration, a 4% lattice-mismatch exists between ZnSe and Si, which presents a rather serious problem with regard to the growth of high-quality heterostructures.

The purpose of this paper is to essentially review the work carried out by the author and his former colleagues at 3M Canada Inc. on the growth of ZnSe on both Ge and Si substrate materials by molecular beam epitaxy. This work was done with a view to expanding the areas of application of ZnSe.

Ge AND Si SUBSTRATE PREPARATION

Since layer quality is critically affected by the nature of the substrate surface, particular attention was paid to the preparation of Ge and Si substrate surfaces prior to ZnSe epitaxy. Of particular concern was to develop a preparation technique which did not involve either acid etching or high-temperature treatment. In the case of Ge, it was found that both C and O contamination could be removed from solvent-washed, as-loaded Ge wafer surfaces by a room-temperature argon-ion sputtering process. Subsequent thermal annealing at temperatures in the range 350°C to 400°C resulted in highly ordered Ge surfaces which exhibited (2x2) surface reconstruction as observed by reflection high energy electron diffraction (RHEED). Thermal treatment alone for Ge resulted in oxygen desorption at temperatures around 450°C; however, carbon contamination persisted even at temperatures >750°C. Room-temperature argon-ion sputtering, in the case of Si, was not successful in removing the oxygen contamination from solvent-washed, as-loaded Si wafer surfaces. However, a simultaneous 'sputter and anneal' treatment, which consisted of the Si surface being argon-ion sputtered while the wafer temperature was held at around 400°C, provided contamination-free Si surfaces. Highly ordered Si surfaces exhibiting (2x1) surface reconstruction were observed following the above treatment. Subsequent annealing at higher temperatures was not required to produce Si surfaces suitable for epitaxy. The Si substrate cleaning temperature of 400°C employed here is much lower than the temperatures normally used in the case of thermally-cleaned Si, (>850°C).

The ZnSe layers grown in this investigation were deposited epitaxially on Ge and Si substrates which were prepared using the respective cleaning techniques described above.

ZnSe GROWTH ON Ge AND Si SUBSTRATES

Since the initial stages of heteroepitaxial growth are crucial to the subsequent orderly development of layer growth, considerable attention was given to the observation of ZnSe growth initiation on atomically-clean and ordered Ge and Si substrate surfaces. These observations will be discussed in the following sections. Also discussed below will be ZnSe/Ge and ZnSe/Si layer characterization. Photoluminescence (PL) and Double-Crystal Rocking Curve (DCRC) x-ray analyses were principally employed in order to optically and structurally characterize the layers. Transmission Electron Microscopy (TEM) in both planar and cross-sectional modes was also used.

ZnSe/Ge Heterostructures

On initiation of growth under typical growth conditions, which are described in detail in reference 3, the streaky RHEED pattern, indicative of a smooth (2x2) reconstructed Ge surface, gave way to a diffuse spotty pattern

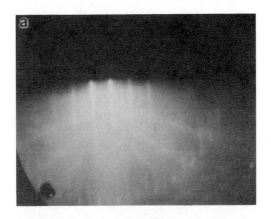

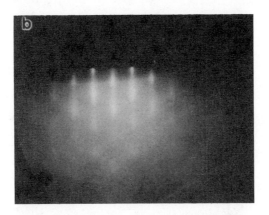

Fig. 1. Evolution of RHEED patterns observed during the initial
stages of growth of ZnSe on (100) Ge. Pattern (a) was
recorded from a sputtered and annealed (100) Ge surface
while patterns (b) and (c) were recorded following 80Å
and 400Å of ZnSe deposition, respectively.

which suggested a three-dimensional island type growth mechanism. In time, the spots became elongated and eventually full streaking was observed. The evolution of the RHEED pattern during the initial growth stages is illustrated in Fig. 1. It is interesting to note that planar growth was only indicated following several 100Å of ZnSe deposition. This can be contrasted to the ZnSe/GaAs case in which planar growth is established normally within tens of angstroms of ZnSe deposition unless growth is initiated on an MBE-grown GaAs surface, in which case, planar growth takes place immediately.[4]

PL analysis of ZnSe/Ge layers produced similar results to those obtained previously from ZnSe/GaAs layers, namely that under 'optimum' growth conditions, the PL spectra are dominated by a donor-bound exciton (DBE) emission

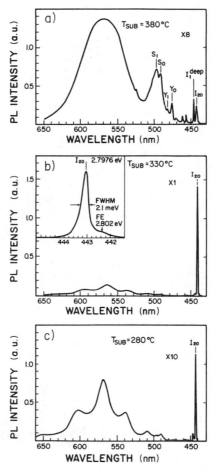

Fig. 2. Typical 4.2K photoluminescence spectra recorded from ZnSe layers grown on (100) Ge using a Zn to Se beam pressure ratio of unity at substrate temperatures of (a) 380°C, (b) 330°C, and (c) 280°C.

peak[5]. As can be seen from the PL spectra illustrated in Fig. 2, the energy
of the dominant DBE peak recorded from ZnSe/(100)Ge layers is 2.7976eV.
Although the linewidth is somewhat broader (2meV compared with 1meV, typically), this emission peak is almost certainly the same as that observed from
ZnSe/(100)GaAs layers grown under the same growth conditions. It is the
author's contention that the emission peak is impurity related and that the
impurity is incorporated during growth either from the background environment
or the source materials.

As evidenced from the spectra shown in Fig. 2 (note the scaling factors),
the intensity of the dominant DBE peak is a sensitive function of the growth
temperature, being a maximum at a temperature around 330°C. This observation
does not imply, however, that the donor concentration is maximized at this
growth temperature. A more appropriate conclusion, since the radiative
efficiency of excitonic transitions is extremely sensitive to layer stoichiometry, is that ZnSe layers grown at this substrate temperature and flux
ratio indicated have close to the stoichiometric composition. Actually,
excitonic peak linewidths are a more reliable indicator of material quality
and, as can be seen from Fig. 3, the minimum DBE peak linewidth and the maximum peak intensity coincide at a growth temperature of 330°C, confirming
330°C to be the 'optimum' growth temperature. Similar data plotted in Fig. 4
suggests that a Zn to Se flux ratio during growth of around unity for a substrate temperature of 330°C is 'optimum'. The data illustrated in Figs. 3
and 4 suggests that the homogeneity range for ZnSe is rather narrow and that,
as expected, the linewidth minimum and intensity maximum of excitonic peaks
are coincident in growth space and indicate near-stoichiometric unintention-
ally-doped ZnSe material.

ZnSe layers were also grown under identical growth conditions on (211)
Ge and (100) GaAs substrates and DBE peak linewidths recorded from these
layers are plotted in Fig. 5 as a function of growth temperature. The DBE
linewidth data from ZnSe/(100)Ge layers is also shown in the figure. Based
on the above discussion, the data illustrated in Fig. 5 suggests that layer
stoichiometry is independent of substrate orientation and polarity with
'optimum' stoichiometry resulting for a fixed set of growth conditions,
regardless of substrate material.

The nature of the substrate material does, however, have a strong influence on the type and density of dislocations present in ZnSe epitaxial
layers. It has been shown that, for a given set of growth conditions, both
the DBE and DCRC linewidths recorded from ZnSe layers grown on Ge substrates
are typically a factor of two larger than those recorded from ZnSe/GaAs
layers of the same thickness[6]. Such data suggests ZnSe/Ge layers to be more
heavily dislocated than ZnSe/GaAs layers. Planar TEM observations have shown
that the larger PL and x-ray peak linewidths recorded from ZnSe/Ge layers
are, in fact, due to the presence of dislocations associated with low-angle
grain boundaries found to exist in ZnSe/Ge layers but which are not present
in ZnSe/GaAs layers[7]. A planar TEM micrograph, recorded from a typical
ZnSe/Ge sample, is shown in Fig. 6, which indicates the layer to have a cell-
type structure. The cell walls are, in fact, composed of planar defects
having very small rigid body displacements as deduced by observing very weak
or no fringe contrast under a variety of diffraction conditions. The dislocation content in ZnSe/Ge layers is associated with these boundaries with
dislocations, which are formed by relaxation processes within the boundary
planes, actually being confined to these planes. It is worth noting that
the cell boundaries are not anti-phase boundaries, which was verified using
convergent beam electron diffraction techniques; but rather are general
displacement boundaries. One hypothesis is that these boundaries delineate

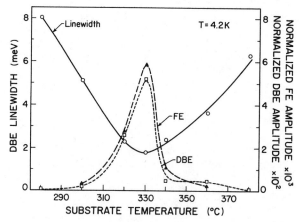

Fig. 3. Donor-bound exciton (DBE) and free-exciton (FE) emission
peak amplitudes together with DBE linewidths plotted as
a function of substrate temperature. The data was
obtained by 4.2K PL analysis of ZnSe/Ge layers grown
using a Zn to Se beam pressure ratio of unity.

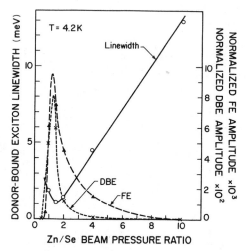

Fig. 4. Donor-bound exciton (DBE) and free-exciton (FE) emission
peak amplitudes as well as DBE linewidths plotted as a
function of Zn to Se beam pressure ratio for ZnSe/Ge
layers grown at a substrate temperature of 330°C. The
data was obtained by 4.2K PL analysis using the 365nm
line from a Hg(Xe) lamp as the source of exciton.

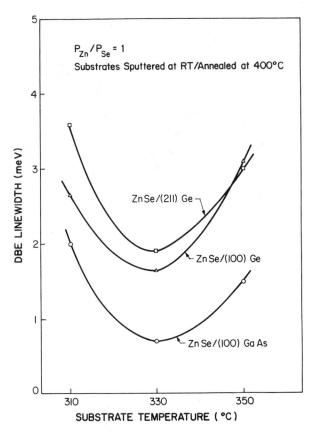

Fig. 5. Linewidths of the dominant donor-bound exciton (DBE)
emission peaks obtained from ZnSe layers grown at three
different substrate temperatures on a variety of substrates, namely, (100)Ge, (211)Ge and (100)GaAs. The
growth conditions were identical in each case. The data
was obtained by 4.2K PL analysis.

regions which are slightly misorientated with respect to each other resulting
as a consequence of island growth during the initial stages of epitaxy. The
RHEED observations of the initial stages of ZnSe growth on Ge, as discussed
above, would seem to support this postulate at least in part.

The DCRC analysis of ZnSe/Ge heterostructures revealed another interesting structural phenomenon associated with this system, namely, that the
epi-layer planes are misorientated, or tilted, with respect to the equivalent
substrate planes. This observation was first discussed in reference 6, in
which tilt angles as large as 1,000 arc sec. were reported. Further studies
have revealed that the magnitude of the tilt angle is a function of both the
ZnSe layer thickness (and consequently, layer stress) and also the substrate
surface orientation. A full discussion of this phenomenon can be found in
reference 8.

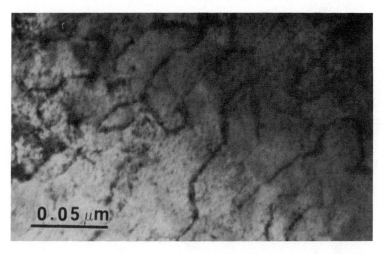

Fig. 6. Planar transmission electron microscopy image of a typical ZnSe layer grown on (100)Ge. Dark lines illustrate dislocations.

ZnSe/Si Heterostructures

Although not studied to the same extent, the striking structural features observed in ZnSe/Ge heterostructures, namely, low angle grain boundaries and epi-layer tilting, were also characteristic of ZnSe/Si heterostructures. However, two major differences were observed between the ZnSe/Ge and ZnSe/Si systems, which most probably are a consequence of the rather severe lattice-mismatch of approx. 4% between ZnSe and Si. Firstly, the ZnSe/Si layers were more heavily dislocated, which was apparent immediately from surface morphology observations and secondly, single-crystal ZnSe growth could not be initiated under normal growth conditions. The second point is discussed in detail in reference 9. Briefly though, RHEED observations indicated that the structural nature of ZnSe layers grown directly on Si substrates depended strongly on the relative magnitudes of the incident Zn and Se fluxes at the growth temperature. Layers grown with the constituent element fluxes set to provide a typical layer growth rate of approx. 0.6μm/h on initiation of growth were polycrystalline, the crystallites being strongly oriented in the [111] direction. However, single-crystal layers were produced by the development of a so-called 'growth-rate ramping' technique. This technique consisted of initiating layer growth at an extremely slow rate by exposing the Si substrate to a normal Zn flux level together with an extremely low Se flux level, the substrate being held typically at around 330°C. The Se flux was then allowed to slowly increase over approximately a 15-minute period to a level equivalent to the Zn flux level. Since, in the MBE-growth of ZnSe, the growth rate is controlled by the arrival rate of the minority specie, the above technique was actually a growth-rate ramping process.

Since growth-rate ramping is not particularly convenient, efforts were made to develop a process which would eliminate the need for this technique. An obvious direction to take, given our previous experience in the growth of

ZnSe on Ge substrates was to grow a Ge buffer layer on Si and subsequently deposit ZnSe on the Ge buffer. Such an approach did, in fact, prove successful in this regard[10]. Prior to ZnSe deposition, Ge epi-layers were grown to a thickness typically around 0.5μm by evaporation of 6N's purity Ge from a resistively-heated PBN crucible in the ZnSe MBE-growth chamber, the Si substrate temperature being around 330°C, which is the 'optimum' temperature for ZnSe growth. The Ge epi-layer growth rate was around 0.33μm/h and, despite the rather low substrate temperature, good structural quality Ge epi-layers were grown on Si as evidenced by RHEED observations. However, although ZnSe layers could be grown directly on the Ge/Si 'substrates' at regular layer growth rates, ZnSe material quality was not noticeably improved over that achieved by growing ZnSe directly on Si. In both cases, DBE peak linewidths in the 6 to 8meV range were obtained, indicative of rather poor quality ZnSe material. Clearly a method of blocking the propagation of extended defects was required in order to improve ZnSe material quality.

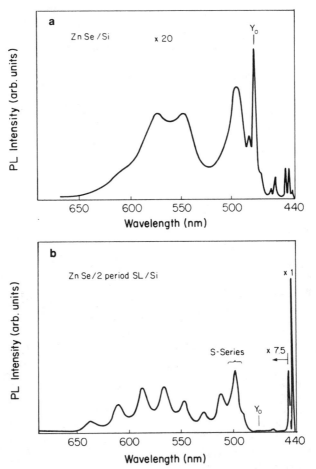

Fig. 7. Photoluminescence spectra recorded at 4.2K from (a) a ZnSe layer grown directly on Si and (b) a ZnSe layer grown on Si with a two-period ZnSe/Ge superlattice buffer structure incorporated. The ZnSe layer thickness in both cases was 2μm.

To this end, an exploratory investigation was carried out into the incorporation of ZnSe/Ge superlattices as buffer structures between ZnSe epilayers and Si substrates. The results of this study are detailed in reference 10. In summary, the incorporation of a two-period ZnSe/Ge superlattice structure consisting of alternate ZnSe and Ge layers, each layer having a thickness of approx. 300Å, resulted in a dramatic improvement in ZnSe epilayer quality. All layers, including the superlattice layers (both ZnSe and Ge), were grown with the substrate temperature maintained at 330°C; pneumatically operated, oven shutters permitting controlled exposure of the substrate to either Ge or Zn and Se fluxes. A material quality improvement was indicated by the observation of a far superior surface morphology and also a greatly improved PL response.

Figs. 7(a) and (b) illustrate 4.2K PL spectra recorded from 2μm thick ZnSe layers grown (a) directly on Si, and (b) with a two-period superlattice buffer incorporated. As can be seen from the figure, the spectrum obtained from the ZnSe/Si layer is dominated by the defect-related Y_o emission peak in contrast to the dominant DBE emission recorded from the layer grown on the superlattice buffer. Furthermore, the defect-related Y_o peak was barely detectable from the layer with the buffer structure incorporated. These results suggested the material to contain a significantly reduced concentration of extented defects by virtue of the superlattice presence. DBE peak linewidths as small as 2meV were recorded from layers grown on Si with a two-period ZnSe/Ge superlattice buffer, which indicates material of comparable quality to the better quality ZnSe layers grown on Ge substrates.

The structural quality of the multilayer arrangement was assessed by cross-sectional TEM analysis, and a TEM micrograph of such a structure is shown in Fig. 8. As can be seen from the micrograph, the individual layers are clearly observable and the interfaces are quite abrupt, suggesting minimal layer interdiffusion, which is not surprising given the low growth

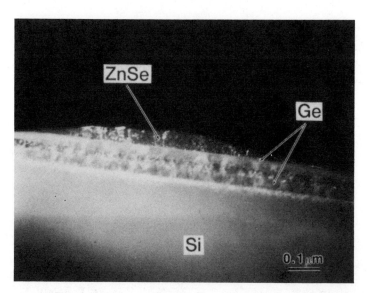

Fig. 8. Cross-sectional transmission electron microscopy image of a ZnSe/(ZnSe/Ge) superlattice/Si structure grown by MBE at a substrate temperature of 330°C. In this case, a two-period superlattice buffer was incorporated.

temperature employed, namely 330°C. A rather surprising result, however, was that layer quality did not improve with the incorporation of an increased number of superlattice periods in the buffer structure. In fact, ZnSe layer quality was observed to deteriorate with an increased number of superlattice periods. This phenomenon was attributed to a breakdown in the structural quality of the buffer structure multi-layers themselves, as evidenced by RHEED observations following the deposition of a critical number of alternate ZnSe and Ge layers. Clearly, further study would be required in order to understand this novel material system. In particular, a theoretical investigation of the system from a mechanical point of view would be of great value in helping to understand the results discussed above.

CONCLUSIONS

The primary difference between ZnSe layers grown on non-polar substrates compared to those grown on polar substrates (in particular, GaAs) is the presence, in the case of the former, of low-angle grain boundaries. Such grain boundaries are thought to form as a consequence of the coalition of slightly misoriented grains formed during an initial island growth process. Dislocations confined in the grain boundaries provide for strain-relief in these layers. Consequently, the nature and density of dislocations in ZnSe/Ge layers and ZnSe/GaAs layers are different. In the case of ZnSe/GaAs layers, classical misfit dislocations are the predominant structural defects responsible for strain relief. Particularly high concentrations of misfit dislocations, in the case of ZnSe/Si layers, significantly add to, and probably dominate, the low-angle grain boundary dislocations present in these layers.

The growth of device-quality ZnSe layers on non-polar substrates would require the elimination of the low-angle grain boundaries and, in the case of Si, the incorporation of a suitable buffer structure to block the propagation of extended defects. It is the author's contention that both of these criteria could be met with further study.

ACKNOWLEDGMENTS

The work of the author's former colleagues at 3M Canada Inc., namely, Drs. H. Mar and J. Kleiman, in this field is gratefully acknowledged. Also, the planar TEM work and analysis of Dr. K. Rajan was invaluable.

This work was supported by 3M Canada Inc. and by the Defense Advanced Research Projects Agency under Office of Naval Research Contract Number N00014-85-C-0552.

REFERENCES

1. G. D. Studtmann, R. L. Gunshor, L. A. Kolodziejski, M. R. Melloch, J. A. Cooper, Jr., R. F. Pierret, D. P. Munich, C. Choi, and N. Otsuka, Pseudomorphic ZnSe/n-GaAs doped-channel field-effect transistors by interrupted molecular beam epitaxy, Appl. Phys. Lett. 52:1249 (1988).
2. N. Peyghambarian, S. H. Park, S. W. Koch, A. Jeffrey, J. E. Potts, and H. Cheng, Room temperature excitonic optical non-linearities of molecular beam epitaxially grown ZnSe thin films, Appl. Phys. Lett. 52:182 (1988).
3. R. M. Park and H. A. Mar, Growth and photoluminescence characterization of ZnSe layers grown on (100)Ge by molecular beam epitaxy, J. Mater. Res. 1:543 (1986).

4. R. L. Gunshor, L. A. Kolodziejski, M. R. Melloch, M. Vaziri, C. Choi, and N. Otsuka, Nucleation and characterization of pseudomorphic ZnSe grown on molecular beam epitaxially grown GaAs epilayers, Appl. Phys. Lett. 50:200 (1987).

5. R. M. Park, H. A. Mar, and N. M. Salansky, Molecular beam epitaxy growth of ZnSe on (100) GaAs by compound source and separate source evaporation: a comparative study, J. Vac. Sci. Technol. B3: 676 (1985).

6. R. M. Park, J. Kleiman, and H. A. Mar, Molecular beam epitaxial growth of ZnSe on (100) GaAs and (100) Ge: a comparative study of material quality, Growth of compound semiconductors, Proc. SPIE vol. 796, Ed. R. L. Gunshor and H. Morkoc, 86 (1987).

7. S. B. Sant, J. Kleiman, M. Melech, R. M. Park, G. C. Weatherly, R. W. Smith, and K. Rajan, Defect characterization of MBE-grown ZnSe/GaAs and ZnSe/Ge heterostructures by cross-sectional and planar transmission electron microscopy, Inst. Phys. Conf. Ser. 87:129 (1987).

8. J. Kleiman, R. M. Park, and H. A. Mar, On epi-layer tilt in ZnSe/Ge heterostructures prepared by molecular beam epitaxy, To be published in J. Appl. Phys. (July '88).

9. R. M. Park and H. A. Mar, Molecular beam epitaxial growth of high-quality ZnSe on (100) Si, Appl. Phys. Lett., 48:529 (1986).

10. R. M. Park, H. A. Mar, and J. Kleiman, ZnSe and ZnSe/Ge epi-layers grown on (100) Si by molecular beam epitaxy, J. Crystal Growth, 86:335 (1988).

ATOMIC LAYER EPITAXY OF ZnSe/ZnS$_x$Se$_{1-x}$ SUPERLATTICES

Hugo J. Cornelissen

Philips Research Laboratories
P.O.Box 80000
5600 JA Eindhoven, The Netherlands

D.A. Cammack and R.J. Dalby[a]

Philips Laboratories
North American Philips Corporation
Briarcliff Manor
New York 10510, USA

Introduction

Several methods exist to grow wide gap II-VI semiconductor materials by molecular beam epitaxy (MBE). For instance, in the case of ZnSe growth one can use either a single source loaded with the compound, or one can use two sources loaded with the elements. An advantage of the latter set-up is that the flux ratio of the elements impinging on the substrate can be varied more freely. In a sense an extreme case is when only one element at a time is admitted to the substrate, which is referred to as the atomic layer epitaxy (ALE) mode. In ALE of ZnSe one cycles through two phases: in each phase enough atoms of one constituent are allowed to impinge on the substrate to reach one full monolayer coverage. Excess atoms are assumed to reevaporate, which seems plausible in view of the high vapor pressure of the elements. Sometimes a pause between the two phases is introduced. Several groups[1,2] have reported good results in growing ZnSe and ZnTe using this ALE growth mode, for instance, the thickness of the layers is simply determined by the number of shutter cycles. Other advantages of ALE may lie in selective dopant incorporation and in possibilities to lower the growth temperature while maintaining good crystal growth. Finally it has to be remarked that the growth rate during ALE is an order of magnitude less than during MBE, which may be disadvantageous because of higher impurity incorporation, but on the other hand may be beneficial for the crystal growth. We have previously reported on the growth of ZnSe/ZnS$_x$Se$_{1-x}$ superlattices by MBE,[3,4] and in the present paper we report on a brief study of the growth and properties of ZnSe/ZnS$_x$Se$_{1-x}$ superlattices in which the ZnSe is grown by ALE.

Experiments

Superlattice structures have been grown as shown schematically in Fig. 1. They are deposited on an MBE grown buffer layer of ZnSe of approximately 1 μm thick, grown with

a) present address: McDonnell Douglas Opto-Electronics Center, 350 Executive Boulevard, Elmsford, NY 10523, USA

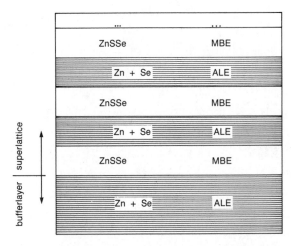

Fig. 1. Schematic drawing of the superlattice structures grown for the present study. They are deposited on a ZnSe buffer layer of $\simeq$ 1 μm thickness of which the main part is grown by MBE and the final part by ALE. The superlattices consist of several tens of layer pairs.

a rate of 0.5 μm/hr at a substrate temperature of 350 °C. The Zn to Se flux ratio is such that both the c(2x2) and the (2x1) surface reconstruction are observed simultaneously in reflection high energy electron diffraction (RHEED). The substrate is n-type (001)GaAs that receives a "standard" preparation procedure prior to growth.[3] The final part of the buffer layer is grown by ALE, where the Zn and the Se shutters are opened alternately for 10 s each, without an intermission.

During the Zn phase of the ALE shutter cycle the c(2x2) surface reconstruction is observed within a few seconds.[4] The Zn phase of ALE is characterized by a relatively high RHEED specular beam intensity as observed in the [100] direction. After closing the Zn shutter and opening the Se shutter the (2x1) reconstruction emerges, again within a few

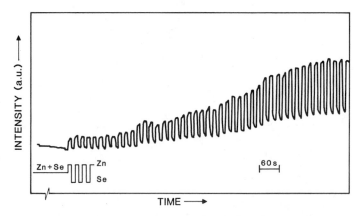

Fig. 2. RHEED specular beam intensity during the growth of the ZnSe buffer layer. Azimuth of the electron beam is [100]. After approximately 1 μm of ZnSe by MBE the ALE mode is started. The specular beam intensity is seen to increase dramatically, indicating a decrease in surface step density on the ZnSe. The shutters were open 10 s each, with no pause in between, as is schematically indicated in the figure (from Ref. 4).

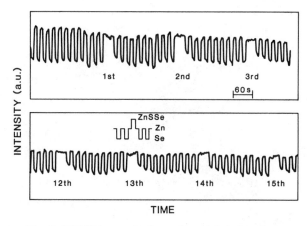

Fig. 3. RHEED specular beam intensity during the superlattice growth. Azimuth of the electron beam is [100]. The chart trace starts in the upper left corner with the ALE of the ZnSe buffer layer. Indicated in the figure are the subsequent $ZnS_{0.2}Se_{0.8}$ superlattice layers (1-st, 2-nd, etc.). In the lower part of the figure the shutter sequence is schematically indicated.

seconds, and the specular beam intensity drops to a lower value. In Fig. 2 a recording of the specular beam intensity is shown as we change from the MBE mode to the ALE mode. The azimuth of the electron beam was [100] but similar behavior was observed for other azimuths. The specular beam intensity under either Zn or Se flux is observed to increase after several tens of shutter cycles, and a saturation of the intensity levels is reached after $\simeq 50$ cycles. Since the specular beam intensity is correlated with the surface step density we conclude that by growing in the ALE mode the surface step density decreases relative to what is found during the MBE mode.

The ALE grown part of the ZnSe buffer layer ends on Zn after 100 shutter cycles. The first ZnS_xSe_{1-x} layer of the superlattice is then grown by closing the Zn shutter and simultaneously opening the shutter of a cell containing $ZnS_{0.5}Se_{0.5}$. Two superlattices have been grown: one with ZnSSe layer thicknesses of 2.5 nm and ZnSe ALE layers of 10 shutter cycles, and a second one with 5.1 nm ZnSSe layer thicknesses and ZnSe ALE layers of 20 shutter cycles. Thicknesses are measured by transmission electron microscopy (TEM). The composition of the ternary layers is $ZnS_{0.2}Se_{0.8}$ as determined by Raman spectroscopy.[5]

The behavior of the specular beam intensity during growth of a superlattice is shown in Fig. 3. During ZnSSe growth no change in the specular beam intensity is observed as can be seen in Fig. 3 where these layers are indicated by 1st, 2nd, 3rd, etc. This is as expected, since from previous work we know that ZnSSe grown from a compound source grows with a Zn-rich surface, which is exactly the surface we end the ALE growth of the ZnSe layers on. Both the initial and the final phase of the ALE ZnSe superlattice layer is the Zn phase. The thickness of the layers is measured to be 1.8 and 3.0 nm for 10 and 20 cycles, respectively. On the average we arrive at a growth rate of 0.6 monolayer per cycle. In the first few shutter cycles of the ALE ZnSe, the modulation in intensity between the Zn and the Se phase is somewhat decreased, but it increases and returns to the modulation value observed prior to the ZnSSe layer growth. This indicates a difference in the deposition of the first few ZnSe monolayers, and a recovery of the RHEED intensity patterns after a few monolayers. These RHEED intensity patterns are observed for tens of $ZnSe/ZnS_{0.2}Se_{0.8}$ layer pairs, but a gradual decrease in total RHEED intensity is observed during the growth of the complete superlattice.

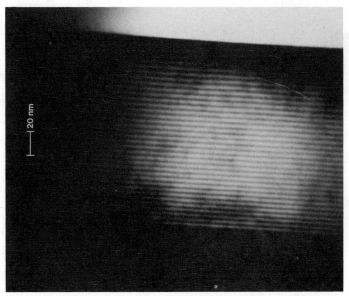

Fig. 4. TEM image of a superlattice structure. The total superlattice thickness is 150 nm, consisting of 36 layer pairs. The ZnSe ALE layer thickness is 1.8 nm, the $ZnS_{0.2}Se_{0.8}$ MBE layer thickness is 2.5 nm. The interfaces are very sharp, and no defects are observed at the interfaces.

The grown structures have been studied with TEM, and in Fig. 4 a TEM image is shown of the superlattice with the thinnest layers and part of the buffer layer. The transition from MBE to ALE during the buffer layer growth can not be observed. The total thickness of the superlattice is 150 nm, consisting of 36 layer pairs. The superlattice layers appear very sharp in this image, and no defects are observed at the interfaces. Also, in the Raman spectroscopy study mentioned above,[5] the ALE grown superlattices can not be distinguished from MBE grown superlattices. From this we conclude that these superlattices are similar to the best superlattices that we grew entirely by MBE.[3] We note, however, that in order to observe the RHEED specular beam intensity the substrate was not rotating during growth, which leads to non-uniformity in thicknesses of MBE grown layers across the wafer. ALE grown layers should be less sensitive to this effect but we did not investigate the non-uniformity in detail.

Conclusions

Judged from the RHEED specular beam intensity during growth of the ZnSe buffer layer, the ALE growth mode results in an atomically smoother ZnSe surface than the MBE mode. A more quantitative conclusion cannot be drawn at the moment, this will require additional analysis. The RHEED intensity during the ALE growth of the ZnSe superlattice layers shows a transient behavior, which might indicate a roughening of the surface during the MBE growth of the $ZnS_{0.2}Se_{0.8}$ superlattice layers, and a subsequent recovering of the surface during the ALE of ZnSe. The fact that the observed growth rate is less than 1 monolayer per ALE shutter cycle may be related to the roughening effects conjectured above, or to non-optimal growth conditions, e.g., a reevaporation of Se during the Se phase of the ALE. This might be prevented by lowering the growth temperature. TEM studies indicate that the superlattices show sharp interfaces and little defects, and that in this respect they are similar to the best superlattices grown entirely by MBE.

Acknowledgment

John Petruzzello is gratefully acknowledged for providing the TEM images.

References

(1) T. Yao and T. Takeda, J. Appl. Phys. **62** (1987) 3071.
(2) M. Kobayashi, N. Mino, H. Katagiri, R. Kimura, M. Konagai, and K. Takahashi, Appl. Phys. Lett. **48** (1986) 296.
(3) D.A. Cammack, R.J. Dalby, H.J. Cornelissen, and J. Khurgin, J. Appl. Phys. **62** (1987) 3071.
(4) H.J. Cornelissen, D.A. Cammack, and R.J. Dalby, J. Vac. Sci. Technol. **B6** (1988) 769.
(5) D.J. Olego, K. Shahzad, D.A. Cammack, and H.J. Cornelissen, Phys. Rev. B September 1988.

GROWTH OF II-VI SEMIMAGNETIC SEMICONDUCTORS

BY MOLECULAR BEAM EPITAXY

*R. N. Bicknell-Tassius, N. C. Giles and J. F. Schetzina

Department of Physics
North Carolina State University
Raleigh, NC 27695-8202, USA

INTRODUCTION

Two dimensional semiconductor systems in the form of quantum well structures and superlattices have attracted considerable attention over the past several years because of the many interesting properties that such layered multilayers exhibit. One of the reasons for the current interest in layered structures is that many of the optical and electronic properties can be tailored through the choice of materials and layer thickness. These structures are prepared by either molecular beam epitaxy or organometallic chemical vapor deposition and are usually composed of lattice-matched III-V compounds such as GaAs-AlAs and GaAs-GaAlAs.

Recently, quantum well structures and superlattices composed of II-VI compounds such as CdMnTe and ZnMnSe have also been grown by molecular beam epitaxy [1-3] and studied by various techniques. Bulk crystals of CdMnTe can be grown in the zincblende structure with manganese concentration of up to x = 0.7. The variation of lattice constant and band gap of CdMnTe with x-value is shown in Fig. 1. CdMnTe is one of a class of materials referred to as semimagnetic or dilute magnetic semiconductors (DMS). Interest in DMS materials stems from the interesting magnetic and magneto-optic properties which they exhibit due to the presence of the transition element Mn. Manganese ions have a total spin S = 5/2 and interact among themselves via an antiferromagnetic exchange interaction. In addition, the Mn ions also interact strongly with conduction band electrons and valence band holes via a spin-exchange interaction. It is this interaction which is new to semiconductor physics and which gives rise to interesting phenomena including giant Faraday rotation , very large electronic g-factors [5], giant negative magnetoresistance [6], and strongly circularly polarized band-edge luminescence.

EXPERIMENTAL DETAILS

In the present work, the CdMnTe layers (Eg = 1.85-2.1 eV) serve as barrier layers between CdTe quantum wells. Superlattices have been prepared consisting of from 14 to 240 CdMnTe-CdTe periods with quantum well thickness ranging from 460 A to 18 A. The SL samples were grown in

---

*Present address: Physikalisches Institut der Universitat Wurzburg, 8700 Wurzburg, FRG

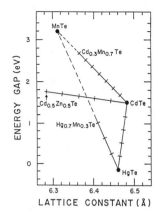

Fig. 1.   Energy gap vs lattice constant for various II-VI compounds

and provided a stable source of Cd and Te during the growth of the CdTe
layers. These two sources also provided a stable source of Cd during
the deposition of the CdMnTe layers.   The third cell contained Mn, and
the fourth cell contained a substitutional dopant, either In or Sb. For
growth of CdMnTe-CdTe multilayers, the Mn source shutter was opened and
closed while maintaining a constant Cd flux from the CdTe sources. The
x-value of the CdMnTe layers was controlled by the choice of the operating temperatures of the Mn source. When such superlattice structures are
grown it results in a periodic variation in the electronic band structure
along the growth direction. This is shown schematically in Fig. 2.

CdMnTe-CdTe Superlattices

To date, many interesting phenomena have been observed in these new

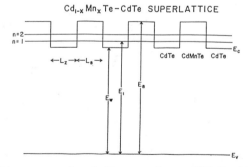

Fig. 2.   Schematic energy band diagram for a CdMnTe-CdTe superlattice.

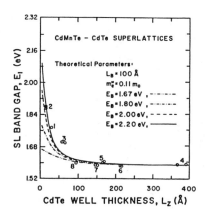

Fig. 3.   Energy gap vs well thickness for CdMnTe-CdTe superlattices

semimagnetic semiconductor superlattices. The photoluminescence spectra
of the superlattices is dominated by a narrow near-edge peak. This near-
edge luminescence is extremely intense - nearly two orders of magnitude
brighter than that typically observed from bulk CdTe[7]. The position
of the photoluminescence edge peak was found to shift to higher energies
as the quantum well layer thickness of the superlattice decreased as
shown in Fig. 3.

Other phenomena which are particular to the semimagnetic semiconductor
superlattices have also been observed. These include strong Zeeman
splitting of the valence and conduction band states, and strongly
circularly polarized band edge emission[8].

CdMnTe-CdTe Laser Structures

To study the potential of these materials as optical emitters, photoluminescence spectra were taken under varying pump powers[9]. The results
of these experiments are summarized in Fig. 4. At low excitation powers,
the spectrum consists of one smoothe peak. Where the pump power is
raised above threshold, several sharp laser modes appear superimposed on
the broad feature observed at lower pump powers. As the pump power is
increased further, these modes increase superlinearly with pump power
and mode Fabry-Perot cavity modes appear. Note that the emission from
this MQW structure is in the visible portion of the spectrum between
665-670nm.

The magnetic field dependence of the laser emission has also been
studied[10]. Figure 5 four typical spectrum taken at 2,3,4 and 5 T. The
cleaved cavity modes are clearly visible in all spectra. Note the
easily visible shift of the peak position to lower energy with increasing
magnetic field, as is expected in a DMS material. Surprisingly, however,
the field dependence is linear and shows no signs of saturation. The
observed shift is 3.4 meV/T.

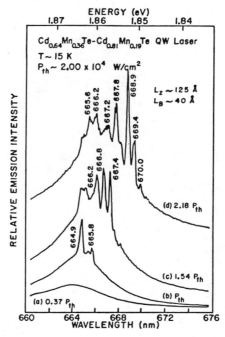

Fig. 4.   Stimulated emission spectra of a CdMnTe-CdTe quantum
well structure.

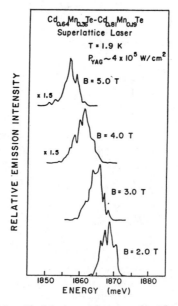

Fig. 5.   Magnetic field dependence of stimulated emission.

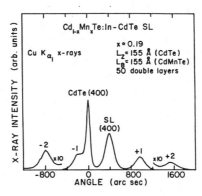

Fig. 6.   Double crystal rocking curve for a CdMnTe-CdTe superlattice.

Recently a new growth technique, Photoassisted MBE, has been employed to grow highly conducting CdTe epilayers for the fabrication of CdTe p-n junctions, and MESFET's[12]. This technique has also been employed to grow highly conducting CdMnTe epilayers and superlattices. Figure 6 shows a double crystal rocking curve of a CdMnTe-CdTe modulation doped superlattice. The main features in the rocking curve is the diffraction peak from the CdTe substrate and shifted be 390 arc seconds is the diffraction from the superlattice. This directly shows that these superlattices are strained-layer-superlattices. In this sample two orders of satellites are observed center around the zero order superlattice peak.

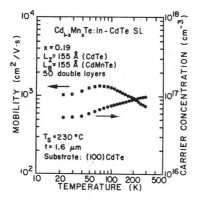

Fig. 7.   Carrier concentration and mobility vs temperature for a n-type CdMnTe-CdTe superlattice.

267

The carrier concentration and mobility for this indium doped sample is shown in Fig. 7. The room temperature carrier concentration for this sample is 1×10^{17} and decreases to 5×10^{16} at 20 K. The mobility at room temperature in this sample is 780 cm^2/V s and increases with decreasing temperature to greater than 1300 cm^2/V s at about 110K. Below 110 K mobility decreases slowly. This behaviour is dramatically different than that observed in single layers of CdTe where the mobility drops significantly at low temperatures.

SUMMARY

Molecular beam epitaxy has been employed to grow many new and interesting dilute magnetic semiconductor structures. They are very efficient optical emitters and are of high structural quality. The new results, with the photoassisted MBE technique, have opened many new avenues for future research in the areas of two-dimensional physics and device fabrication.

ACKNOWLEDGEMENTS

The work at NCSU was supported by army Research Office contract DAAG29-84-K-0039, DARPA/ARO contract DAAL03-86-K-0146 and National Science Foundation grant DMR83.13036.

REFERENCES

1. R.N. Bicknell, R.W. Yanka, N.C. Giles-Taylor, D.K. Blanks, E.L. Buckland and J.F. Schetzina, Appl. Phys. Lett. 45, 92 (1984)
2. L.A.Kolodziejski, R.L. Gunshor, T.C. Bonsett, R. Venkatasub-ramanian, S. Datta, R.B. Bylsma, W.M. Becker and N. Otsuka, Appl. Phys. Lett. 47, 169 (1985).
3. S. Venugoplan, L.A. Koldziesjki, R.L. Gunshor and A.K. Ramdas, Appl. Phys. Lett. 45, 975 (1984)
4. J.A. Gaj, R.R. Galazka, S. Nagata and P.H. Keesom, Phys. Rev. B 22, 334 (1980)
5. D.L. Peterson, A. Petrou, M. Datta, A.K. Ramdas and S. Rodriguez, Solid State Commun. 43, 667 (1982)
6. J.K. Furdyna, J. Vac. Sci. Technol. 21, 220 (1982)
7. R.N. Bicknell, N.C. Giles-Taylor, D.K. Blanks, R.W. Yanka, E.L. Buckland and J F Schetzina, J Vac. Sci. Technol. B 3, 709 (1985)
8. J. Warnock, A. Petrou, R.N. Bicknell, N.C. Giles-Taylor, D.K. Blanks and J.F. Schetzina, Phys. Rev. B 32, 8116 (1985).
9. R.N. Bicknell, N.C. Giles-Taylor, N.G. Anderson, W.D. Laidig and J.F. Schetzina, J. Vac. Sci. Technol. A 4, 2126 (1986)
10. E.D. Issacs, D. Heiman, J.J. Zayhowski, R.N. Bicknell and J.F. Schetzina, Appl. Phys. Lett. 48, 275 (1986)
11. R.N. Bicknell, N.C. Giles and J.F. Schetzina, J. Vac. Sci. Technol. A 5, 3059 (1987)
12. D.L. Dreifus, R.M. Kolbas, K.A. Harris, R.N. Bicknell, R.L. Harper and J.F. Schetzina, Appl. Phys. Lett. 51, 931 (1987)

EXCITON SELF-TRAPPING IN ZnSe/ZnTe SUPERLATTICE STRUCTURES

L. A. Kolodziejski* and R. L. Gunshor

School of Electrical Engineering
Purdue University
West Lafayette, Indiana 47907 USA

A. V. Nurmikko

Division of Engineering and Department of Physics
Brown University
Providence, Rhode Island 02912 USA

N. Otsuka

Materials Engineering
Purdue University
West Lafayette, Indiana 47907 USA

INTRODUCTION

The growth by molecular beam epitaxy (MBE) and the optical and structural characterization of a variety of wide bandgap II-VI compound semiconductors have attracted much attention in the last few years; a review of these new II-VI compound semiconductor superlattices can be found in reference 1. Multiple quantum well and superlattice structures incorporating layers of the diluted magnetic (or semimagnetic) semiconductors (DMS) provide bandgap modulation while exhibiting novel phenomenon arising from the presence of the magnetic ion. Subtlties arising from the exchange interaction between the magnetic ions and band electrons in an external magnetic field provide additional insight into the spatial distribution of carrier wavefunctions in structures of atomic dimensions. In one example, this novel diagnostic tool allows determination of valence band offset in the II-VI DMS strained-layer superlattice structures.

Epitaxial layers, superlattices, and multiple quantum well structures involving wide bandgap semiconductors such as ZnSe and ZnTe are potentially important for optical devices operating in the visible portion of the spectrum. Alloying ZnSe with Te to form Zn(Se,Te) permits adjustment of the lattice constant, a lowering of the bandgap (roughly from 2.8 to 2.4 eV), and provides opportunities for amphoteric doping. Of particular interest is the observation that the photoluminescence yield of Zn(Se,Te) can be significantly enhanced over that of bulk ZnSe crystals[2,3] and epitaxial layers, above liquid helium temperatures due to localization of excitons in the radom alloy. In this paper we will emphasize the identification of self-trapping of excitons at Te sites which dominates the recombination process in ZnTe/ZnSe superlattices[4] composed of thin (1-2 monolayers) ZnTe alternated with thicker ZnSe layers. Te represents an isoelectronic center in ZnSe, whereas Se is an isoelectronic center in ZnTe; both situations will be addressed below. It is expected that the self-trapping phenomenon plays a role in the radiative processes of a variety of structures containing the ZnSe/ZnTe heterointerface.

EPITAXIAL GROWTH OF ZnTe-CONTAINING HETEROSTRUCTURES

The growth of the Zn(Se,Te) mixed crystal by molecular beam epitaxy is complicated by a difficulty in controlling the composition. In the work reported by Yao et al.[5], over the entire range of Te fraction, a three to ten overpressure of Te was required. In our laboratory we have grown a number of Zn(Se,Te) epilayers with varying fractions of Te; a particular difficulty was encountered when a small fraction of Te was desired (<10%), resulting in widely varying compositions under what appeared to be similar growth conditions. To circumvent the problems associated with controlling the alloy concentrations, we have designed ZnSe-based structures consisting of ultrathin layers (or 'sheets') of ZnTe spaced by appropriate dimensions to approximate a Zn(Se,Te) mixed crystal with low or moderate Te composition. In one case, such a "pseudo-alloy" was grown as an epilayer consisting of a 100 period superlattice with each period containing one monolayer of ZnTe (~3 Å) separated by 100 Å ZnSe. Alternatively, either one monolayer or two monolayers of ZnSe have been layered between 100 Å of ZnTe to form a superlattice containing 50 periods. In a second prototype structure, the pseudo-alloy was used to modify the ZnSe quantum well in a ZnSe/$Zn_{0.8}Mn_{0.2}$Se multiple quantum well. In such multiple quantum well (MQW) structures, either one or two ZnTe sheets were placed in the center of each well; the wells had thicknesses ranging from 44 to 130 Å. A third set of structures which were synthesized consisted of epitaxial layers of (Zn,Mn)Se, with Mn mole fractions ranging from 1% to 20%, in which was placed the ultrathin 'sheets' of ZnTe with spacings of 100 Å; in this case the ZnTe ultrathin sheets were either 1, 2, or 4 monolayers thick. The lattice mismatch between bulk ZnSe and ZnTe is approximately 7.4%; in the aforementioned structures the ZnTe sheets are likely to accommodate most of the strain. (Because of the significant amount of strain which existed in the ZnTe/(Zn,Mn)Se heterostructures, the structure containing four monolayers of ZnTe was spaced by six monolayers of $Zn_{0.8}Mn_{0.2}$Se; the lattice parameter of ZnSe increases as Mn is incorporated.)

Figure 1 illustrates schematically the various heterostructures which were designed and fabricated using a combination of molecular beam epitaxy and atomic layer epitaxy (ALE) growth techniques. GaAs (100) substrates were used and were chemically prepared in the conventional manner[6] for the structures which were composed of primarily ZnSe or (Zn,Mn)Se. Structures which were based in ZnTe, with ultrathin sheets of ZnSe, were grown on GaSb (100) substrates. The GaSb substrates were chemically etched in a 0.12% solution of Br_2/methanol for 2 minutes; oxide desorption occurred around 550 °C in an antimony flux. During the growth of the ZnSe material, the fluxes of Zn and Se were set to be equal as measured by a quartz crystal monitor and the substrate temperature for all growths was maintained at 320 °C. Following the deposition of a 1-2 μm thick buffer layer of ZnSe (or ZnTe), a multiple quantum well (with (Zn,Mn)Se barrier layers) or a superlattice of ZnSe/ZnTe was grown by the MBE growth technique, but interrupted for the ALE growth of the ZnTe (or ZnSe) monolayer sheets. In an effort to optimize the interface abruptness of the ZnTe (or ZnSe) monolayer, the ALE of ZnTe (or ZnSe) was performed on a recovered ZnSe (or ZnTe) surface. The ALE growth involved the following sequence of shutter operation (which included appropriate desorption times after each deposition): 1) Zn, 2) Te (or Se), 3) Zn, and 4) Se (or Te). The remainder of the structure, both prior and following the ALE growth

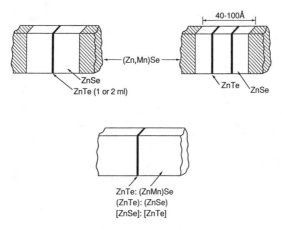

Fig. 1. Schematic drawing of the various ZnTe-containing heterostructures: (top) ZnTe is placed within the ZnSe quantum wells of the ZnSe/(Zn,Mn)Se MQW structures, and (bottom) one period of the 'comb' superlattices is illustrated consisting of ZnTe(1 ml):ZnSe, ZnSe (1 ml):ZnTe, and ZnTe (1 or 4 ml):(Zn,Mn)Se.

of the ultrathin sheets, was grown by molecular beam epitaxy. Reflection high energy electron diffraction (RHEED) was used to monitor the surface structure during growth. After recovery of each Zn-, Te-, or Se-surface, we observed the expected Zn-stabilized, Te-stabilized, or Se-stabilized reconstructed surface, respectively, in agreement with the observations reported by Yao and Toshihiko[7].

The pseudo-alloy structures have been examined by cross-sectional transmission electron microscopy (TEM). Clear images of ultrathin ZnTe sheets were observed in dark field images of the (100) reflection. These ultrathin sheets appeared as narrow bright lines with widths of about 5 Å in the dark field images, indicating the existence of continuous ZnTe monolayers in ZnSe crystals. Although Xe ions were used for the final thinning of cross-sectional samples, superlattice areas have suffered some damage. Because of the damage and the lower resolution of the dark field imaging technique, details of ZnSe/ZnTe heterointerfaces could not be seen in the images. However, the TEM observation has clearly shown that ZnTe sheets have formed coherent interfaces with ZnSe layers.

The first reported MBE-grown multiple quantum well (MQW) structures involving ZnSe employed the DMS material (Zn,Mn)Se as the wider bandgap barrier layer. (Iron, another magnetic transition metal ion, has recently been incorporated into ZnSe to form the first epitaxial (Zn,Fe)Se alloy grown by MBE[8].) The band structure of the host II-VI semiconductor (ZnSe) is not directly modified by the presence of Mn since the two s electrons of the outer shell replace those of Zn, and become part of the band electrons in extended states. The five electrons in the unfilled 3d shell of Mn, however, give rise to localized magnetic moments which are partially aligned in an external magnetic field. The resultant magnetic moment interacts with the band electrons causing a Zeeman splitting which is orders of magnitude larger than for the host II-VI semiconductor. The presence of the magnetic ion (with the associated Zeeman shifts) provides a unique and useful feature to the superlattices and multiple quantum well structures in which Mn is incorporated. The magnetic field-induced changes[9] in optical transition energies in superlattices incorporating the DMS material provide additional insight into excitonic behavior and valence band offset in strained-layer structures in general. Detailed studies of the magneto-optical behavior of the MQW structures indicated that the magnetic-field-induced band shifts are primarily due to exchange interactions associated with the penetration of the hole wavefunctions into the (Zn,Mn)Se barrier layers. Concluding that much of the hole envelope wavefunction resides in the (Zn,Mn)Se barrier layer implied that the strained valence band offset is small and is likely to be less than 20 meV[9].

Figure 2 shows the photoluminescence excitation spectrum near the n=1 exciton ground state from a (100)-oriented MQW sample of ZnSe/Zn$_{1-x}$Mn$_x$Se (x=0.23), with a well width of 67 Å[9]. For such typical parameters, only the n=1 transition is seen, consisting of the light-hole (LH) and heavy-hole (HH)

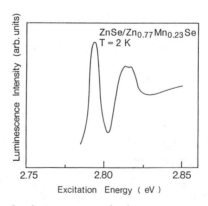

Fig. 2.   Photoluminescence excitation spectrum for a ZnSe/(Zn,Mn)Se MQW structure (see text). The n=1 light and heavy hole resonances dominate the spectrum.

excitonic resonances. The identification of the resonances was made through magnetooptical studies[9] where both LH and HH transitions are found to exhibit large Zeeman splittings due to finite exciton wavefunction overlap with the (Zn,Mn)Se barrier layers.

Once captured in a ZnSe quantum well, the photoenergetic electrons and holes relax by optical phonon emission to the n=1 confined particle states. This relaxation step is fast (probably well below one psec in such polar material). The subsequent exciton formation and further energy relaxation (localization by quantum well width fluctuations) is a slower process which can be time-resolved through the use of picosecond pulsed laser excitation and a streak camera[10]. Recombination lifetime is, of course, also obtained from the data (~200 psec); some details of this can be found in reference[11].

OPTICAL PROPERTIES of HETEROSTRUCTURES CONTAINING ULTRATHIN ZnTe SHEETS

In striking contrast to the photoluminescence from a ZnSe/(Zn,Mn)Se MQW (Fig. 3(a)), the photoluminescence resulting when the ZnTe sheets are inserted into each quantum well is shown in Figure 3(b). For this structure the ZnSe well width was approximately 44 Å as measured by TEM. The luminescence in this instance exhibits two broad emission bands with half widths on the order of 50 meV and 100 meV with peak positions approximately 100 and 300 meV below the band edge region, respectively. In the following we denote the two bands as S_1 and S_2, respectively, following the notation of reference 3. The spatially integrated luminescence yield at T=2K for quantum wells with and without the ZnTe sheets was comparable; however, that for the sample without ZnTe decreased precipitously with temperature. The location of the absorption edge in this quantum well sample was established through photoluminescence excitation spectroscopy, a relavant portion of which is shown in Figure 3(c). The spectral position of the distinct n=1 light hole exciton resonance indicates that the addition of the ultrathin ZnTe layers does not significantly shift the position of this lowest energy transistion; this is readily verified by comparing the excitation spectra of the MQW sample without ZnTe (see Fig. 2). With band offset information about the ZnTe/ZnSe heterojunction only now under study theoretically[12], electron affinity data suggests that the ZnTe layer would provide a narrow additional potential well in the valence band of depth as much as 1 eV, while contributing to a thin barrier in the conduction band of some 0.6 eV in height. Similar spectra (showing the S_1 and S_2 bands) were also observed for the "pseudo-alloy" structure composed of a ZnSe/ZnTe superlattice with one period having a one monolayer sheet of ZnTe spaced by 100 Å of ZnSe.

We have also examined the temperature dependence (Fig. 4) of the luminescence spectra in the pseudo-alloy structures. We note that for the sample of Fig. 4, two distinct temperature regions were observed. Starting from a low temperature the emission band S_1 is first quenched, this occurs quite rapidly in a fairly narrow temperature range of T=70-90 K. Second, at temperatures beyond T=120 K or so, a strong spectral red shift is seen for the remaining S_2 luminescence but with decreasing quantum efficiency. This shift is much larger than a simple bandgap reduction and indicates that the partial quenching of the emission band S_2 brings yet additional lower energy band(s) into view.

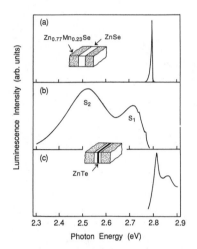

Fig. 3. Comparison of photoluminescence spectra at T=2K of (a) a ZnSe/(Zn,Mn)Se MQW sample (L_w=67 Å) with (b) that of a similar structure but with the insertion of monolayer sheets of ZnTe in the middle of the quantum well (L_w=44 Å). (The amplitude of emission in (a) has been reduced to bring the peak to scale.) The photoluminescence excitation spectra of sample (b) is shown in the bottom panel (c). In both cases the yellow Mn-ion internal emission was significantly below that from the blue-green region.

These results can be compared with recent work which has shown how exciton self-trapping is a very efficient process in bulk Zn(Se,Te) alloys, where it is characterized by broad emssion lines and large Stokes shifts relative to the fundamental absorption edge[3]. In particular, the S_1 and S_2 emission bands have been associated with trapping and strong accompanying local lattice relaxation at single and double Te sites, respectively. In a bulk crystal of low Te concentration (x<0.05) exciton recombination from the S_1 band dominates at low temperature but is partially quenched at higher

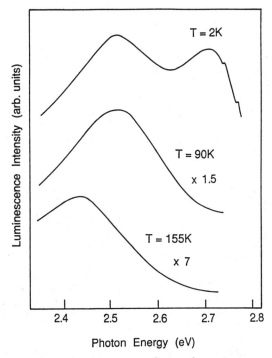

Fig. 4.   Photoluminescence from the
ZnSe/(Zn,Mn)Se MQW sample with ZnTe
sheets shown as a function of temperature.

temperatures.   This occurs as a consequence of excitons in traps associated with emission S_1 first obtaining sufficient thermal energy to reach over a potential barrier for their localization, the process being activated in a narrow temperature range similar to that of the MBE samples.   However, the initial dominance of the S_2 band at low temperatures and the emergence of other red-shifted features in heterostructures, suggest a more complex trapping process in the ZnSe/ZnTe pseudo-alloy than encountered in the case of a dilute bulk Zn(Se,Te) mixed crystal.

In the modified quantum well structures considered here, the process leading to recombination spectra, such as in Fig. 3(b), is then interpreted as follows. Photoexcitation of electron-hole pairs is followed by rapid thermalization of the carriers within the ZnSe quantum wells. However, before any substantial exciton recombination across the superlattice bandgap takes place, exciton capture at the ultrathin ZnTe layers at the middle of the wells occurs with simultaneous local lattice relaxation and subsequent strong radiative recombination at Te sites. Whereas the typical exciton lifetimes of ~200 psec. for the weakly localized Wannier excitons in the ZnSe/(Zn,Mn)Se quantum wells is measured, the self-trapped excitons at Te sites result in observed lifetimes of ~10-50 nsec, depending on the particular structure. The lifetime increase is a consequence of the reduction in electron-hole wavefunction overlap in the structures containing isoelectronic centers. As discussed above, although the electron is strongly confined in the ZnSe well of ZnSe/(Zn,Mn)Se MQWs, the small valence band offset results in significant penetration of the hole wavefunction in the Mn-containing barrier layers; this results in substantial Zeeman splitting in an external magnetic field. The strong localization[13] of the hole at Te sites in the quantum well is further confirmed by an absence of Zeeman splitting in ZnTe-containing ZnSe/(Zn,Mn)Se multiple quantum wells; the hole wavefunction no longer penetrates into the barrier regions.

A question which is raised by these observations concerns the sharpness of the ZnSe/ZnTe interface, i.e., to what degree is interdiffusion of the anions necessary to provide a finite mixed crystal region compatible for the trapping process described above. Alternatively, approaching the question from the perspective of the two dimensional growth aspect in atomic layer epitaxy, one may ask how much do the ZnTe deviate from an ideally complete monolayer. Since the largest contribution to luminescence at low temperature is by the emission band S_2 (and not S_1 as in a dilute bulk mixed crystal), and since there is the additional clear emergence of lower emission bands above 120 K, we can conclude that pairs of Te isoelectronic traps, as well as more complicated clusters, are more numerous than single isolated Te lattice sites. Starting from an initially perfect monolayer of ZnTe, the dominance of the S_2 feature implies that interdiffusion is unlikely to extend over much more than a molecular monolayer (~3Å) in our structures; otherwise, within the resulting inhomogeneous Zn(Se,Te) interface region the density of single Te sites would dominate. Fig. 5 compares the photoluminescence of two similar ZnSe/(Zn,Mn)Se multiple quantum well structures having a one monolayer ZnTe sheet (Fig. 5(a)) and having two contiguous monolayers of ZnTe (Fig. 5(b)) centered in the ZnSe well. In the latter the S_1 band (single Te sites) has essentially disappeared and the clear dominance of a slightly red-shifted S_2 band (doubly-coordinated Te sites) is apparant; the majority of the Te is now bonded to Zn atoms in a manner approaching that of a ZnTe lattice. The implication of these observations is that increasing the ZnTe sheet thickness to two monolayers results in excitons that are strongly trapped in the layer plane; additional increases in layer thickness should result in band offsets characteristic of a Type II superlattice, with holes confined to the ZnTe.

SUMMARY

We have fabricated and evaluated a number of structures which show quite dramatically the strong exciton capture at a ZnSe/ZnTe interface. The self-trapping of excitons at Te sites is likely to dominate the recombination

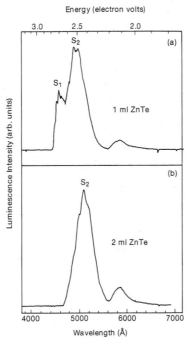

Fig. 5: Low temperature (8 K) luminescence originating from two ZnTe-containing ZnSe/(Zn,Mn)Se multiple quantum wells. In (a) both S_1 and S_2 bands are observed in the case of the ZnSe well perturbed by the insertion of one monolayer (ml) of ZnTe, whereas in (b) only the S_2 band is observed for the case of two contiguous monolayers of ZnTe placed in the center of the ZnSe well.

process in ZnSe/ZnTe superlattices[14,15] composed of thin (1 or 2 monolayer) ZnTe alternated with thicker ZnSe layers. Furthermore, it is expected that the self-trapping phenomenon plays a role in the radiative processes of a variety of structures containing the ZnSe/ZnTe heterointerface. An important observation is the increase (from a few hundred picoseconds to several tens of nanoseconds) in lifetime associated with the exciton trapping at isoelectronic centers. The capture process discussed here may be useful in enhancing the quantum efficiency in semiconductor structures at non-cryogenic temperatures where defects might otherwise limit luminescence from free or weakly bound excitons.

ACKNOWLEDGEMENT

The authors would like to gratefully acknowledge substantial contributions from the following students and co-workers: J. M. Gonsalves, Q. Fu, D. Lee, M. Vaziri, C. Choi, D. Lubelski, S. Rajagopalan, M. Kobayashi, S. Durbin, and D. Mathine. The research at Purdue and Brown is supported by a joint Defense Advanced Research Projects Agency/Office of Naval Research University Research Initiative Program N00014-86-K0760. Additional support was provided to Purdue by Office of Naval Research Contract N00014-82-K0653, Air Force Office of Scientific Research Grant 85-0185, and National Science Foundation-MRG Grant DMR-8520866. Additional support to Brown was provided by Office of Naval Research Contract N00014-83-K0638.

REFERENCES

*current address: Department of Electrical Engineering and Computer Science, Massachusetts Institute of Technology, Cambridge, Massachusetts 02139

1. R. L. Gunshor, L. A. Kolodziejski, A. V. Nurmikko, and N. Otsuka, Semimagnetic and Magnetic Semiconductor Superlattices, in: "Annual Review of Materials Science," R. A. Huggins, ed., Annual Reviews, Inc., Palo Alto, CA (1988).
2. A. Reznitsky, S. Permogorov, S. Verbin, A. Naumov, Yu. Korostelin, and S. Prokov'ev, Solid State Comm. 52:13 (1984).
3. D. Lee, A. Mysyrowicz, A. V. Nurmikko, and B. J. Fitzpatrick, Phys. Rev. Lett. 58:1475 (1987).
4. L. A. Kolodziejski, R. L. Gunshor, Q. Fu, D. Lee, A. V. Nurmikko, J. M. Gonsalves, and N. Otsuka, Appl. Phys. Lett. 52:1080 (1988).
5. T. Yao, Y. Makita, and S. Maekawa, J. Crystal Growth 45:309 (1978).
6. L. A. Kolodziejski, R. L. Gunshor, N. Otsuka, S. Datta, W. M. Becker, and A. V. Nurmikko, IEEE Trans. Quantum Electronics QE-22:1666 (1986).
7. T. Yao and Y. Toshihiko, Appl. Phys. Lett. 48:160 (1986).
8. B. T. Jonker, J. J. Krebs, S. B. Qadri, and G. A. Prinz, Appl. Phys. Lett. 50:848 (1987).
9. Y. Hefetz, J. Nakahara, A. V. Nurmikko, L. A. Kolodziejski, R. L. Gunshor, and S. Datta, Appl. Phys. Lett. 47:989 (1985).
10. Y. Hefetz, W. C. Goltsos, A. V. Nurmikko, L. A. Kolodziejski, and R. L. Gunshor, Appl. Phys. Lett. 48:372 (1986).
11. Y. Hefetz, W. C. Goltsos, D. Lee, A. V. Nurmikko, L. A. Kolodziejski, and R. L. Gunshor, Superlattices and Microstructures 2:455 (1986)

12. Y. Rajakarunanayake, R. H. Miles, G. Y. Wu, and T. C. McGill, <u>Phys. Rev. B</u> 37:10212 (1988).

13. A. V. Nurmikko, L. A. Kolodziejski, and R. L. Gunshor, to appear in the Proceedings of the Fifth International Conference on Molecular Beam Epitaxy, August 28-September 1, 1988, Sapporo, Japan.

14. H. Fujiyasu, K. Mochizuki, Y. Yamazaki, M. Aoki, and A. Sasaki, <u>Surf. Sci.</u> 174:542 (1986).

15. M. Kobayashi, N. Mino, H. Katagiri, R. Kimura, M. Konagai, and K. Takahashi, <u>Appl. Phys. Lett</u>. 48:296 (1986).

RESONANT RAMAN SCATTERING BY LO PHONONS IN II-VI COMPOUNDS AND DILUTED MAGNETIC SEMICONDUCTORS

W. Limmer, H. Leiderer, and W. Gebhardt

Faculty of Physics
University of Regensburg
D-8400 Regensburg, FRG

1. INTRODUCTION

Resonant Raman scattering by LO phonons is an optical method which provides insight into the relations between lattice-dynamical and electronic properties of crystals. A resonant enhancement of the Raman scattering efficiency is observed whenever the energy of the incident photon approaches a critical point in the combined density of states of electronic interband transitions. By analyzing the energy dependence and the absolute values of the Raman efficiency detailed information is obtained about the electron-phonon interactions, involved in the scattering process, and about band-structure parameters as gap energy E_0, gap broadening η and the optical deformation potential D_{opt}.

In this paper we report on resonant Raman scattering by LO phonons near the E_0 gap in the binary and ternary II-VI semiconductors $ZnSe$, $Zn_{0.9}Mn_{0.1}Se$, $Cd_{0.75}Mn_{0.25}Te$ and $Cd_{0.55}Mn_{0.45}Te$. The substitution of the group II elements Zn and Cd by the magnetic transition metal Mn leads to the interesting class of ternary compounds known as diluted magnetic semiconductors (DMS). Typical properties are the dependence of band parameters on Mn-concentration x and a giant Zeeman splitting of valence and conduction bands in external magnetic fields [1,2]. We have investigated the influence of this Zeeman splitting on the resonance behaviour of the Raman scattering efficiency in $Cd_{0.75}Mn_{0.25}Te$.

Inelastic scattering of light by phonons occurs via virtual transitions of the electronic system. The intermediate electronic states are excitons. In the polar semiconductors considered here there are two different mechanisms of electron-phonon interaction [3]. The first one is caused by the deformation potential which leads to the usual selection rules of dipole allowed scattering. The second one, the so-called Fröhlich interaction, is due to the macroscopic electric field accompanying the LO phonons (dipole forbidden scattering). For the Fröhlich interaction an impurity-induced scattering mechanism has been proposed in addition to the intrinsic one [3,4]. Here, the electron-hole pair, created by the incident photon in the Raman process, scatters twice, once inelastically due to the electron-phonon interaction and once elastically due to an electron-impurity interaction. A comparison of experimental and theoretical results has shown that even in high purity materials the impurity-induced mechanism is the dominant one [5].

We have carried out model calculations for the Raman scattering efficiency where an intrinsic and an impurity-induced mechanism are considered for the Fröhlich interaction and to our knowledge for the first time also for the deformation-potential interaction. The contributions of light and heavy hole valence bands are taken explicitly into account and the case of Zeeman-splitted subbands is considered. It should be emphasized that we used in our calculations for the different scattering mechanisms one consistent formalism with the same adjustable parameters for gap energy, gap broadening and impurity concentration.

2. THEORY

For a quantitative comparison of experimental and theoretical results it is convenient to use the Raman scattering efficiency $dS/d\Omega$ as a measure of the scattered Raman intensities [6]. This quantity has the dimensions of an inverse lenght and represents the ratio between scattered and incident power for a unit path length within the solid. It is defined as

$$\frac{dS}{d\Omega} := \frac{dP_S}{d\Omega} \cdot \frac{1}{P_L \cdot L} \qquad (1)$$

where $dP_S/d\Omega$ denotes the scattered power per solid angle $d\Omega$ inside the crystal, P_L the incident power and L the scattering length. In a microscopic description of the scattering process the Raman efficiency $dS/d\Omega$ is related to the scattering amplitude W_{fi} which describes the transition from the initial state i to a final state f [7]

$$\frac{dS}{d\Omega} := \frac{\omega_S}{\omega_L} \cdot \frac{n_L}{c} \cdot \frac{1}{d\Omega} \cdot \frac{2\pi}{\hbar} \cdot \sum_f |W_{fi}|^2 \cdot \delta(\hbar\omega_i - \hbar\omega_f) \qquad (2)$$

where ω_L and ω_S are the frequencies of the incident and scattered photon, respectively, n_L is the index of refraction and c the speed of light in vacuum.

For intrinsic Stokes scattering by one LO phonon the scattering amplitude can be evaluated in third order perturbation theory, leading to a sum of six different terms. The most resonant one is given by [8,9]

$$W_{fi}(1LO; int) = \sqrt{\bar{n}+1} \cdot \sum_{\alpha,\beta} \frac{<0|H_{eR}^+(\hat{e}_S)|\beta><\beta|H_{eL}|\alpha><\alpha|H_{eR}^-(\hat{e}_L)|0>}{(E_\beta - \hbar\omega_S) \cdot (E_\alpha - \hbar\omega_L)} \qquad (3)$$

where $\bar{n}(\Omega_{LO}, T)$ is the phonon occupation number, $|0>$ the electronic ground state, with all valence bands occupied and all conduction bands empty, $H_{eR}^-(\hat{e}_L)$ and $H_{eR}^+(\hat{e}_S)$ are the electron-photon interaction operators for the absorption and emission of the incident and scattered photon, respectively, $\hat{e}_L$ and $\hat{e}_S$ are the corresponding light polarization vectors and H_{eL} is the electron-phonon interaction operator. The sum runs over all intermediate electronic states α and β. Scattering by two LO phonons is described in fourth order perturbation theory, assuming an iteration of the electron-phonon interaction. Impurity-induced scattering by one LO phonon is very similar to the two-LO-phonon scattering, with one electron-phonon interaction replaced by an electron-impurity interaction H_{eI} [3]. The corresponding amplitude for Stokes scattering is given by

$$W_{fi}(1LO, imp) = \sqrt{n+1} \times$$

$$\times \sum_{\alpha,\beta,\gamma} \left\{ \frac{<0|H_{eR}^+(\hat{e}_S)|\gamma><\gamma|H_{eI}|\beta><\beta|H_{eL}|\alpha><\alpha|H_{eR}^-(\hat{e}_L)|0>}{(E_\gamma + \hbar\Omega_{LO} - \hbar\omega_L) \cdot (E_\beta + \hbar\Omega_{LO} - \hbar\omega_L) \cdot (E_\alpha - \hbar\omega_L)} \right.$$

$$\left. + \frac{<0|H_{eR}^+(\hat{e}_S)|\gamma><\gamma|H_{eL}|\beta><\beta|H_{eI}|\alpha><\alpha|H_{eR}^-(\hat{e}_L)|0>}{(E_\gamma + \hbar\Omega_{LO} - \hbar\omega_L) \cdot (E_\beta - \hbar\omega_L) \cdot (E_\alpha - \hbar\omega_L)} \right\} \tag{4}$$

where Ω_{LO} is the LO-phonon frequency. In polar semiconductors the electron-phonon interaction can be separated into a short- and a long-range part.

$$H_{eL} = H_{DP} + H_F \tag{5}$$

The short-range part is the deformation-potential interaction [10,11,12]

$$H_{DP} = \sum_l \vec{V}_{DP}(\vec{r}_l) \cdot \vec{u}_{rel}(\vec{r}_l) \tag{6}$$

where the sum runs over all electrons in the crystal. The relative displacement $\vec{u}_{rel}$ of the two sublattices is given by [3]

$$\vec{u}_{rel}(\vec{r}) = u_0 \cdot \hat{e}_{LO} \cdot e^{-i\vec{q}\cdot\vec{r}} \cdot \left\{ d_{\vec{q}}^+ + d_{-\vec{q}} \right\}; \qquad u_0 = \sqrt{\frac{\hbar}{2NM^*\Omega_{LO}}} \tag{7}$$

where $\vec{q}$ is the LO-phonon wave vector, N the number of primitive cells in the crystal, M^* their reduced mass, $d_{\vec{q}}^+$ and $d_{-\vec{q}}$ are phonon creation and destruction operators, respectively, and $\hat{e}_{LO}$ is a unit polarization vector of the LO phonon. We define the optical deformation potential D_{opt} as

$$D_{opt} := \frac{1}{\sqrt{3}} \cdot <X | a_0 \cdot V_{DP,z} | Y >; \tag{8}$$

where X and Y are the p_x and p_y valence band wave functions at the Γ-point of the Brillouin zone [13] and a_0 is the lattice constant.

The long-range part H_F is the Fröhlich interaction, which is induced by the macroscopic electric field accompanying the longitudinal optical phonons in polar materials [3]

$$H_F = \sum_l \frac{iC_F}{\sqrt{V}} \cdot \frac{\vec{q} \cdot \hat{e}_{LO}}{q^2} \cdot e^{-i\vec{q}\vec{r}} \cdot \left\{ d_{\vec{q}}^+ + d_{-\vec{q}} \right\} \tag{9}$$

where V is the scattering volume. The Fröhlich constant C_F is given by

$$C_F := e \cdot \sqrt{2\pi\hbar\Omega_{LO} \left(\frac{1}{\epsilon_\infty} - \frac{1}{\epsilon_0} \right)} \tag{10}$$

where e is the free-electron charge, ϵ_0 the static and ϵ_∞ the high-frequency dielectric constant.

For the electron-impurity interaction we consider the Fourier transformation of a screened Coulomb potential [3]

$$H_{eI} = \sum_l \frac{4\pi e^2}{V\epsilon_0(q^2 + q_F^2)} \cdot e^{-i\vec{q}\vec{r}_l} \tag{11}$$

where $q_F = 2/\lambda$, λ being the mean distance between impurities. To simplify the calculations of the scattering amplitudes we consider only uncorrelated electron-hole pairs as intermediate states. In this approximation an excited electronic state $|\alpha>$ can be written in second quantization as

$$|\alpha> = c^+_{n,\vec{k}} v_{m,\vec{k}-\vec{\mathcal{K}}} |0> \tag{12}$$

where $c^+_{n,\vec{k}}$ is the creation operator of an electron with wave vector $\vec{k}$ in conduction band n and $v_{m,\vec{k}-\vec{\mathcal{K}}}$ the destruction operator of an electron with wave vector $\vec{k} - \vec{\mathcal{K}}$ in valence band m. In order to give an example of our theoretical results, we present an expression for the Raman scattering efficiency for one-LO-phonon scattering in the case of the intrinsic deformation-potential :

$$\frac{dS}{d\Omega}(1LO, int; DP) = \left(\frac{\omega_S}{\omega_L}\right)^2 \cdot \frac{n_S}{n_L} \cdot \left(u_0 \cdot \frac{D_{opt}}{a_0}\right)^2 \cdot \left(\frac{P^2}{m_0}\right)^2 \cdot \left(\frac{\mu_{lh}\mu_{hh}}{m_0^2}\right)^2 \times$$

$$\times \frac{e^4 V m_0^2(a^*)^2}{c^4\hbar^8 3(2\pi)^2} \cdot [\bar{n}+1] \cdot |\hat{e}_S^* \cdot \begin{pmatrix} d & ie & 0 \\ ie & -d & 0 \\ 0 & 0 & 0 \end{pmatrix} \cdot \hat{e}_L|^2 ; \tag{13}$$

with

$$d := R_v(-1; -3, 1) + R_v(1; -1, 3) - R_v(1; 3, -1) - R_v(-1; 1, -3); $$
$$e := R_v(-1; -3, 1) + R_v(1; -1, 3) + R_v(1; 3, -1) + R_v(-1; 1, -3); \tag{14}$$

$$R_v(n; m', m) := \frac{1}{\frac{1}{2}(se_m + se_{m'})x} \cdot arctan\left[i\frac{\frac{1}{2}(se_m + se_{m'})x}{x2_{n,m'} + x1_{n,m}}\right] \tag{15}$$

where P is the interband matrix element of linear momentum [6] and m_0 the free-electron mass. The indices m, m' and n for the valence and conduction subbands, respectively, are related to the z-component of the total angular momentum by $m, m', n = 2 \cdot m_J$ (Fig.1). The other quantities are defined as

$$se_m = \frac{m_e}{m_e + m_{h,m}}; \quad sh_m = \frac{m_{h,m}}{m_e + m_{h,m}}; \quad \frac{1}{\mu_m} = \frac{1}{m_e} + \frac{1}{m_{h,m}}; \tag{16}$$

$$a^* := \frac{\hbar}{\sqrt{2m_0\hbar\Omega_{LO}}}; \quad x := a^* \cdot q; \quad x_F := a^* \cdot q_F; \tag{17}$$

$$xj_{n,m}^2 = \left(\frac{\mu_m}{m_0}\right) \cdot \frac{1}{\hbar\Omega_{LO}} \cdot [\hbar\omega_L - E_g(n,m) + i\eta(n,m) - (j-1) \cdot \hbar\Omega_{LO}]$$
$$- \delta_{j,2} \cdot se_m \cdot sh_m \cdot x^2; \quad \{j = 1, 2\}; \tag{18}$$

where m_e is the effective mass of the conduction band electron and $m_{h,m}$ the effective mass of an electron in the valence band m. For zero magnetic field the Raman tensor reduces to the well known form

284

$$\ddot{R}_{DP}(1LO; int) = \begin{pmatrix} d & ie & 0 \\ ie & -d & 0 \\ 0 & 0 & 0 \end{pmatrix} \quad \xrightarrow[\longrightarrow]{B \to 0\,T} \quad \begin{pmatrix} 0 & a_{DP} & 0 \\ a_{DP} & 0 & 0 \\ 0 & 0 & 0 \end{pmatrix} \quad (19)$$

Intrinsic one-LO-phonon scattering by the Fröhlich interaction leads to the following Raman tensor

$$\ddot{R}_F(1LO; int) = \begin{pmatrix} a & ic & 0 \\ -ic & a & 0 \\ 0 & 0 & b \end{pmatrix} \quad \xrightarrow[\longrightarrow]{B \to 0\,T} \quad \begin{pmatrix} a_F & 0 & 0 \\ 0 & a_F & 0 \\ 0 & 0 & b_F \end{pmatrix} \quad (20)$$

Impurity-induced one-LO-phonon scattering is described by Raman tensors similar to those in Eq. (19) and (20). However, the corresponding Raman efficiencies exhibit a quite different resonance behaviour. The outgoing resonance is much stronger than the ingoing one, in contrast to the intrinsic one-LO-phonon scattering. This behaviour results from double resonances in the denominator of the scattering amplitude W_{fi} of Eq. (4).

In the case of two-LO-phonon scattering we consider a process which is described by an iteration of the intrinsic Fröhlich interaction [14]. For degenerate valence and conduction subbands the Raman tensor is diagonal and the corresponding Raman efficiency exhibits an outgoing resonance at $\hbar\omega_L = E_o + 2\hbar\Omega_{LO}$.

Diluted magnetic semiconductors

One of the most striking properties of DMS is the giant Zeeman splitting of the valence and conduction bands in external magnetic fields. This behaviour is a result of the exchange interaction between the localized $3d^5$ electrons of the Mn^{2+}-ions and itinerant band electrons [2]. In our calculations Zeeman splitting of the valence and conduction bands is taken into account following J.A.Gaj et.al. [15,16]. These authors have considered the exchange interaction within the framework of the $\vec{k} \cdot \vec{p}$ -method, neglecting Landau quantization.

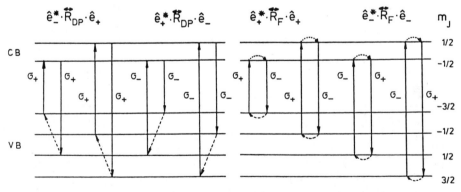

Fig.1. Schematic diagram of the Zeeman splitting of Γ_8 valence and Γ_6 conduction band. The allowed transitions for Raman scattering by the deformation potential $(\hat{e}_S^* \cdot \ddot{R}_{DP} \cdot \hat{e}_L)$ and Fröhlich interaction $(\hat{e}_S^* \cdot \ddot{R}_F \cdot \hat{e}_L)$ are shown. In the first case only hole scattering occurs.

Fig.1 shows schematically the Zeeman splitting of the electronic bands at the Γ-point of the Brillouin zone. Allowed transitions in the Raman process are displayed for a backscattering configuration where the directions of incident and scattered light are perpendicular to the (001) face of the crystal and parallel to the magnetic field $\vec{B}$ (Faraday configuration). The polarization vectors $\hat{e}_+$ and $\hat{e}_-$ for circular polarized light are defined as

$$\hat{e}_+ := 1/\sqrt{2} \cdot (1, i, 0); \qquad \hat{e}_- := 1/\sqrt{2} \cdot (1, -i, 0) \tag{21}$$

Applying magnetic fields up to a few Tesla the splitting of the electronic bands becomes comparable to phonon energies and thus significant changes in the resonance behaviour of the Raman scattering efficiency are expected.

3. EXPERIMENTAL DETAILS

The samples $Cd_{0.75}Mn_{0.25}Te$, $Cd_{0.55}Mn_{0.45}Te$ and $Zn_{0.9}Mn_{0.1}Se$ were grown at the crystal growth laboratory of this faculty by H.H.Otto from the vapour phase using iodine as transport agent. The Mn-concentrations x were established by X-ray powder diffraction analysis (Guinier method) from the known x-dependence of the lattice constant a_0 [17,18]. The $ZnSe$ and Si samples were obtained from M.Cardona and W.Kauschke, Max-Planck-Institut für Festkörperforschung Stuttgart.

The Raman measurements were performed with a cw jet-stream dye laser operating with DCM, rhodamine 6G and stilbene 3. The dyes were pumped with all visible or ultraviolet lines of a cw Ar^+ laser. The laser beam was focused onto the sample with a cylindrical lens and the power density was kept below 10 Wcm^{-2}. The detection system was computer controlled and consisted of a Jarrel-Ash 1m double-grating monochromator with a RCA 31034 photomultiplier tube connected to a photon-counting system. The measured intensities were stored in the computer system which enabled us to remove the background in the Raman spectra by appropriate computer programs. Absolute values of the Raman scattering efficiency were obtained by using a sample substitution method [6]. For $Cd_{0.75}Mn_{0.25}Te$ and $Cd_{0.55}Mn_{0.45}Te$ (red spectral range) we used $ZnSe$ as a reference [19]. In the case of $ZnSe$ and $Zn_{0.9}Mn_{0.1}Se$ (blue spectral range) Si was used [20]. The measured scattering rates R of sample and reference outside the crystal are related to the Raman efficiency by [6,9]

$$\frac{dS}{d\Omega}(\omega_L)_{sample} = \frac{K(\omega_L, \omega_S)_{reference}}{K(\omega_L, \omega_S)_{sample}} \cdot \frac{dS}{d\Omega}(\omega_L)_{reference} \cdot \frac{R(\omega_L)_{sample}}{R(\omega_L)_{reference}} \tag{22}$$

where

$$K(\omega_L, \omega_S) := \frac{(1 - r_L)(1 - r_S) \cdot \{1 - exp[-(\alpha_L + \alpha_S) \cdot L]\}}{n_S^2 \cdot (\alpha_L + \alpha_S)} \tag{23}$$

The frequency-dependent factors $K(\omega_L, \omega_S)$ must be used to correct the measured scattering rates with respect to absorption (α), reflectivity (r) and refraction (n). The quantities referring to the incident and scattered photon energies have subscripts L and S, respectively. The absorption coefficients of $Cd_{0.55}Mn_{0.45}Te$ in the range 1.80 - 2.03 eV and of $Zn_{0.9}Mn_{0.1}Se$ in the range 2.55 - 2.77 eV were obtained from absorption measurements, performed by K.Hochberger at our institute. All other data for optical constants were taken from the literature. The measurements on $Cd_{0.55}Mn_{0.45}Te$,

$Cd_{0.75}Mn_{0.25}Te$, $Zn_{0.9}Mn_{0.1}Se$ and $ZnSe$ were carried out at the temperatures 300 K, 2 K, 80 K and 105 K, respectively. The $Cd_{0.75}Mn_{0.25}Te$ sample, together with $ZnSe$ as a reference, was inserted in a liquid-He-bath cryostat within a superconducting split coil magnet system, which enabled us to apply external magnetic fields up to 5 Tesla. All the other samples and the corresponding references were mounted on the cold finger of a N_2 cryostat (80 - 300 K). All measurements were performed in a backscattering configuration with the direction of incident and scattered light perpendicular to the (001) faces of sample and reference. The directions x, y and z were chosen along the crystal axes [100], [010] and [001], respectively. The orientation of the samples was carried out with the help of an optical two-circle reflection goniometer and the Laue method for final checking.

According to the Raman tensors given in Sec. 2, the efficiencies for deformation-potential and Fröhlich scattering can be determined separately by choosing an appropriate scattering configuration.

$$\frac{dS}{d\Omega}(1LO) \quad \propto \quad \begin{cases} |a_{DP}|^2 + |a_{DPi}|^2 & \text{for} \quad z(xy)\bar{z} \quad \text{and} \quad z(yx)\bar{z} \\ |a_F|^2 + |a_{Fi}|^2 & \text{for} \quad z(xx)\bar{z} \quad \text{and} \quad z(yy)\bar{z} \end{cases}$$

$$\frac{dS}{d\Omega}(2LO) \quad \propto \quad |a_{2F}|^2 \qquad \text{for} \quad z(xx)\bar{z} \quad \text{and} \quad z(yy)\bar{z} \tag{24}$$

4. RESULTS AND DISCUSSION

In this section we present a quantitative comparison of experimental results and theoretical calculations for the Raman scattering efficiencies near the E_0 gap. The two-LO-phonon resonances were fitted first, yielding accurate values of the E_0 gap and its broadening η from position and width of the resonance. For all samples the values of E_0 coincided exactly with those obtained from luminescence spectra. The impurity concentrations n_i were obtained from a fit of one-LO-phonon scattering by the impurity-induced Fröhlich mechanism. Thereby, all adjustable parameters which determine the qualitative behaviour of the theoretical curves were fixed. The optical deformation potential D_{opt} appears as a constant factor in the corresponding fit curves and was obtained by fitting the curve of the intrinsic deformation-potential mechanism to the experimental values. All other parameters, needed for the calculation of the fit curves, were taken from the literature. Thus, no further adjustable parameters were used for the fits of the one-LO-phonon scattering by the intrinsic Fröhlich mechanism and the impurity-induced deformation-potential mechanism. In the following subsections only a few of the measured resonance Raman spectra are shown as typical examples. The numbers in brackets denote the ratios of experimental and theoretical values of the Raman scattering efficiency in the maximum of the resonance.

I. ZnSe

Fig.2(a) shows the resonance behaviour of the Raman efficiency for scattering by two LO phonons in the configuration $z(x,x)\bar{z}$. A maximum appears, whenever the energy of the scattered photon equals the gap energy E_0 (outgoing resonance), i.e. for $\hbar\omega_L = E_0 + 2\hbar\Omega_{LO}$ where $\hbar\Omega_{LO} = 31.5$ meV. The solid line is a fit to the experimental data assuming an iteration of the intrinsic Fröhlich mechanism as pointed out in Sec. 2.

The best qualitative agreement with the experimental results is obtained if we assume that the gap energy E_0 is 2.777 eV and its broadening η is 0.2 meV. Further parameters, used in these calculations, are taken from the literature. Scattering by one LO phonon in the same configuration is displayed in Fig.2(b). Again a maximum occurs whenever the energy of the scattered photon equals the gap energy, i.e. for $\hbar\omega_L = E_0 + \hbar\Omega_{LO}$. The solid line is a fit representing the sum of intrinsic (int.) and impurity-induced (imp.) Fröhlich mechanism.

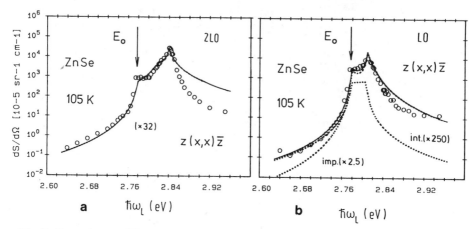

Fig.2. Experimental Raman efficiencies for scattering by (a) two LO phonons and (b) one LO phonon in ZnSe at 105 K for the scattering configuration $z(x,x)\bar{z}$. The solid lines are fits assuming (a) an iteration of the intrinsic Fröhlich mechanism, (b) an intrinsic (int.) and an impurity-induced (imp.) mechanism for the Fröhlich interaction.

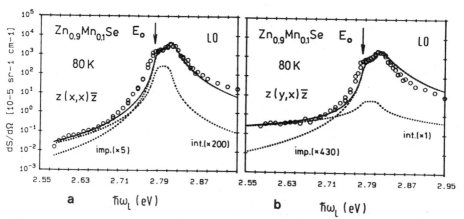

Fig.3. Experimental Raman efficiencies for scattering by one LO phonon at 80 K in $Zn_{0.9}Mn_{0.1}Se$. The results are shown for (a) Fröhlich interaction and (b) deformation-potential interaction. The solid lines are fits assuming for both interactions an intrinsic and an impurity-induced mechanism.

288

It can clearly be seen, that the behaviour of the Raman scattering efficiency near the maximum of the resonance is dominated by the impurity-induced mechanism. From the fit in Fig.2(b) we obtain an impurity concentration n_i of $3.5 \cdot 10^{18} cm^{-3}$.

II. $Zn_{0.9}Mn_{0.1}Se$

The results for one-LO-phonon scattering by the Fröhlich interaction are plotted in Fig.3(a). Similar to the case of $ZnSe$ (Fig.2(b)) the experimental data are well discribed by the impurity-induced Fröhlich mechanism. For the impurity concentration n_i we obtain the same value as in the case of $ZnSe$. The gap energy $E_0 = 2.785$ eV and its broadening $\eta = 4$ meV were determined from a fit to the two-LO-phonon resonance. Fig.3(b) shows the resonance behaviour of the Raman efficiency for one-LO-phonon scattering by the deformation-potential interaction. According to the form of the corresponding Raman tensor in Eq. (19), scattering by the deformation-potential interaction is observed in the configuration $z(x,y)\bar{z}$ or $z(y,x)\bar{z}$, i.e. for $\hat{e}_L$ and $\hat{e}_S$ perpendicular to each other. The Raman efficiency shows an incoming resonance at $\hbar\omega_L = E_0$ and a much more pronounced outgoing resonance at $\hbar\omega_L = E_0 + \hbar\Omega_{LO}$. In the energy range 2.55 - 2.70 eV the experimental data are well described by the model calculations for the intrinsic scattering mechanism. From the fit we obtain for the deformation potential D_{opt} a value of 30 eV. Near the resonance maximum the intrinsic mechanism fails to describe the experimental behaviour. In this energy range a much better qualitative agreement is obtained by the impurity-induced mechanism. However, the ratio between experimental and theoretical values of the Raman efficiency is about 430.

III. $Cd_{0.55}Mn_{0.45}Te$

The Raman spectra of $Cd_{1-x}Mn_xTe$ show two LO-phonon lines denoted by LO_1 and LO_2. The LO_1-phonon is a vibration of the "CdTe" compound of the mixed crystal while the LO_2-phonon belongs to the "MnTe" component. In Fig.4 the Raman efficiency for scattering by LO_1-phonons is shown where $\hbar\Omega_{LO1} = 19.2$ meV. By fitting the two-LO_1-phonon resonance (Fig.4(a)) we obtain the gap energy $E_0 = 2.04$ eV and its broadening $\eta = 25$ meV. The results for the one-LO_1-phonon scattering by the Fröhlich interaction are given in Fig.4(b). The fit curves for the intrinsic and the impurity-induced mechanism are plotted with their resonance maximum at the same height to demonstrate the qualitative difference between the two scattering mechanisms. For the impurity concentration n_i we obtain a value of $1.5 \cdot 10^{19} cm^{-3}$. Fig.4(c) shows the Raman efficiency for the deformation-potential interaction. Similar to the case of $Zn_{0.9}Mn_{0.1}Se$ (Fig.3(b)) an excellent qualitative agreement of experimental and theoretical results is obtained if we assume that the impurity-induced mechanism dominates near resonance. However, for photon frequencies far below the gap energy E_0 the intrinsic mechanism seems to be the more important one. We obtain for the optical deformation potential $D_{opt} = 25$ eV.

IV. $Cd_{0.75}Mn_{0.25}Te$

We have investigated the influence of external magnetic fields on the resonant Raman scattering in $Cd_{0.75}Mn_{0.25}Te$. Normalized Raman intensities were measured for σ_+ and σ_- incident light without field, at 1.75 T and at 4 T. Fig.5 shows the results for

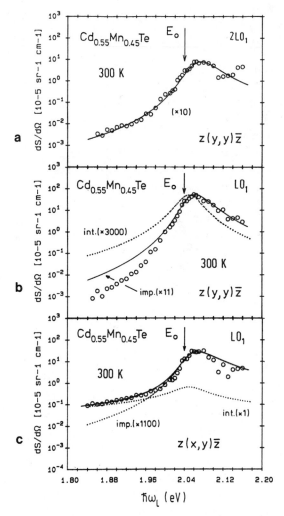

Fig.4. Experimental Raman efficiencies for scattering by LO_1 phonons at 300 K in $Cd_{0.55}Mn_{0.45}Te$. (a) Two-LO_1-phonon scattering in the configuration $z(y,y)\bar{z}$, (b) one-LO_1-phonon scattering by the Fröhlich interaction and (c) one-LO_1-phonon scattering by the deformation-potential interaction. The solid lines are fits assuming (a) an iteration of the intrinsic Fröhlich interaction and (b),(c) an intrinsic and an impurity-induced scattering mechanism for both Fröhlich and deformation-potential interaction.

scattering by one LO_1 phonon in the Faraday configuration. The lines are drawn as a guide to the eye. For σ_+ incident light the maximum of the resonance shifts to lower energies with increasing magnetic field. For σ_- incident light two local maxima appear, one is shifted to higher energies and the other to lower energies.

The measured intensities are not corrected for absorption. Therefore a strong decrease in intensity is observed for increasing energy. The experimental results are in qualitative agreement with our model calculations which lead to the Raman tensors in Eq. (19) and (20). A quantitative comparison of the experimental and theoretical results will be given elsewhere.

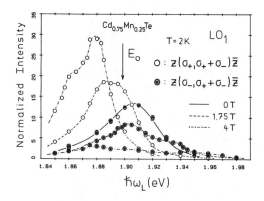

Fig.5. Normalized Raman intensities for scattering by one LO_1 phonon with external magnetic fields in $Cd_{0.75}Mn_{0.25}Te$ at 2 K.

5. CONCLUSION

We have investigated the absolute Raman efficiencies for LO phonon scattering in $ZnSe$, $Zn_{0.9}Mn_{0.1}Se$, $Cd_{0.75}Mn_{0.25}Te$ and $Cd_{0.55}Mn_{0.45}Te$ as a function of incident photon energy $\hbar\omega_L$ near the E_0 gap. Using appropriate configurations, we observed scattering through the deformation-potential and Fröhlich interaction separately. The maxima of the Raman efficiencies occur for both scattering configurations whenever the energy of the scattered photon equals the gap energy E_0. The experimental data are in qualitative agreement with theoretical calculations, assuming uncorrelated electron-hole pairs as intermediate states. Scattering by one LO phonon is shown to be dominated by the impurity-induced mechanism. However, a quantitative comparison shows that the experimental values of the Raman efficiency are always higher than the calculated ones. The application of an external magnetic field in $Cd_{0.75}Mn_{0.25}Te$ causes a large Zeeman splitting of the valence and conduction bands. As a result the maximum of the resonance curve shifts to higher or lower energies for increasing field, depending on the polarization of incident light.

6. ACKNOWLEDGEMENT

We wish to thank M. Cardona and W. Kauschke, Max-Planck-Institut für Festkörperforschung Stuttgart, who enabled us to carry out parts of the measurements on $Zn_{0.9}Mn_{0.1}Se$ at their institute.

7. REFERENCES

[1]   J.K. Furdyna, J. Appl. Phys. **53**, 7637 (1982)

[2]   N.B. Brandt and V.V. Moshchalkov, Advances in Physics **33**, 193 (1984)

[3]   J. Menendez and M. Cardona, Phys. Rev. B **31**, 3696 (1985)

[4]   A.A. Gogolin and E.I. Rashba, Solid State Commun. **19**, 1174 (1976)

[5]   W. Kauschke and M. Cardona, Phys. Rev. B **33**, 5473 (1986)

[6]   M. Cardona, in *Light Scattering in Solids II*, Vol. 50 of *Topics in Applied Physics*, edited by M. Cardona and G. Güntherodt (Springer, Berlin 1982)

[7]   R. Loudon, *The Quantum Theorie of Light* (Clarendon Press, Oxford, 1973)

[8]   R. Loudon, Proc. Roy. Soc. A **275**, 218 (1963)

[9]   W. Richter, in *Solid State Physics*, Vol. 78 of *Springer Tracts in Modern Physics*, edited by G. Höhler (Springer, Berlin 1976)

[10] G.L. Bir and G.E. Pikus, Soviet Physics Solid State **2**, 2039 (1961)

[11] A.K. Ganguly and J.L. Birman, Phys. Rev. **162**, 806 (1967)

[12] A. Pinczuk and E. Burstein, in *Light Scattering in Solids I*, Vol. 8 of *Topics in Applied Physics*, edited by M. Cardona (Springer, Berlin, 1975), p. 23

[13] E.O. Kane, in *Semiconductors and Semimetals*, edited by R.K. Willardson and A.C. Beer (Academic Press, New York, 1966), Vol. 1, p. 75.

[14] R. Zeyher, Phys. Rev. B **9**, 4439 (1974)

[15] J.A. Gaj, J. Ginter and R.R. Galazka, Phys. Stat. Sol. (b) **89**, 655 (1978).

[16] J.A. Gaj, R. Planel and G. Fishman, Solid State Commun. **29**, 435 (1979)

[17] J. Bak, U. Debska, R.R. Galazka, G. Jasiolek, E. Mizera and B. Bryza, II Intern. Congress Crystall, Warszawa, 1978, Coll. Abstracts 245

[18] L.A. Kolodziejski, R.L. Gunshor, R. Venkatasubramanian, T.C. Bonsett, R. Frohne, S.Datta, N. Otsuka, R.B. Bylsma, W.M. Becker and A.V. Nurmikko, J. Vac. Sci. Technol. B **4**, 583 (1986)

[19] J.M. Calleja, H. Vogt and M. Cardona, Philosophical Magazine A **45**, 239 (1982)

[20] J. Wagner and M. Cardona, Solid State Commun. **48**, 301 (1983)

RESONANCE RAMAN SCATTERING IN [111]-ORIENTED CdTe/CdMnTe SUPERLATTICES

L Vina, L L Chang*, M Hong*, J Yoshino*, F Calle[+],
J M Calleja[+] and C Tejedor[+]

Instituto de Ciencia de Materiales, Universidad de
Zaragoza-C.S.I.C., 50009 Zaragoza, Spain
*IBM Thomas J Watson Research Center, PO Box 218,
Yorktown Heights, NY 10598, USA
[+]Universidad Autonoma de Madrid, Cantoblanco, 28049,
Madrid, Spain

INTRODUCTION

$Cd_{1-x}Mn_xTe$ alloys can be grown in zinc-blende structure up to a
composition of $x \sim 0.7$. The band gap of these crystals can be continuously
tuned increasing the Mn composition.[1] The difference between the band
gaps of CdTe and the ternary alloy provides the potential for the
electronic confinement in superlattices (SL's) and quantum wells. High
quality SL's based on $Cd_{1-x}Mn_xTe$ have been grown in the last years using
epitaxial techniques.[2-4] The optical properties of these SL's have been
studied extensively in the literature.[5] Resonance Raman Scattering (RRS)
provides the link between electronic and lattice-dynamical properties:
the energy dependence of the Raman efficiency is determined by the
electron-phonon coupling and by band-structure parameters. Recent
reviews deal with vibronic[6] and electronic[7] properties of semiconductor
SL's studied with RRS. The phonon spectra of $Cd_{1-x}Mn_xTe$ bulk crystals[8]
and $CdTe/Cd_{1-x}Mn_xTe$ superlattices[9-12] have been studied previously in the
literature; these studies will be used here to compare new results in
the SL's. Usually, $CdTe/Cd_{1-x}Mn_xTe$ SL's are grown on GaAs substrates.
Two orientations of CdTe epitaxial films have been observed: (100)-CdTe
parallel to (100)-GaAs[13] or (111)-CdTe parallel to (100)-GaAs.[14] Marked
diferences between [100] and [111] $CdTe/Cd_{1-x}Mn_xTe$ SL's, attributed to the
quality of the interfaces and to the presence of strain have been reported
in the literature.[3,9,15] Here we present a RRS study on [111] SL's which
shows a similar behaviour to that observed in [001] superlattices.[9]

EXPERIMENTAL DETAILS

The samples used in our study have been grown by molecular beam
epitaxy on (100)-oriented GaAs substrates. A thick (between 0.15μm and
0.2μm (111)-oriented CdTe buffer layer was followed by a $CdTe/Cd_{1-x}Mn_xTe$
SL. Samples with x ranging from 0.10 to 0.21, with well widths $d_1=86$Å
and 100Å, and barrier thicknesses, d_2, from 20Å to 100Å were investigated.
The period $d=d_1 + d_2$ was repeated 25 times for the samples with the
widest barrier-width and 100 times for those with the thinnest one. For

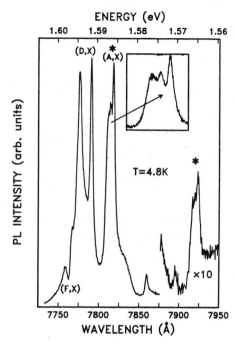

Fig. 1 Excitonic PL spectrum at 4.8K for a (111)-CdTe layers on GaAs
The inset shows fine structure corresponding to acceptor-bound
excitons (A,X). The structures labelled with * in the low-
energy range correspond to one-phonon replica of (A,X).
Power density was 0.3W/cm^2.

the samples with intermediate barrier thicknesses, the period was chosen
so as to keep the total alloy thickness approximately constant. Details
of the growth technique have been described in a previous publication.[4]
The samples were mounted on a variable temperature crystat and kept in a
He-gas atmosphere. Raman spectra were obtained using discrete lines of
a Kr$^+$-ion laser, as well as with excitation from an LD700 dye-laser
pumpted by the Kr$^+$ laser. The dye-laser was also used to obtain photoluminescence (PL) and PLE spectra. Power densities were kept below
1W/cm^2 for PL and PLE and below 20W/cm^2 for the Raman spectra. The light
was analysed with a double grating monochromator and detected by a
Peltier-cooled GaAs photomultiplier using standard photon counting
techniques.

RESULTS AND DISCUSSION

    Figure 1 shows a PL spectrum, excited at 1.68eV, for a (111)-CdTe
film grown on (100)-GaAs. Sharp excitonic peaks, with full width at half
maximum smaller than 1meV, are clearly resolved in the spectrum. This
spectrum, comparable to those reported for high purity bulk-CdTe,[16,17]
reflects the high crystal quality obtainable by molecular beam epitaxy
of CdTe on GaAs. The peaks at 7760Å and 7767Å are attributed to the
upper- and lower-branch of the free-exciton (F,X) polariton, respectively.[18]

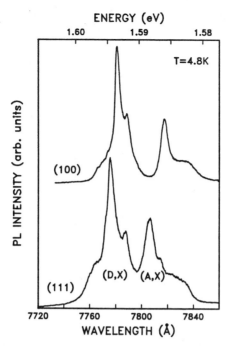

Fig. 2. Comparison of typical PL spectra for (100)-CdTe and (111)-CdTe
layers on GaAs. Power density was 0.1W/cm².

The peaks at 7776Å and 7790Å correspond to donor-bound excitons (D,X),[19]
while those between 7800Å and 7850Å are assigned to acceptor-bound
excitons (A,X).[19] The inset in the figure shows fine structure of the
(A,X) recombination. The peaks at the low-energy side of the spectrum (*)
depict the one-LO-phonon replica of (A,X), since their energy separation,
21.3meV, corresponds to the longitudinal phonon energy in CdTe.[8]

Figure 2 compares typical PL spectra for (100)-CdTe and (111)-CdTe
layers on GaAs. The spectra were obtained with excitation at 1.61eV. The
quality of these layers is slightly worse than that of Fig. 1, as can be
inferred from the width of the excitonic peaks and from the absence of
free-exciton recombination. By comparison with Fig. 1, we assign the
high-energy structures to donor-bound excitons (D,X) and the low-energy
ones to acceptor-bound excitons (A,X). The figure demonstrates the
similar quality of (100)- and (111)-oriented CdTe layers obtained
typically. There is only a small blue shift (∿1.5meV) of the PL of the
(111) sample with respect to that of the (100), maybe due to the presence
of strain in the samples.

The PL (dashed line) and the PLE (solid line) spectra of a 25-periods
SL's with 100Å-wide wells and barriers are shown in Fig. 3. A small Stokes
shift (8meV) between the PL and the PLE is clearly seen in the figure.
This shift has been attributed to well-width fluctuations and/or impurity
effects.[11] Three broad structures at 1.636, 1.653 and 1.695eV are seen in
the PLE spectrum. To assign these structures a calculation of the levels in
the superlattice was performed in the envelope function approximation.

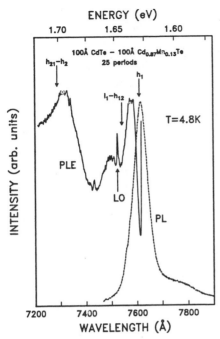

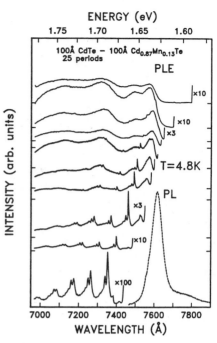

Fig. 3. PL (dashed line) and PLE solid line) spectra for a CdTe/$Cd_{1-x}Mn_xTe$ superlattice with 100Å-wide wells and barriers. The arrows show the result of a calculation in the envelope function approximation.

Fig. 4. PL (dashed line) and PLE spectra recorded at different energies of the same SL as in Fig. 3. The spectra are shifted vertically for clarity and enhancement factors are shown on the right-hand-side.

A free-exciton dependence on Mn concentration[20] $E(x)=1.5976+1.564x(eV)$, a 90-100% rule for the conduction- and valence-band discontinuity[21] and the following effective masses:[22] $m_o^*=0.096$, $m_h^*=0.4$ and $m_1^*=0.1$ were used for the calculations. The vertical arrows show the results of the calculations. The peaks at $\sim1.63eV$ and $\sim1.65eV$ correspond to the first heavy-hole exciton (h_1) and to the first light-hole exciton (1_1), respectively, and finally the peak at $\sim1.70eV$ corresponds to the exciton associated with the second conduction and valence band levels (h_2). Close to 1_1 and h_2 are also h_{12} and h_{21}, where the first (second) subindex indicates conduction (valence) band level.

Sharp peaks, which correspond to first and higher order LO-CdTe phonons, are also seen in the PLE spectrum of Fig. 2. These phonons and those of the $Cd_{1-x}Mn_xTe$ barriers become clearer when the spectrometer in the PLE spectra is placed at higher energies. Figure 4 shows PLE spectra of the same sample as in Fig. 3 for different detection energies. As the detection energy is moved toward higher values, sharp periodic structures become resolvable. The energy distance from the first two peaks to the detection energy is 21.3meV and 24.3meV, which correspond to the CdTe LO-phonon and to an interface phonon (IF), respectively. The latter peak was assigned previously[11] to a MnTe-like LO-phonon (LO_2) of the barriers, however, more recent results[12] favour the present assignment. The additional structures, with a period between the main peaks of 21.3meV, correspond to higher order phonons (LO and IF) and combinations

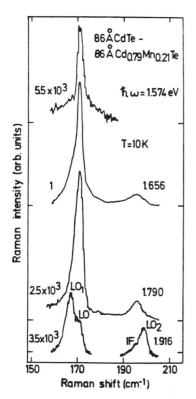

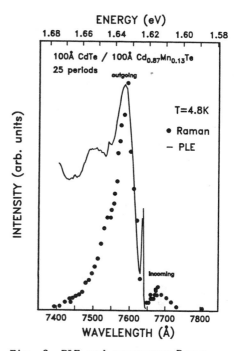

Fig. 5. Raman spectra at 10K for a
86Å-CdTe/86A-Cd$_{0.79}$Mn$_{0.21}$
Te SL.

Fig. 6. PLE and resonance Raman
for the sample of Fig. 3.

of them. Oscillatory structure in the PLE spectrum of bulk CdTe has been
observed previously.[23] Hot-exciton formation and hot-electron effects
can be ruled out as responsible for the periodic structure, since phonon
lines can be observed above and below the h$_1$ exciton and a second
periodicity, slightly higher than the LO frequency, is absent in our
spectra.[23]

Unpolarized Raman spectra in the vicinity of h$_1$ are shown in Fig.5
for a sample with 86Å-wells and 86Å-Cd$_{0.79}$Mn$_{0.21}$Te barriers. The energy
of the incident photon is indicated on the right-hand-side of the
spectra. Far above from resonance (1.916eV), the Raman spectrum is
dominated by the CdTe- and MnTe-like modes in the barriers, at 166cm^{-1}
(LO$_1$) and 198cm^{-1}(LO$_2$), respectively. Two additional structures are seen
in the spectrum at 171cm^{-1}, which correspond to the LO-phonon in CdTe,[8]
and at 195cm^{-1}. Due to its resonance behaviour, the latter structure is
assigned to a MnTe-like interface mode (IF). As the laser energy
approachs h$_1$, LO and IF are resonantly enhanced, while LO$_2$, which is
confined in the barriers, is strongly reduced. The LO$_1$ mode, which
can propagate in the wells and barriers,[9] is observed in the spectra as
a shoulder in the low-energy side of LO. These spectra of [111] CdTe/
Cd$_{1-x}$Mn$_x$Te SL's resemble those obtained in [100] superlattices.[9]

A comparison between the resonant behaviour of the LO phonon and
the excitons observed in PLE is shown in Fig. 6 for a sample with 100Å
wells and barriers and 13% Mn. The Raman points have been shifted by
one-LO phonon towards lower energies, in order to compare directly the
resonance of the outgoing channel with the h$_1$ exciton. Impurity-induced
LO-phonon scattering presents a maximum for scattered light at the
energy of the electronic excitation[24] (outgoing resonance). The

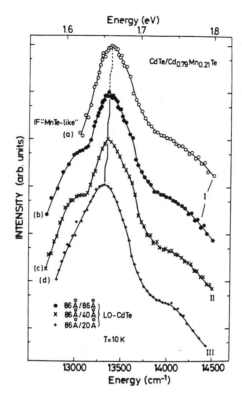

Fig. 7 Resonance Raman intensities of the MnTe-like IF- and LO-phonons in three SL's with 86Å-wide wells and 86Å (I), 40A (II) and 20Å (III) wide barriers. The lines are a guide to the eye.

asymmetry in the line shape of the outgoing resonance also points to impurity effects, although the influence of bound excitons as inter-mediate electronic states cannot be ruled out.[10,25] The incoming resonance is also seen in Fig. 6 at an energy of 1.615eV, one LO-phonon below h_1. A much weaker enhancement of the Raman cross-section is also observed at the energy of l_1; the incoming and outgoing channels show a comparable behaviour, in agreement with the results of Ref. 10.

We have assigned, in a previous work,[11] a phonon at 196cm^{-1} to the MnTe-like LO_2-phonon. However, the Raman spectrum at 1.916eV (Fig. 5), together with the similar resonant behaviour of this mode and of the LO CdTe-phonon depicted in Fig. 7 (curves a and b), suggests an interface character of the phonon labelled IF.[26] Figure 7 also shows the resonance of the LO-CdTe phonon for three SL's with 86Å-wide wells and 86Å (I), 40Å (II) and 20Å (III) barriers, with 21% Mn, as a function of incident laser energy. The $\sim$25cm^{-1} shift between the resonances of IF and LO corresponds to the energy difference of their frequencies. The enhancements around 13100cm^{-1} and 13300cm^{-1} are due to the incoming and outgoing resonances with h_1, respectively. The shoulder at $\sim$13900cm^{-1} corresponds to h_2.

The formation of minibands can be hinted from a comparison between curves b), c) and d) in Fig. 7. An increasing broadening of the resonance together with a small red shift, is observed as the barrier width is reduced. With respect to sample I, the measured broadenings are $\Delta\Gamma=(0(+10)$cm^{-1} (sample II), $\Delta\Gamma=70(+10)$cm^{-1} (sample III). The corresponding shifts amount to $\Delta E=10(+10)$cm^{-1} and $\Delta E=40(+15)$cm^{-1} for samples II and III, respectively. A crude estimation, using a Kronig-Penney model, gives a broadening (shift) of $30cm^{-1}$ ($15cm^{-1}$) and $100cm^{-1}$ ($55cm^{-1}$) for samples II and III, respectively.

In conclusion, we have shown that good quality [111] CdTe films and CdTe/$Cd_{1-x}Mn_xTe$ SL's can be grown by molecular beam epitaxy on GaAs substrates. Confined, propagating and interface longitudinal optical phonons in [111] SL's are similar to those observed in [100] SL's. Differences between [100] and [111] CdTe/$Cd_{1-x}Mn_xTe$ SL's, such as a dependence of the phonon frequencies on laser energies in [111] samples,[9,12] disappear when the samples are subjected to thermal cycling.[12,27] The results of these experiments enable the direct observation of the formation of a miniband in a superlattice.

ACKNOWLEDGEMENTS

We gratefully acknowledge the assistance of H Munekata for help in growing the superlattices.

REFERENCES

1. P. Lautenschlager, S. Logothetidis, L. Viña and M. Cardona, Phys. Rev. B 32, 3811 (1985), and references therein.
2. R.N. Bickell, R.W. Yanka, N.C. Giles-Taylor, D.K. Blanks, E.L. Buckland, and J.F. Schetzina, Appl. Phys. Lett. 45, 92 (1984).
3. L.A. Kolodziejski, R.L. Gunshor, N. Otsuka, X.-C. Zhang, S.-K. Chang and A.V. Nurmikko, Appl. Phys. Lett. 47, 882 (1985).
4. J. Yoshino, H. Munekata, and L.L. Chang, J. Vac. Sci. Technol. B 5, 683 (1987).
5. See, for example, A.V. Nurmikko, R.L. Gunshor, and L.A. Kolodziejski, IEEE J. Quant. Electron. QE-22, 1785 (1986)
6. M.V. Klein, IEEE J. Quantum Electron. QE-22, 1760 (1986).
7. G. Abstreiter, R. Merlin, and A. Pinczuk, IEEE J. Quantum Electron. QE-22, 1771 (1986).
8. S. Venugopalan, A. Petrou, R.R. Galazka, A.K. Ramdas, and S. Rodriguez, Phys. Rev. B 25, 2681 (1982).
9. E.-K. Suh, D.U. Bartholomew, A.K. Ramdas, S. Rodriguez, S. Venogupalan, L.A. Kolodziejski, and R.L. Gunshor, Phys. Rev. B 36, 4316 (1987)
10. S.-K. Chang, H. Nakata, A.V. Nurmikko, L.A. Kolodziejski, and R.L. Gunshor, Journal de Physique C5, 345 (1987)
11. L. Viña, L.L. Chang and J. Yoshino, Journal de Physique C5, 317 (1987)
12. L. Viña, F. Calle, J.M. Calleja, C. Tejedor, M. Hong, and L.L. Chang, in Proceedings of the 19th International Conference on the Physics of Semiconductors, Warsaw 1988, edited by J. Kossut (to be published).
13. R.N. Bicknell, R.W. Yanka, N.C. Giles-Taylor, J.F. Schetzina, T.J. Magee, C. Leung, and H. Kawayoshi, Appl. Phys. Lett. 44, 313 (1984).
14. L.A. Kolodziejski, T. Sakamoto, R.L. Gunshor, and S. Datta, Appl. Phys. Lett. 44, 799 (1984).
15. S.-K. Chang, A.V. Nurmikko, L.A. Koldziejski, and R.L. Gunshor, Phys. Rev. B 33, 2589 (1987)
16. J.E. Espinsoa, J.M. Gracia, H. Navarro, A. Zehe, and R. Triboulet, J. Lumin. 28, 163 (1983).
17. N.C. Giles-Taylor, R.N. Bicknell, D.K. Blanks, T.H. Myers, and J.F. Schetzina, J. Vac. Sci. Technol. A 3, 76 (1985).
18. S. Suga, K. Cho, P. Hiesinger, and T. Koda, J. Lumin. 12-13, 109 (1976)
19. Z.C. Feng, A. Mascarenhas, and W.J. Choyke, J. Lumin. 35, 329 (1986) and references therein.
20. D. Heiman, P. Becla, R. Kershaw, D. Ridgley, K. Dwight, A. Wold, and R. Galazka, Phys. Rev. B 34, 3961 (1986)
21. S.-K. Chang, A.V. Nurmikko, J.-W. Wu, L.A. Kolodziejski, and R.L. Gunshor, Journal de Physique C5, 367 (1987).

22. Landolt-Bornstein Tables, Vol.17, ed. by O. Madelung, M. Schulz and H. Weiss (Springer-Verlag, New York, 1982).
23. A. Nakamura and C. Weisbuch, Solid-State Electron. 21, 1331 (1978)
24. J. Menendez and M. Cardona, Phys. Rev. B 31, 3696 (1985).
25. P.J. Colwell and M.V. Klein, Solid State Commun. 8, 2095 (1970)
26. A.K. Sood, J. Menendez, M. Cardona and K. Ploog, Phys. Rev. Lett. 54, 2115 (1985)
27. L. Viña, M. Hong, L.L. Chang, F. Calle, J.M. Calleja, and C. Tejedor (unpublished).

MAGNETO-OPTICAL ABSORPTION IN II-VI SEMICONDUCTORS:

APPLICATION TO CdTe

J.T. Devreese[*] and F.M. Peeters[0]

University of Antwerp (U.I.A.)
Department of Physics
Universiteitsplein 1
B-2610 Antwerpen(Wilrijk)

The linear magneto-optical absorption theory is introduced. We start from the Kubo formula for the frequency dependent conductivity and apply the memory function formalism. The electron-LO-phonon interaction and the band non-parabolicity are incorporated. The theory is used to analyse recent experimental data on $CdTe$. For the first time, data have become available for $\omega_c > \omega_{LO}$ (ω_c is the cyclotron frequency and ω_{LO} is the long-wavelength LO-phonon frequency). These new experimental results provide a quantitative test of the theory of large polarons. For the full magnetic field range ($B : 0 \rightarrow 20.5T$) theory and experiment agree very well and the proposed band mass $m_b/m_e = 0.0920$ is consistent with the electron-phonon coupling constant $\alpha = 0.3$.

1. Introduction

Cyclotron resonance is a standard technique for the determination of the mass of conduction electrons and holes in solids. A magnetic field is applied which quantises the motion of the electrons perpendicular to the magnetic field into Landau orbits. The separation between these levels is $\hbar\omega_c^*$, where $\omega_c^* = eB/m^*c$ is the cyclotron resonance frequency, B the magnetic field strength, m^* the cyclotron mass of the electrons, e the elementary charge, and c the velocity of light. In the *Faraday active mode configuration* circular polarized light is radiated parallel to the magnetic field. This induces transitions between different Landau levels if the energy of the radiation($\hbar\omega$) equals the energy separation between the Landau levels($\hbar\omega_c^*$). Thus from a sweep of the magnetic field it is experimentally possible to determine the cyclotron mass of the electrons.

The electron-phonon interaction will change the above picture in three different ways[1-3]: 1) For small magnetic fields a *polaron*(=electron + lattice distortion) will move through the crystal. This leads to: a) a shift in the energy: $\Delta E = -\alpha\hbar\omega_{LO}$ for $\alpha << 1$, with α the electron-phonon coupling constant; b) a mass renormalization: $m^*/m_b = 1 + \alpha/6$ for $\alpha << 1$, with m_b the electron band mass. The latter implies that the separation between the Landau levels will be $\omega_c^* = eB/m^*c$ which is smaller than for the non-interacting problem where the cyclotron frequency would be $\omega_c = eB/m_bc$. If the electron-phonon coupling constant is not small, values for ΔE and m^*/m_b are obtained[4] which are larger than as given above. 2) When the magnetic field is such that $\omega_c^* \sim \omega_{LO}$ an anti-crossing of energy levels is observed and found theoretically. For this magnetic field value the energy of the first Landau level which

Table I. Physical parameters for some polar materials

Material	α	$\hbar\omega_{LO}(meV)$	$E_g(eV)$	m_b/m_e	m^*/m_e
$CdTe$	0.3	21.1	1.606	0.0920	0.0965
$InSb$	0.02	24.4	0.235	0.0135	0.0135
$GaAs$	0.068	36.75	1.520	0.0661	0.0669
$AgBr$	1.54	17.3	2.7	0.2108	0.2818
$AgCl$	1.84	24.3	3.3	0.2805	0.3988

has an energy $\Delta E + \frac{3}{2}\hbar\omega_c$ becomes degenerate with the level $\Delta E + \hbar\omega_{LO} + \frac{1}{2}\hbar\omega_c$. Due to the electron-phonon interaction the degeneracy of these levels will be lifted and one observes a *splitting* of the cyclotron resonance peak into two different peaks. 3) In the large magnetic field limit, i.e. $\omega_c >> \omega_{LO}$, one of the energy levels is *pinned* to $\hbar\omega_{LO} + \Delta E$. The cyclotron resonance peak appears in the continuum.

Few experiments exist to provide a quantitative test of the validity of currently accepted polaron theories. The pioneering studies of high-frequency cyclotron resonance in III-V and II-VI semiconductors[5-8] showed the power of an experiment in which large polarons are probed spectroscopically at a frequency approaching that of the longitudinal optical phonon.

Pinning and splitting of the cyclotron resonance spectrum has been observed first in the material $InSb$ by Dickey *et al*[5]. $InSb$ is a very weak polar semiconductor($\alpha = 0.02$) and consequently the polaron effects are very small. Furthermore band non-parabolicity is very important in this material. Away from the resonant field where $\omega_c^* \sim \omega_{LO}$ band non-parabolicity has a very similar effect on the electron effective mass as the polaron effect, both effects lead to a linear increase of m^* with B.

More suitable candidates for a detailed quantitative test of the theory of large polarons exist within the alkali and silver halides[9-11]. Here the wide band gap ensures that the rigid-band states exhibit little deviation from a parabolic conduction band while the highly ionic lattice produces polar coupling constants in the range from 1 to 5. Recently[11] we were able to find a good agreement between cyclotron measurements in $AgBr$ and $AgCl$ for magnetic fields in the range $B : 0 \to 16T$. We found important polaron contributions to the cyclotron mass which could not be explained within traditional pertubation-type theories. A calculation[12] based on a generalization[13] of the Feynman polaron model was necessary in order to explain the data.

These alkali and silver halides exhibit polaron masses in low magnetic fields which ranges from $\sim 0.3m_e$ in $AgBr$ to $\sim 1.0m_e$ in $RbCl$. These large band masses makes it impracticable with present steady magnetic fields to make the cyclotron frequency exceed the frequency of the longitudinal optical phonon. Therefore measurements had to concentrate on cyclotron resonance for $\omega_c/\omega_{LO} < 0.5$.

In the present paper we study cyclotron resonance in the material $CdTe$. This material has a relative small band mass and a low LO-phonon frequency(see Table I) which allows experimentally for the possibility to reach the resonance condition $\omega_c^* \approx \omega_{LO}$ at a magnetic field of $B \sim 17T$ which is within the present day range of attainable steady magnetic fields.

Often[1-3,14-19] the cyclotron resonance mass is obtained from the position of the Landau levels(E_n) in the following way: $m^* = m_b\omega_c/\omega_c^*$ (m_b is the electron band mass and $\omega_c = eB/m_b c$ is the cyclotron resonance frequency for a non-interacting electron) where the cyclotron resonance frequency is determined by $\omega_c^* = (E_1 - E_0)/\hbar = \omega_c - (\Delta E_1 - \Delta E_0)/\hbar$ with ΔE_n the electron-phonon correction to the position of the nth Landau level.

In this paper we will advocate a different approach and calculate the *magneto-optical absorption spectrum* which, in a cyclotron resonance experiment, is the experimental measured quantity.

2. The energy levels of a polaron in a magnetic field

A polaron in a magnetic field is described by the Fröhlich Hamiltonian

$$H = \frac{1}{2m_b}\left(\mathbf{p} + \frac{e}{c}\mathbf{A}\right)^2 + \sum_{\mathbf{k}}\hbar\omega_{LO}(a_{\mathbf{k}}^+ a_{\mathbf{k}} + \frac{1}{2}) + \sum_{\mathbf{k}}(V_k a_{\mathbf{k}}e^{i\mathbf{k}\cdot\mathbf{r}} + V_k^* a_{\mathbf{k}}^+ e^{-i\mathbf{k}\cdot\mathbf{r}}) \quad , \quad (1)$$

where $\mathbf{A}$ is the vector potential, $a_k(a_k^+)$ the annihilation (creation) operator of an LO phonon with momentum $\hbar\mathbf{k}$ and energy $\hbar\omega_{LO}$. The interaction coefficient is

$$V_k = i\frac{\hbar\omega_{LO}}{k}\sqrt{\frac{4\pi\alpha}{V}}\sqrt{\frac{\hbar}{2m_b\omega_{LO}}} \quad ,$$

V is the volume of the crystal and the electron-phonon coupling constant is defined in the following way

$$\alpha = \frac{1}{2}\frac{e^2}{\sqrt{\hbar/2m_b\omega_{LO}}}\left(\frac{1}{\epsilon_\infty} - \frac{1}{\epsilon_0}\right)\frac{1}{\hbar\omega_{LO}} \quad , \quad (2)$$

where ϵ_∞ and ϵ_0 are respectively the high frequency and the static dielectric constant of the material.

Within second-order perturbation theory the energy shift of the nth Landau level due to the electron-phonon interaction is given by

$$\Delta E_n = -\sum_{m=0}^{\infty}\sum_{\mathbf{q}}\frac{|M_{n,m}(\mathbf{q})|^2}{D_{n,m}} \quad , \quad (3)$$

where

$$M_{n,m}(\mathbf{q}) = <m, k_z; \mathbf{q}|H_I|n, 0; \mathbf{0}> \quad , \quad (4)$$

is the matrix element of the electron-phonon interaction H_I. The energy denominator in Eq.(3) is given by

$$D_{n,m} = \hbar\omega_{LO} - \Delta_n + \frac{\hbar^2 q_z^2}{2m_b} + \hbar\omega_c(m - n) \quad , \quad (5)$$

where the choice: (1) $\Delta_n = 0$ leads to Rayleigh-Schrödinger perturbation theory (RSPT) which gives accurate results for ΔE_n if $\omega_c \ll \omega_{LO}$; (2) $\Delta_n = \Delta E_n$ results in Wigner-Brillouin perturbation theory (WBPT) which can account for the *splitting* of the degenerate energy levels; and (3) $\Delta_n = \Delta E_n - \Delta E_n^{RSPT}$ gives an "improved" Wigner-Brillouin theory (IWBPT) as was developed in Ref.15 and further discussed in Ref.16. IWBPT gives the correct *pinning* behavior for small α.

For a parabolic conduction band it is possible to perform the summation over the intermediate Landau levels in Eq.(3) explicitly. In Ref.19 we obtained an expression where only one integral had to be done numerically.

3. The cyclotron resonance absorption spectrum

Theoretically there are two different viewpoints in calculating the cyclotron resonance frequency. First, and the most often used in polaron physics because of its apparent simplicity, relies on a calculation of the position of the Landau levels. This approach was elaborated on in the forgoing section. An alternative approach which we want to advocate is by calculating the *cyclotron resonance absorption spectrum* itself. It is this quantity which is measured experimentally. The position of the cyclotron resonance peak then gives the cyclotron resonance frequency.

Within linear response theory the cyclotron resonance absorption spectrum in the Faraday active mode configuration is proportional to the real part of the conductivity tensor

$$Re\sigma_+(\omega) \sim \frac{Im\Sigma(\omega)}{(\omega - \omega_c - Re\Sigma(\omega))^2 + (Im\Sigma(\omega))^2} \quad , \tag{6}$$

where $\Sigma(\omega)$ is the memory function which is a function of the cyclotron resonance frequency ω_c and of the electron-phonon coupling constant. The memory function[20] is in essence a force-force correlation function where the time evolution is restricted to the Liouville space which is perpendicular to the velocity operator $\dot{x}+i\dot{y}$ (the magnetic field is taken along the z-axis). In order to calculate the memory function we use second-order perturbation theory and i) decouple the electron-phonon system, ii) the phonons are taken at thermal equilibrium and iii) the electron system will be calculated within different approximations. To first order in the electron-phonon coupling constant the memory function has the form

$$\Sigma(\omega) = \frac{1}{\omega} \int_0^\infty dt(1 - e^{i\omega t})ImF(t) \quad , \tag{7}$$

with

$$F(t) = -\frac{1}{\hbar m_b} \sum_{\mathbf{k}} (k_x^2 + k_y^2)|V_k|^2 < [b_{\mathbf{k}}(t), b_{\mathbf{k}}^+(0)] > \quad , \tag{8}$$

where we defined the operator $b_{\mathbf{k}}(t) = a_{\mathbf{k}}(t)e^{i\mathbf{k}\cdot\mathbf{r}(t)}$. Within the above approximation

$$< b_{\mathbf{k}}(t)b_{\mathbf{k}}^+(0) > = (1 + n(\omega_{LO}))e^{-i\omega_{LO}t}S^*(-\mathbf{k}, t) \quad , \tag{9}$$

where $S(\mathbf{k}, t)$ is the space Fourier transform of the electron density-density correlation function.

At low temperatures one has $Im\Sigma(\omega) \simeq 0$ when $\omega < \omega_{LO}$ and consequently the cyclotron resonance frequency is given by the non-linear equation

$$\omega - \omega_c - Re\Sigma(\omega) = 0 \quad , \tag{10}$$

From the solution $\omega = \omega_c^*$ the cyclotron resonance mass is defined as $m^* = m_b\omega_c/\omega_c^*$. The physical meaning of Eq.(10) is apparent: $Re\Sigma(\omega_c^*)$ is the shift in the cyclotron resonance peak due to the electron-phonon interaction. Note also that theoretically the cyclotron mass is always referred to the electron band mass which is not a directly meausurable quantity.

In most bulk polar systems it is possible to apply the one particle approximation in calculating polaron effects. The electron density is typically of the order of $n_e \sim 10^{13}cm^{-3}$ which leads to a Fermi energy $E_F \ll k_bT$ and consequently Boltzmann statistics is valid. This is different from the presently very important GaAs-heterostructures where a one-electron theory does not give the correct quantitative results[21]. In a typical GaAs-heterostructure the electron density is $n_e \sim 10^{11} - 10^{12}cm^{-2}$ which for a typical electron layer thickness of $100\mathring{A}$ results in an effective 3D electron density which is 4-5 orders of magnitude larger than for typical bulk systems. The resulting Fermi energy satisfies $E_F \gg k_bT$ and consequently the

304

electrons will obey Fermi-Dirac statistics. The occupation probabilities of the Landau states is important in such systems and screening of the electron-phonon interaction has been found to influence the results.

Therefore the electron density-density correlation function will be calculated within the *one* electron approximation. In the weak electron-phonon coupling limit Eq.(8) reduces to the following expression

$$F(t) = -\sum_{\mathbf{k}} \frac{(k_x^2 + k_y^2)}{\hbar m_b} |V_k|^2 < e^{i\mathbf{k}\cdot\mathbf{r}(t)} e^{-i\mathbf{k}\cdot\mathbf{r}(0)} > e^{-i\omega_{LO}t} \quad , \tag{11}$$

where for simplicity we have limited ourselves to the zero temperature case. The results for the non-zero temperatue situation is given in Ref.12. The electron density-density correlation function is given by

$$S(\mathbf{k}, t) = < e^{i\mathbf{k}\cdot\mathbf{r}(0)} e^{-i\mathbf{k}\cdot\mathbf{r}(t)} >$$

$$= \exp\left[i\frac{\hbar k_z^2}{2m_b}t - \frac{\hbar(k_x^2 + k_y^2)}{2m_b\omega_c}(1 - e^{-i\omega_c t})\right] \quad . \tag{12}$$

If Eq.(12) is inserted into Eq.(11) one finds for the memory function (7) $\Sigma(\omega) = \Sigma^M(\omega) - \Sigma^M(-\omega)$. The real part of this function is[12]

$$Re\Sigma^M(\omega) = \frac{\omega_c}{2\omega} \sum_{n=0}^{\infty} \left[\frac{1}{\sqrt{1 + n\omega_c}} - \frac{\Theta(1 + n\omega_c - \omega)}{\sqrt{1 + n\omega_c - \omega}}\right]$$

$$- \frac{1}{2\omega} \sum_{n=0}^{\infty} \frac{1}{n!} \left[\sqrt{\omega_c}\left(G_{n+\frac{1}{2}}(\frac{1 + n\omega_c}{\omega_c}) - G_{n+\frac{1}{2}}(\frac{1 + n\omega_c - \omega}{\omega_c})\right)\right.$$

$$\left. - \sqrt{1 + n\omega_c}G_n(\frac{1 + n\omega_c}{\omega_c}) - \sqrt{1 + n\omega_c - \omega}G_n(\frac{1 + n\omega_c - \omega}{\omega_c})\Theta(1 + n\omega_c - \omega)\right], \tag{13}$$

where $\Theta(x) = 1(x > 0), 0(x < 0)$ is the well-known theta function and we defined

$$G_\nu(x) = P \int_0^\infty \frac{t^\nu e^{-t}}{t - x} dt \quad , \tag{14}$$

where $P(\cdot)$ stands for the principal value. The imaginary part is less involved and is given by

$$Im\Sigma(\omega) = -\frac{\omega_c}{2\omega} \sum_{n=0}^{\infty} \frac{\Theta(\omega - (1 + n\omega_c))}{\sqrt{\omega - (1 + n\omega_c)}}$$

$$+ \frac{1}{2\omega} \sum_{n=0}^{\infty} \frac{1}{n!} \sqrt{\omega - (1 + n\omega_c)} E_n(\frac{\omega - (1 + n\omega_c)}{\omega_c})\Theta(\omega - (1 + n\omega_c)) \quad , \tag{15}$$

with $E_n(x) = G_n(-x)$ for $x > 0$. The generalization of the above expressions (13) and (15) to non-zero temperature is given in Ref.12. When finite electron-phonon coupling effects are important the electron density-density correlation function has to be calculated beyond the non-interacting limit. This was done in Ref.12 where $S(k, t)$ was calculated for the generalized Feynman polaron model in a magnetic field. As shown in Ref.22 this leads to significant quantitative corrections to the cyclotron polaron mass when $\alpha > 0.5$.

In Table II the numerical results for the polaron cyclotron mass are given as function of the magnetic field for $\alpha = 0.3$ and for different approximations. We compare the results as found from the IWBPT theory[19] with the above weak coupling results(the third column in Table II as indicated by $v = w$; v and w are the variation parameters which define the Feynman polaron model) as based on a memory function formalism. If we go beyond the

Table II. Polaron cyclotron mass for the lower branch when $\alpha = 0.3$

ω_c/ω_{LO}	IWBPT	Memory function formalism	
		$v = w$	$v \neq w$
0.0	1.0458	1.0505	1.0523
0.1	1.0496	1.0549	1.0573
0.2	1.0543	1.0610	1.0633
0.3	1.0604	1.0680	1.0705
0.4	1.0678	1.0767	1.0794
0.5	1.0771	1.0879	1.0907
0.6	1.0889	1.1023	1.1052
0.7	1.1044	1.1214	1.1243
0.8	1.1253	1.1471	
0.9	1.1541	1.1817	
1.0	1.1933	1.2277	
1.1	1.2450	1.2861	
1.2	1.3093	1.3559	

weak electron-phonon coupling limit, slightly larger (about 0.3% for $\omega_c/\omega_{LO} = 0.5$) cyclotron masses are found as indicated by the last column. Note that the IWBPT results give much lower values for the polaron mass which shows the inadequacy of this type of pertubation calculation for $\alpha = 0.3$. In Ref.22 it was found that when we calculate the memory function within second-order pertubation theory results for the polaron cyclotron mass are obtained which are accurate for values of α larger than 0.1. The results of Table II comfirm that even for $\alpha = 0.3$ such a calculation gives very accurate results. Therefore we will limit ourselves to the $v = w$ results for the material $CdTe$.

The numerical results for the cyclotron mass as defined from the position of the cyclotron resonance peak in the continuum are given in Table III. Results are listed for the IWBPT theory where we have taken the principal value of the right-hand side of Eq.(3). In this case the frequency ω of the radiation is larger than the LO-phonon frequency and an appreciable broadening of the cyclotron resonance peak occurs which implies $Im\Sigma(\omega) \neq 0$. Theoretically it is no longer possible to use the non linear equation (10) in order to determine the cyclotron resonance frequency. The imaginary part of the memory function has to be included. The numerical results of Table III show that this pushes the cyclotron resonance peak to higher frequencies which results in a smaller cyclotron mass. This shift may be as large as 7% for $\omega_c = \omega_{LO}$.

Table III. Polaron cyclotron mass for the upper branch when $\alpha = 0.3$

ω_c/ω_{LO}	IWBPT	Memory function formalism	
		$Im\Sigma = 0$	$Im\Sigma \neq 0$
0.9	0.8233	0.8350	0.7627
1.0	0.8539	0.8629	0.8048
1.1	0.8750	0.8822	0.8365
1.2	0.8903	0.8965	0.8602
1.3	0.9020	0.9074	0.8784
1.4	0.9110	0.9162	0.8917
1.5	0.9183	0.9234	0.9036

4. Comparison with experiment

In order to make a detailed quantitative comparison between the experimental measured cyclotron resonance mass and the theoretical calculated values it is necessary to incorporate the electron band non-parabolicity. The band non-parabolicity will be taken into account via the assumption of a *local parabolic band* approximation. In this approach the Landau levels in the absence of polaron coupling are determined from the two-level Kane theory. In the zero temperature limit the electron momentum along the magnetic field will be zero in the low density limit and we find

$$\epsilon_{np}(n) = \frac{E_g}{2}\left(-1 + \sqrt{1 + \frac{\hbar\omega_c}{E_g}(4n + 2)}\right) \quad , \tag{16}$$

where E_g is the band gap(see Table I for numerical values for some typical polar materials).

Due to the band non-parabolicity the cyclotron resonance frequency will be different from ω_c even in the absence of any polaron effects. The shifted cyclotron resonance frequency is given by $\hbar(\omega_c)_{np} = \epsilon_{np}(n = 1) - \epsilon_{np}(n = 0)$. In order to incorporate polaron effects together with the band non-parabolicity, we insert $(\omega_c)_{np}$ as obtained from above, into the memory function. This results in the non linear equation

$$\omega - (\omega_c)_{np} - Re\Sigma(\alpha, (\omega_c)_{np}, \omega) = 0 \quad , \tag{17}$$

where the solution $\omega = \omega_c^*$ equals the theoretical cyclotron mass which is affected both by band non-parabolicity and by polaron effects.

Note that in the present approximation the correct behaviour for either vanishing electron-phonon coupling, or parabolic energy band limit, is obtained. Also in the limit of zero magnetic field the correct behaviour is obtained because we included *all* the intermediate states. Furthermore the polaron effect and the band non-parabolicity are not considered to be additive.

For $CdTe$ $\hbar\omega_{LO}/E_g \sim 0.013$ and as a consequence band non-parabolicity effects are very weak. This is in contrast with e.g. $InSb$ where $\hbar\omega_{LO}/E_g \sim 0.1$ and band non-parabolicity is of paramount importance. For $CdTe$ the band non-parabolicity correction to the cyclotron resonance frequency can be accurately approximated by $\hbar(\omega_c^*)_{np} = \hbar(\omega_c^*)_p[1 - 2\hbar(\omega_c^*)_p/E_g + ...]$ where $(\omega_c^*)_p$ is the cyclotron resonance frequency after polaron corrections as determined from Eq.(10) with a parabolic conduction band.

In Fig. 1 our theoretical results are compared with the available experimental results for $CdTe$. The early results from Mears *et al*[7] and from Waldman *et al*[8] are shown together with the recent data from Oberti *et al*[23]. A very close agreement is found for a coupling constant of $\alpha = 0.3$ and a band mass of $m_b/m_e = 0.0920$.

In earlier theoretical calculations of the polaron cyclotron resonance mass by Larsen[14] it turned out to be necessary to take $\alpha = 0.4$ in order to fit the experimental results of Waldman *et al*[8]. The reason is that in Ref.8 the polaron effects are underestimated and as a consequence one needed a higher α in order to compensate for this[24].

Note that the analysis of Ref.24 and the present study resolves the inconsistency between the value of α as found from the definition (Eq.(2)) and from a comparison between the experimental cyclotron data and the theoretical calculation. From the definition of α we find $\alpha = 0.28 \pm 0.06$ if we take $m_b/m_e = 0.0920$ and assume the following values for the dielectric constants[8,25] $\epsilon_0 = 9.6 \pm 0.2$ and $\epsilon_\infty = 7.13$. The results quoted in the Landolt-Börnstein tables[26] are in the range $\epsilon_0 = 10.6 - 9.4$ and $\epsilon_\infty = 7.4 - 7.1$ which results in a coupling constant $\alpha = 0.22 - 0.35$. Thus even in the most favorable case one is not able to get a

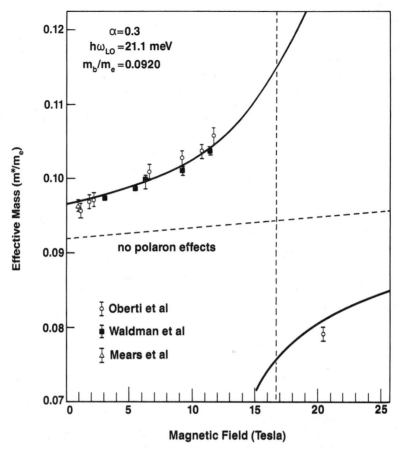

Fig. 1. The polaron cyclotron resonace mass as function of the magnetic field for $CdTe$. Different experimental data are compared with our present theoretical result(solid curve) which is based on a memory function formalism. The result without polaron effects is shown by the dashed line. The vertical dashed line indicates where the cyclotron resonance frequency is equal to the LO-phonon frequency.

coupling constant as high as $\alpha = 0.4$. Note that a similar situation was encountered in the silver halides[11] $AgBr$ and $AgCl$ where a theory was needed which went beyond second-order perturbation theory and which calculated the full cyclotron resonance spectrum. This theory was able to provide a value for α consistent with the value as calculated based upon the definition of α.

In conclusion we were able to find a consistent description of the polaron contribution to the cyclotron mass in $CdTe$ within the magnetic field range $B : 0 \rightarrow 20.5T$. The recent experimental data by Oberti et al could be explained by our theory.

Acknowledgements

This work was sponsored by F.K.F.O.(Fonds voor Kollektief Fundamenteel Onderzoek, Belgium), project No. 2.0072.80.

References

* Also at R.U.C.A., B-2020 Antwerp and University of Technology, NL-5600 MB Eindhoven.

0 Senior Research Associate of the National Fund for Scientific Research (Belgium).

1 *Polarons and Excitons*, edited by C.G. Kuper and G.D. Whitfield (Oliver and Boyd, Edinburgh, 1963).

2 *Polarons in Ionic Crystals and Polar Semiconductors*, edited by J.T. Devreese (North-Holland, Amsterdam, 1972).

3 *Polarons and Excitons in Polar Semiconductors*, edited by J.T. Devreese and F.M. Peeters (Plenum, New York, 1984).

4 R.P. Feynman, Phys. Rev. **97**, 660 (1955).

5 D.H. Dickey, E.J. Johnson and D.M. Larsen, Phys. Rev. Lett. **18**, 599 (1967).

6 C.J. Summers, P.G. Harper and S.D. Smith, Solid Stat. Commun. **5**, 615 (1967).

7 A.L. Mears and R.A. Stradling, Solid Stat. Commun. **7**, 1267 (1969).

8 J. Waldman, D.M. Larsen, P.E. Tannenwald, C.C. Bradley, D.R. Cohn and B. Lax, Phys. Rev. Lett. **23**, 1033 (1969).

9 J.W. Hodby, J. Phys. **C4**, L8 (1971).

10 J.W. Hodby, J.G. Crowder and C.C. Bradley, J. Phys. **C7**, 3033 (1974).

11 J.W. Hodby, G.P. Russell, F.M. Peeters, J.T. Devreese and D.M. Larsen, Phys. Rev. Lett. **58**, 1471 (1987).

12 F.M. Peeters and J.T. Devreese, Phys. Rev. **B34**, 7246 (1986).

13 F.M. Peeters and J.T. Devreese, Phys. Rev. **B25**, 7281 (1982).

14 D.M. Larsen, Phys. Rev. **142**, 428 (1966).

15 D.M. Larsen and E.J. Johnson, J. Phys. Soc. Jpn., Suppl.. **21**, 443 (1966).

16 G. Lindemann, R. Lassnig, W. Seidenbusch and E. Gornik, Phys. Rev. **B28**, 4693 (1983).

17 D.M. Larsen, in *Polarons in Ionic Crystals and Polar Semiconductors*, Ref. 2, p.237

18 D.M. Larsen, J. Phys. **C7**, 2877 (1974).

19 F.M. Peeters and J.T. Devreese, Phys. Rev. **B31**, 3689 (1985).

20 F.M. Peeters and J.T. Devreese, Phys. Rev. **B28**, 6051 (1983).

21 M.A. Hopkins, R.J. Nicholas, M.A. Brummell, J.J. Harris and C.T. Foxon, Phys. Rev. **B36**, 4789 (1987); F.M. Peeters, X. Wu and J.T. Devreese, Solid Stat. Commun. **65**, 1505 (1988); C.J.G.M. Langerak, J. Singleton, P.J. van der Wel, J.A.A.J. Perenboom, D.J. Barnes, R.J. Nicholas, M.A. Hopkins and C.T.B. Foxon (to be published).

22 J.T. Devreese and F.M. Peeters, Solid Stat. Commun. **58**, 861 (1986).

23 Oberti *et al*, (to be published).

24 F.M. Peeters and J.T. Devreese, Physica **127B**, 408 (1984).

25 A. Manabe, A. Mitsuishi, H. Yoshinaga, Jpn. J. Appl. Phys. **6**, 593 (1967).

26 *Landolt-Börnstein tables*, editor O. Madelung (Springer-Verlag, Berlin, 1982), Vol. **17b**, p. 225.

OPTICALLY-DETECTED MAGNETIC RESONANCE OF II-VI COMPOUNDS

J. John Davies

Department of Applied Physics
University of Hull
Hull HU6 7RX, U.K.

INTRODUCTION

The first ODMR experiments on II-VI compounds were reported in 1975[1,2] and since that time the technique has been used with considerable success to investigate recombination processes in these and other semiconductors, especially those with the wider bandgaps. Because of the much greater spectral resolution that ODMR provides compared with standard optical spectroscopy, extremely detailed identification of recombination centres is possible, and in this respect studies of II-VI compounds have been particularly successful. Reviews of the early investigations on these materials were given by Cox[3] and by Cavenett[4] and accounts of work up to 1985[5] and up to 1987[6] have been given by the present author. The purpose of the present article is to summarize the type of information that ODMR provides in the context of II-VI materials, to outline both the advantages and the limitations of the technique and finally to discuss the possibilities of extending the investigations to multiple quantum well (MQW) and superlattice structures.

GENERAL PRINCIPLES AND EXPERIMENTAL CONSIDERATIONS

In this article we shall concentrate on ODMR experiments in which the magnetic resonance transitions are detected by monitoring the photo-luminescence (rather than the absorption). The technique relies on the existence of spin selection rules in the luminescence process and is essentially a method of detecting magnetic resonance transitions that occur in the excited state of the system. In this sense, ODMR is complementary to conventional electron spin resonance, which provides information about the ground (or some metastable) state.

The general principle is illustrated in Fig.1. The excited state is usually populated by laser radiation and either the intensity or the polarization of the resulting luminescence is monitored. Because of the spin selection rules, the luminescence properties are sensitive to the distribution of population amongst the different spin levels of the excited state. Magnetic resonance transitions occurring in the excited state alter this spin distribution and hence affect the luminescence: by monitoring the appropriate emission property (intensity or polarization) it is therefore possible to detect the occurrence of these transitions.

FIG.1.

Schematic Diagram
to illustrate the
principle of ODMR

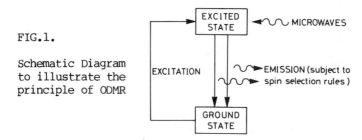

Several different experimental arrangements for ODMR have been described[2,3,7,8], and a typical layout is given in fig.2.   In order to lengthen the spin-lattice relaxation times it is usual to arrange for the specimen to be immersed in superfluid helium (which is allowed to enter the resonator cavity).   The incident microwaves (typically 9 GHz, 23 GHz or 35 GHz) are chopped at a frequency f usually in the range 30 Hz to 50 Hz and changes in the emission that occur at this frequency are detected with a lock-in system.   These changes are monitored as the magnetic field is swept slowly and constitute the ODMR spectrum.   The frequency f is an important parameter since for the greatest ODMR signal it should be of the order of the decay rate of the excited state.   The dynamical response of the luminescence to the application of the microwaves at resonance is governed by a combination of optical decay times, spin-lattice relaxation times and microwave power level and can be complicated[9].

FIG.2a.

Experimental arrangement
for ODMR.   The magnetic
field is not shown.

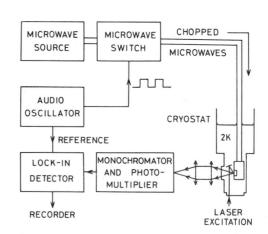

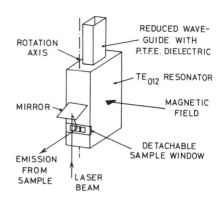

FIG.2b.

Detail of the resonator
cavity of Fig.2a.
Several designs of cavity
have been described[2,3,7,8].

Lasers are usually used for excitation since they enable high power
densities at the specimen and since they are reasonably stable. The
directions of excitation and of observation vary according as to whether
optical pumping is required and whether σ_+, σ_- or π polarized emission
is to be monitored. Since the ODMR spectra are often anisotropic with
respect to the direction of the magnetic field, a mechanism to alter the
specimen orientation is essential.

In order to saturate the magnetic resonance transitions in a time
shorter than the lifetime of the excited state, the microwave source
should usually be of high power and devices with outputs in excess of 1W
are often used. If the magnetic resonance signals are inhomogeneously
broadened, more efficient saturation (and hence enhancement of the ODMR
signal) can be achieved by superimposing a rapid field modulation[10]. It
is also possible to use ODMR in a time-resolved mode[11,12], in which the
laser is pulsed and the luminescence sampled after a specified delay time;
during alternate cycles, a microwave pulse (up to several watts, e.g. from
a travelling-wave tube) is applied and the effect on the gated
luminescence signal is monitored.

THE EXCITED STATE

For II-VI compounds there are two archetypal recombination processes
that have been studied by ODMR: two-centre recombination (e.g. donor/
acceptor, donor/deep centre) and excitonic recombination. Since the
nature of the ODMR spectrum differs according to the type of recombination
process, it is convenient to discuss each process in turn.

Two-centre recombination processes

Such processes are extremely common in II-VI compounds and lead to
strong photoluminescence bands that are often very wide and which cannot
therefore be studied by conventional Zeeman spectroscopy. To illustrate
how ODMR can be applied we consider recombination between a donor and an
acceptor, each imagined to have an effective spin of 1/2 and sufficiently-
well separated for spin-exchange interactions (see later) to be
negligible. The combined spins lead to the four energy levels of the
excited state (i.e. the state before recombination) shown in the upper
part of fig.3. After recombination, a singlet ground state is formed and
we assume that the recombination selection rules are such that spin is

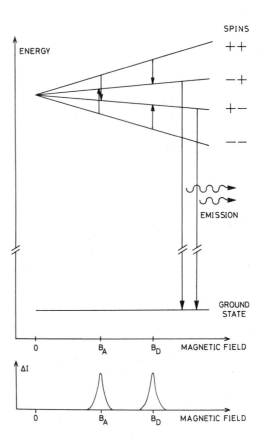

FIG.3.

Energy levels for a pair of
recombining centres (e.g. a
donor and an acceptor).

conserved, the allowed radiative transitions being as indicated. At low
temperatures, spin-lattice relaxation rates are slow and (if all four
excited levels are populated by the laser at equal rates) the steady-state
populations of the levels |++> and |--> will exceed those of |+-> and
|-+>. When the magnetic field is such as to satisfy the magnetic
resonance condition for either the donors or the acceptors, microwave-
induced transitions occur which transfer the excess population to the
strongly-emitting levels; in consequence, the emission intensity
increases, giving rise to the ODMR signals (lower part of fig.3).

An example is shown in figure 4. Here the ODMR spectrum was
obtained by monitoring the intensity of the shallow donor/shallow acceptor
green "edge" emission from CdS[13]. The presence of signals from shallow
donors and from shallow acceptors confirms that the emission is due to
recombination between these centres. In addition, the high resolution of
the technique reveals the presence of two different types of acceptor
centres (due to lithium and sodium). As expected for shallow centres,
the ODMR signals in this case have axial symmetry about the direction of
the crystal c-axis.

In general, the observation of anisotropy in the ODMR spectra can be
very useful, since it enables the local symmetries of the centres to
be established. This has proved particularly important for

314

FIG.4.

ODMR spectrum for
donor–acceptor
recombination in
CdS (23 GHz).

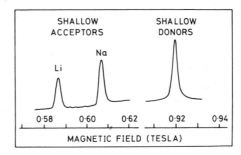

FIG.5. Dependence of deep-centre ODMR line positions on
magnetic field directions obtained by monitoring
the 530 nm green emission in ZnSe that contains
copper.

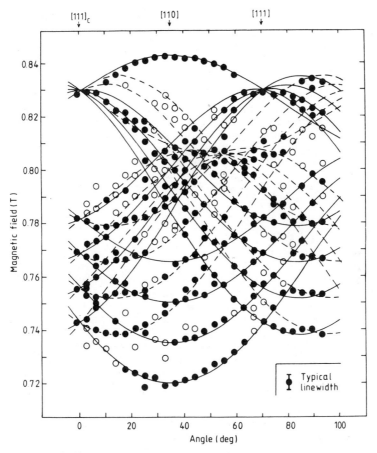

315

recombination involving strongly-localized deep centres, an example
being shown in fig.5, which concerns the angular variation of the
deep-centre ODMR spectra[14] obtained by monitoring the green emission
from ZnSe containing copper.   As in conventional ESR, the observation
of hyperfine structure (in this case, a four-line pattern
corresponding to the nuclear spin of 3/2 for copper) points to the
identity of the participating element, whilst the overall behaviour
indicates a centre whose local symmetry is orthorhombic.   A full
analysis[14] leads to identification of the 530 nm band as due to
recombination between shallow donors and deep centres that consist of
a substitutional copper/interstitial copper pair.   A large number of
such optically-active deep centres in II-VI compounds have now been
identified in this way (see e.g. references in [5] and [6])

So far, we have considered recombining centres whose spatial
separation is large.   When the separation is reduced, spin exchange
interactions become important and marked changes occur in the ODMR
spectra.   Furthermore, because of the Coulomb and other interactions
between the centres, the emission energy also depends on the separation.
Hence, by monitoring the appropriate emission energy it becomes possible
selectively to study recombining pairs of given separation and thus to
measure the spin exchange energy as a function of distance[15,16].   Such
measurements provide useful information on the nature of the wavefunctions
of the recombining centres.

When the intrapair separation becomes sufficiently small, the
exchange energy J will be such that $J \gg (g_1-g_2) \beta B$, where g_1 and g_2 are
the g-values of the two centres, β is the Bohr magneton and B the magnetic
field.   Under these circumstances, two centres of spin 1/2 will form
singlet and triplet combined spin states separated in energy by J.   In
the triplet state the g-value is the mean of g_1 and g_2[17].   Furthermore,
effects such as anisotropic exchange will split the triplet levels even in
zero field.   If the triplet lies beneath the singlet, recombination will
occur from the triplet levels and "triplet state ODMR" may be observable,
as in the following section.

Triplet state recombination

ODMR from triplet states in II-VI compounds has been observed from
strongly-coupled pairs in two-centre recombination processes (see above)
and from deeply-bound two-particle (i.e. one electron and one hole)
excitons.   In figure 6 we show the behaviour of triplet levels in a
magnetic field directed along the symmetry axis of a system that has axial
symmetry.   If the allowed optical transitions are as indicated, magnetic
resonance at either B_1 or B_2 causes an increase in the emission intensity.
More commonly, the emission from the $M_s = \pm 1$ levels is of different
polarization from that from $M_s = 0$: in this case the magnetic resonance
transitions are manifested as changes in polarization of the emitted
light.   Triplet state ODMR spectra are characterized by two signals
separated in field by an amount $\Delta B = 2D/g B$, where D is the "zero field
splitting parameter" (see fig.6).   Striking examples in II-VI compounds
are the zinc vacancy/zinc interstitial Frenkel pairs observed in ZnSe by
Rong and Watkins[18] and the excitons deeply bound at Te_2 isoelectronic
centres in $CdS_{0.98}Te_{0.02}$ studied by Hilton et al[19] (fig.7).

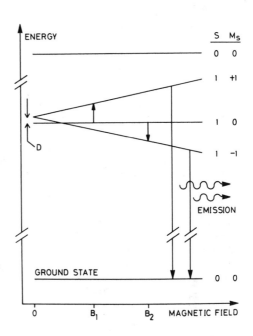

FIG.6.

Energy levels for
triple state recombination

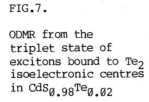

FIG.7.

ODMR from the
triplet state of
excitons bound to Te_2
isoelectronic centres
in $CdS_{0.98}Te_{0.02}$

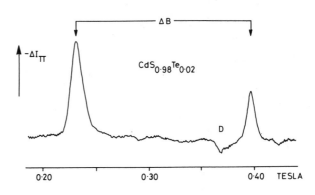

SPECTRAL RESOLUTION, SENSITIVITY AND LIMITATIONS

ODMR linewidths in II-VI compounds are typically of order 5 mT which, for g = 2, corresponds to 6×10^{-7} eV. The width of the optical emission bands is typically in the range 0.01 eV to 0.2 eV. Thus huge improvements in resolution, approaching a factor of one million, are possible. It is this that forms the major advantage of the technique.

In discussing sensitivity we note that the fractional changes in emission intensity when resonance occurs are typically of order 0.1% (though exceptionally this can rise to several per cent). If the noise level is set by statistical variations in the detected photon rate (shot noise) one needs to detect 10^6 photons in the integrating time of the detection system, which we take to be 1s. Allowing for the solid angle subtended by the collection optics ($\sim 10^{-1}$ sr) and for a detection efficiency of 10%, we require the specimen to emit 10^9 photons per second. In the steady state, this would require the number n of emitting centres that are excited to be such that $n\tau^{-1} > 10^9$ s^{-1}, where τ is the decay constant of the excited state. We must also have $n\tau^{-1} = \alpha \xi N$, where N is the number of incident laser photons per second, α is the fraction absorbed and ξ is the quantum efficiency of the luminescence process. At first sight it is possible by increasing the laser power to increase the emission intensity to the required level; in practice, this may not be possible because of specimen heating and also because an excessively high excited carrier concentration can result in a breakdown of the spin selection rules[20]. For these reasons a general calculation of the sensitivity of the technique is not possible. The essential condition is that $n\tau^{-1} > 10^9$ s^{-1}.

An important restriction on the validity of the technique is set by the lifetime of the excited state. In order to saturate the microwave transitions before recombination takes place, one must have $g^2\beta^2 B_\mu^2 T_2 > \tau^{-1}$, where B_μ is the microwave magnetic field and T_2 is the spin-spin relaxation time. In ODMR experiments, B_μ is typically of order 10^{-3} – 10^{-4}T and for II-VI compounds T_2 is of order 10^{-7}s. It therefore becomes difficult to observe microwave-induced changes in emission if the recombination times are less than about 10^{-8}s. It should also be noted that for very short recombination times, lifetime broadening will become important and will reduce both the sensitivity and resolution of the experiment; if τ is less than 10^{-9}s, the broadening (for g = 2) will be in excess of 10mT.

We note finally that care must be taken when analysing ODMR signals, particularly if more than one optically-active species is present in the sample. Competitive recombination processes[21,22], electron transfer processes[23] and energy transfer processes[24,25] (both radiative and non-radiative) can cause major changes in the ODMR spectra, which must then be interpreted with caution.

APPLICATION TO MQWs AND SUPERLATTICES

One of the distinguishing features of MQW structures is the strong photoluminescence due to excitonic recombination, and the effects of quantum confinement on such spectra can be extremely marked. From the previous section we see that for ODMR to be successful requires that the recombination time be longer than about 10^{-8}s. For excitonic recombination in type I structures this appears unlikely. However, for type II MQWs (possible examples are ZnSe/ZnTe and CdS/CdSe) the lifetimes could well be such that excitonic recombination is accessible for ODMR study. Indeed, such experiments have recently been reported for the III-V type II MQW system formed from GaAs/AlAs, for which van Kesteren et al[26] have studied the heavy hole excitons. The recombination time is of order 1 μs and the ODMR signals were observed as changes in the polarization of the emission. The results confirm the energy level scheme attributed to the exciton and enable the exchange splitting and the electron and hole g-values to be determined.

One of the motivations for ODMR studies in MQWs is to obtain further information on the effects of quantum confinement. By monitoring the destruction of spin memory in the optical pumping/recombination emission cycle in a lattice-matched GaInAs/InP single quantum well, Cavenett and coworkers[27,28] have investigated the magnetic resonance of confined electrons; in this particular example the electron g-value appears unaffected by confinement, but this result may not be general and similar experiments on II-VI structures would be of great interest. In suitably-doped layers, quantum confinement effects would be expected on two-centre recombination processes in which one (or both) of the centres is shallow and therefore not too strongly localized.

Even in multiple-layer structures where the thicknesses are too large for noticeable quantum confinement effects, there is still considerable interest because of the strain that is present and which may make possible the observation of magnetic resonance signals from hole states at shallow acceptors or in the valence band. For cubic, zincblende semiconductors such signals are expected[29] to be too wide for observation unless the degeneracy of the valence band is removed by strain, either applied externally or present because of the adoption of the hexagonal, wurtzite form (e.g. CdS). The degeneracy will therefore be removed in strained-layer superlattices and suitably doped ZnS/ZnSe structures are of interest in this respect.

CONCLUSIONS

ODMR is well-established as a technique for investigating recombination processes in II-VI compounds, both in bulk crystalline and epitaxial form. The method has a much higher resolution than optical spectroscopy and under favourable conditions has high sensitivity. The main restriction to its applicability is the need for the recombination times to be in excess of 10^{-8}s. Whilst studies of excitonic recombination in type I superlattices are therefore unlikely to be successful, it may well be possible to investigate such processes in type II structures. For suitably-doped superlattices of both types, ODMR studies of two centre recombination processes would appear to be feasible and the presence of strain in these structures may make possible the observation of signals not obtainable in bulk material.

ACKNOWLEDGEMENTS

I should like to thank Dr. J.E. Nicholls for stimulating discussions and rewarding collaboration and the SERC for support of the work carried out at Hull.

REFERENCES

1. J.R. James, J.E. Nicholls, B.C. Cavenett, J.J. Davies and D.J. Dunstan, Solid State Commun., 17:969 (1975).
2. J.E. Nicholls, J.J. Davies, B.C. Cavenett, J.R. James and D.J. Dunstan, J. Phys. C (Solid State Phys.) 12:361 (1979)
3. R.T. Cox, Rev. Physique Appl. 15:653 (1980).
4. B.C. Cavenett, Advan. Phys. 30:475 (1981).
5. J.J. Davies, J. Crystal Growth 72:317 (1985).
6. J.J. Davies, J. Crystal Growth 86:599 (1988).
7. D. Verity, J.J. Davies, J.E. Nicholls and F.J. Bryant, J. Appl. Phys. 52:737 (1981).
8. K.M. Lee, Rev. Sci. Instr. 53:702 (1982).
9. D.J. Dunstan and J.J. Davies, J. Phys. C (Solid State Phys.) 12:2927 (1979).
10. J.J. Davies, J. Phys. C (Solid State Phys.) 11:1907 (1978).
11. D. Block and R.T. Cox, J. Luminescence 24/25:167 (1981).
12. D. Block, A. Hervé and R.T. Cox, Phys. Rev. B 25:6049 (1982).
13. J.L. Patel, J.E. Nicholls and J.J. Davies, J. Phys. C (Solid State Phys.) 14:1339 (1981).
14. J.L. Patel, J.E. Nicholls and J.J. Davies, J. Phys. C (Solid State Phys.) 14:5545 (1981).
15. R.T. Cox and J.J. Davies, Phys. Rev. B. 34:8591 (1986).
16. F. Rong, W.A. Batty, J.F. Donegan and G.D. Watkins, Phys. Rev. B 37:4329 (1988).
17. J.J. Davies and J.E. Nicholls, J. Phys. C (Solid State Phys.) 15:5321 (1982).
18. F. Rong and G.D. Watkins, Phys. Rev. Letters 56:2310 (1986).
19. C.P. Hilton, J.E. Nicholls, J.J. Davies and O. Goede, to be published.
20. D. Verity, J.E. Nicholls, J.J. Davies and F.J. Bryant, J. Phys. C. (Solid State Phys.) 13:L87 (1980).
21. D. Verity, J.J. Davies and J.E. Nicholls, J. Phys. C. (Solid State Phys.) 14:485 (1981).
22. J.L. Patel, J.J. Davies, J.E. Nicholls and B. Lunn, J. Phys. C (Solid State Phys.) 14:4717 (1981).
23. K.P. O'Donnell, K.M. Lee and G.D. Watkins, J. Phys. C (Solid State Phys.) 16:L723 (1983).
24. J.J. Davies, J. Phys. C. (Solid State Phys.) 16:L867 (1983).
25. M. Godlewski, W.M. Chen and B. Monemar, Phys. Rev. B 37:2570 (1988).
26. H.W. van Kesteren, E.C. Kosman, F.J.A.M. Greidanus, P. Dawson, K.J. Moore and C.T. Foxon, Phys. Rev. B, in press (1988).
27. B.C. Cavenett, G.R. Johnson, A. Kana'ah, M.S. Skolnick and S.J. Bass, Superlattices and Microstructures 2:323 (1986).
28. G.R. Johnson, A. Kana'ah, B.C. Cavenett, M.S. Skolnick and S.J. Bass, Semicond. Sci. Technol. 2:182 (1987).
29. G. Lancaster, "Electron Spin Resonance in Semiconductors", Hilger and Watts, London, p.49 (1966).

ACOUSTOELECTRIC CHARACTERIZATION OF WIDE GAP II-VI SEMICONDUCTORS IN THE

CASE OF BIPOLAR PHOTOCONDUCTION [*]

Josef Rosenzweig

Institut für angewandte Physik der Universität Karlsruhe
Kaiserstrasse 12, D-7500 Karlsruhe, Fed. Rep. of Germany[**]

INTRODUCTION

The purpose of the present article is to demonstrate that acoustoelectric methods may be used to characterize piezoelectric semiconductors
in the case of bipolar photoconduction. The wide gap II-VI semiconductor
CdTe is used as an example for this characterization method.

Measurements of the electric conductivity and of the Hall effect are
not sufficient to determine the densities and mobilities of holes and electrons simultaneously present in a bipolar photoconductor. Therefore,
measurements with other methods should be performed to extract further
relations for the analysis. It is shown that the acoustoelectric effect
may be used as an independent method.

The application of acoustoelectric measurements for determination of
densities and mobilities of two types of charge carriers at bipolar photoconduction was first proposed by Kietis et al.[1-3]. The analysis of these
authors was restriced only to the case in which traps for charge carriers
do not affect the acoustoelectric current. Our previous measurements on
photoconducting CdTe at the ultrasonic frequency of 23 MHz, however, showed
that the influence of traps on the acoustoelectric phenomena cannot be
ignored[4]. Therefore it is necessary to discuss the general case of acoustic
wave propagation in a semiconductor with two types of carriers in the
presence of traps. In the present article this problem is analyzed and the
theoretical results are applied to the explanation of experiments on CdTe.
The detailed analysis is presented in ref.5, the main results were published
in ref.6.

THEORY

If a piezoelectrically active sound wave propagates in a semiconductor,
a wave of longitudinal electric field is generated. This wave generates a
wave of space charge and, therefore, drags charge carriers in the direction
of wave propagation. As a result, a DC current, the acoustoelectric current,

---

[*] Based upon the Habilitation Thesis, University of Karlsruhe (1987).
[**] Present address: Fraunhofer-Institut für Angewandte Festkörperphysik
Eckerstrasse 4, D-7800 Freiburg, Fed. Rep. of Germany

appears in this direction[7,8]. In the case of bipolar conduction electrons and holes travel under the influence of the sound wave. Therefore, the sign of the acoustoelectric current depends on the values of the mobilities and densities of charge carriers. Similarly to the Ohmic Hall effect, it can be imagined in a simplified picture that the acoustoelectric current will disappear if the partial acoustoelectric currents of electrons and holes are equal. In the case of a semiconductor with traps this partial currents may be altered and therefore the total acoustoelectric current may also be changed. In the following general relations for the acousto-electric current in a bipolar semiconductor containing traps will be derived.

The total density of space charge, generated by the acoustic wave, may be divided between free carriers and carriers localized in traps. The trapping can be described by complex trapping factors f_n and f_p (ref.8). These trapping factors are the ratios of the free bunched carrier densities $\tilde{n}$ and $\tilde{p}$, to the total bunched carrier densities $\tilde{n}_s$ and $\tilde{p}_s$ in the space charge wave:

$$\tilde{n} = f_n \tilde{n}_s, \quad \tilde{p} = f_p \tilde{p}_s. \tag{1}$$

The trapping factors can be split into real and imaginary parts

$$f_n = f_{nr} + if_{ni}, \quad f_p = f_{pr} + if_{pi}. \tag{2}$$

In an one dimensional case in the presence of an applied longitudinal electric field E, the current densities of electrons and holes, J_n and J_p respectively, are:

$$J_n = e\mu_n(n + f_n\tilde{n}_s)(E + \tilde{E}) + \mu_n k_B T \, \partial(f_n\tilde{n}_s)/\partial x \,, \tag{3}$$

$$J_p = e\mu_p(p + f_p\tilde{p}_s)(E + \tilde{E}) - \mu_p k_B T \, \partial(f_p\tilde{p}_s)/\partial x \,. \tag{4}$$

Here, e is the elementary charge, μ_n and μ_p are the mobilities of electrons and holes; n and p are the electron and hole densities without the space charge perturbation generated by the acoustic wave; k_B is the Boltzmann constant and T the temperature, $\tilde{E}$ is the wave of longitudinal electric field generated by the piezoelectrically active sound wave.

In the following it is assumed that the recombination times of carriers $\tau_{rec,n}$ and $\tau_{rec,p}$ exceed the period of the sound wave $2\pi/\omega$. In this case the continuity equations for the electron and hole current densities must be satisfied separately:

$$\partial J_n/\partial x = -\partial(-e\tilde{n}_s)/\partial t, \tag{5}$$

$$\partial J_p/\partial x = -\partial(e\tilde{p}_s)/\partial t \,. \tag{6}$$

The Poisson equation for the electric displacement $\tilde{D}$ in the wave is

$$\partial\tilde{D}/\partial x = e(\tilde{p}_s - \tilde{n}_s) \,. \tag{7}$$

Assuming now all alternating quantities to be proportional to $\exp[i(\omega t - kx)]$ and omitting the nonlinear products $\tilde{n}_s\tilde{E}$ and $\tilde{p}_s\tilde{E}$ for small sound intensities, the differential equations may be easily transformed to a system of linear algebraic equations. From this system of equations we obtain the following formulae for the total densities of bunched carriers $\tilde{n}_s$, $\tilde{p}_s$ and the electric displacement $\tilde{D}$ as functions of $\tilde{E}$:

$$\tilde{n}_s = - \frac{\varepsilon\varepsilon_0 k \, \omega_{cn}/\omega}{e(\gamma_n - if_n\omega/\omega_{Dn})} \tilde{E}, \tag{8}$$

$$\tilde{P}_s = \frac{\varepsilon\varepsilon_0 k\omega_{cp}/\omega}{e(\gamma_p - if_p\omega/\omega_{Dp})}\,\tilde{E}, \tag{9}$$

$$\tilde{D} = i\varepsilon\varepsilon_0\left(\frac{\omega_{cn}/\omega}{\gamma_n - if_n\omega/\omega_{Dn}} + \frac{\omega_{cp}/\omega}{\gamma_p - if_p\omega/\omega_{Dp}}\right)\tilde{E}. \tag{10}$$

Here, ε is the dielectric constant of the semiconductor at sound frequency ω, ε_0 is the permittivity of vacuum,

$$\omega_{cn} = en\mu_n/(\varepsilon\varepsilon_0), \qquad \omega_{cp} = ep\mu_p/(\varepsilon\varepsilon_0) \tag{11}$$

are the "dielectric relaxation frequencies" of charge carriers,

$$\omega_{Dn} = (\omega/k)^2 e/(\mu_n k_B T), \qquad \omega_{Dp} = (\omega/k)^2 e/(\mu_p k_B T) \tag{12}$$

are the "diffusion frequencies",

$$\gamma_n = 1 + (k/\omega)f_n\mu_n E = 1 - f_n v_n/(\omega/k),$$
$$\gamma_p = 1 - (k/\omega)f_p\mu_p E = 1 - f_p v_p/(\omega/k), \tag{13}$$

where

$$v_n = -\mu_n E, \qquad v_p = \mu_p E \tag{14}$$

are the drift velocities of electrons and holes.
Formula (10) extends the treatment of Fink and Quentin (ref.9) to the case of carrier trapping.

Eq.(10) has to be solved together with the equations of state for a piezoelectric material

$$\tilde{D} = \varepsilon\varepsilon_0\tilde{E} + \beta\tilde{S}, \tag{15}$$

$$\tilde{T} = C\tilde{S} - \beta\tilde{E} \tag{16}$$

and the equation of motion for the continuum

$$\rho_m\,\partial^2\tilde{u}/\partial t^2 = \partial\tilde{T}/\partial x \tag{17}$$

to get the dispersion relation. Here, $\tilde{u}$ is the displacement, $\tilde{S} = \partial\tilde{u}/\partial x$ the strain and $\tilde{T}$ the stress in the acoustic wave, C the elastic stiffness, β the piezoelectric constant and ρ_m the mass density.

Substitution of $\tilde{D}$ from eq. (10) in eq.(15) yields

$$\tilde{E} = -\beta H\,\tilde{S}/\varepsilon\varepsilon_0, \tag{18}$$

where

$$H = \left[1 - \frac{i\omega_{cn}/\omega}{\gamma_n - if_n\omega/\omega_{Dn}} - \frac{i\omega_{cp}/\omega}{\gamma_p - if_p\omega/\omega_{Dp}}\right]^{-1}. \tag{19}$$

Substitution of $\tilde{E}$ from eq.(18) in eq.(16) yields

$$\tilde{T} = C(1 + K^2 H)\tilde{S} \equiv C'\tilde{S} \tag{20}$$

323

with
$$K^2 = \beta^2/(C\varepsilon\varepsilon_0) \quad \text{and} \quad C' = C(1 + K^2 H), \tag{21}$$

where K is the electromechanical coupling constant.

Eqs. (17) and (20) yield the wave equation

$$\partial^2\tilde{u}/\partial t^2 = (C'/\rho_m)\partial^2\tilde{u}/\partial x^2, \tag{22}$$

and with the wave $\exp[i(\omega t - kx)]$ the following dispersion relation:

$$\omega^2 = (C'/\rho_m)k^2. \tag{23}$$

For a real frequency ω we have a complex wave vector k

$$k(\omega) = \omega(\rho_m/C')^{1/2} = k_r - i\alpha/2 \tag{24}$$

with the absorption coefficient

$$\alpha = -2\omega\rho_m^{1/2} \, \text{Im}(C'^{-1/2}). \tag{25}$$

In all piezoelectric materials $K^2 \ll 1$, therefore

$$\alpha \approx (\omega/v_{so})K^2 \, \text{Im}(H) = k_r K^2 \text{Im}(H), \tag{26}$$

where $v_{so} = (C/\rho_m)^{1/2}$ is the sound velocity without the piezoelectric effect. For $K^2 \ll 1$ the inequality $\alpha \ll k_r$ is valid and therefore

$$k \approx k_r = \omega/v_s \approx \omega/v_{so} = k_o. \tag{27}$$

Eq. (26) together with eq. (19) yields the absorption coefficient in the general case of bipolarity and trapping:

$$\alpha = K^2\frac{\omega}{v_s}\frac{P_1(N_1^2+N_2^2)\omega_{cp}/\omega + N_1(P_1^2+P_2^2)\omega_{cn}/\omega}{(N_1P_1-N_2P_2-P_2\omega_{cn}/\omega - N_2\omega_{cp}/\omega)^2 + \left[(P_2+\omega_{cp}/\omega)N_1+(N_2+\omega_{cn}/\omega)P_1\right]^2}. \tag{28}$$

The real functions N_1, N_2, P_1 and P_2 are given by

$$N_1 = 1 - f_{nr}v_n/v_s + f_{ni}\,\omega/\omega_{Dn},$$

$$N_2 = f_{ni}\,v_n/v_s + f_{nr}\,\omega/\omega_{Dn},$$

$$P_1 = 1 - f_{pr}\,v_p/v_s + f_{pi}\omega/\omega_{Dp}, \tag{29}$$

$$P_2 = f_{pi}\,v_p/v_s + f_{pr}\,\omega/\omega_{Dp}.$$

The acoustoelectric current density J_{AE} can be calculated from the following equation including the trapping of electrons and holes:

$$J_{AE} = \frac{1}{2}\text{Re}\left[(e\mu_n f_n\tilde{n}_s + e\mu_p f_p\tilde{p}_s)\,\tilde{E}^*\right]. \tag{30}$$

This equation extends the result obtained by Greebe[8] for one type of carriers to the case of bipolarity. Only the mobile parts of the space charge waves $f_n\tilde{n}_s$ and $f_p\tilde{p}_s$ contribute to the acoustoelectric current. Substitutions of $\tilde{n}_s$ and $\tilde{p}_s$ from eqs. (8),(9) and $\tilde{E}$ from eq. (18) in eq. (30) yield the expression for the acoustoelectric current density in a bipolar semiconductor containing traps

$$J_{AE} = WK^2 \frac{2\omega}{v_s} \cdot \frac{\mu_p P_3 (N_1^2 + N_2^2)\omega_{cp}/\omega - \mu_n N_3 (P_1^2 + P_2^2)\omega_{cn}/\omega}{(N_1 P_1 - N_2 P_2 - P_2 \omega_{cn}/\omega - N_2 \omega_{cp}/\omega)^2 + [(P_2 + \omega_{cp}/\omega)N_1 + (N_2 + \omega_{cn}/\omega)P_1]^2} . \quad (31)$$

Here, $W = \frac{1}{2}C \, |\tilde{S}|^2$ is the energy density of the sound wave and N_3, P_3 are the following real functions:

$$N_3 = f_{nr} - (f_{nr}^2 + f_{ni}^2)v_n/v_s,$$
$$P_3 = f_{pr} - (f_{pr}^2 + f_{pi}^2)v_p/v_s. \quad (32)$$

Using eqs. (28) and (31) we obtain for the general case the Weinreich relation, expressing the acoustoelectric current density as a function of the absorption coefficient:

$$J_{AE} = \alpha W \frac{\mu_p P_3 (N_1^2 + N_2^2)\omega_{cp}/\omega - \mu_n N_3 (P_1^2 + P_2^2)\omega_{cn}/\omega}{P_1 (N_1^2 + N_2^2)\omega_{cp}/\omega + N_1 (P_1^2 + P_2^2)\omega_{cn}/\omega} . \quad (33)$$

Some special cases

1. One type of carriers, trapping and diffusion are included. If only electrons contribute to conductivity, following reduced expressions for α and J_{AE} can be derived by setting $\omega_{cp} = 0$, $P_1 = 1$ and $P_2 = 0$ in eqs. (28) and (31):

$$\alpha = \frac{K^2}{v_s} \cdot \frac{\omega_{cn}(1 - f_{nr}v_n/v_s + f_{ni}\omega/\omega_{Dn})}{(1 - f_{nr}v_n/v_s + f_{ni}\omega/\omega_{Dn})^2 + (\omega_{cn}/\omega + f_{nr}\omega/\omega_{Dn} + f_{ni}v_n/v_s)^2} , \quad (34)$$

$$J_{AE} = - \frac{WK^2}{v_s} \cdot \frac{\mu_n \omega_{cn}[f_{nr} - (f_{nr}^2 + f_{ni}^2)v_n/v_s]}{(1 - f_{nr}v_n/v_s + f_{ni}\omega/\omega_{Dn})^2 + (\omega_{cn}/\omega + f_{nr}\omega/\omega_{Dn} + f_{ni}v_n/v_s)^2} . \quad (34)$$

These formulae are the same as Southgate and Spector[10] obtained for this special case. The minus sign at f_{ni} in ref.10 is caused by the wave $\exp[i(kx - \omega t)]$ used there.

2. Bipolar conduction without trapping, diffusion is included. The trapping factors in this special case are $f_{nr} = 1$, $f_{ni} = 0$, $f_{pr} = 1$, $f_{pi} = 0$ and therefore

$$\gamma_n = \gamma_{no} \equiv 1 - v_n/v_s = 1 + \mu_n E/v_s,$$
$$\gamma_p = \gamma_{po} \equiv 1 - v_p/v_s = 1 - \mu_p E/v_s. \quad (36)$$

The functions N_1, N_2, N_3, P_1, P_2, P_3 (eqs.(29), (32)) reduce to

$$N_1 = \gamma_{no}, \quad N_2 = \omega/\omega_{Dn}, \quad N_3 = \gamma_{no},$$
$$P_1 = \gamma_{po}, \quad P_2 = \omega/\omega_{Dp}, \quad P_3 = \gamma_{po}. \quad (37)$$

For the absorption coefficient and the acoustoelectric current density we obtain:

$$\alpha = K^2 \frac{\omega}{v_s} \frac{\gamma_{po}(\gamma_{no}^2 + \omega^2/\omega_{Dn}^2)\omega_{cp}/\omega + \gamma_{no}(\gamma_{po}^2 + \omega^2/\omega_{Dp}^2)\omega_{cn}/\omega}{\left[\gamma_{no}\gamma_{po} - \dfrac{\omega^2}{\omega_{Dn}\omega_{Dp}} - \dfrac{\omega_{cn}}{\omega_{Dp}} - \dfrac{\omega_{cp}}{\omega_{Dn}}\right]^2 + \left[(\dfrac{\omega}{\omega_{Dp}} + \dfrac{\omega_{cp}}{\omega})\gamma_{no} + (\dfrac{\omega}{\omega_{Dn}} + \dfrac{\omega_{cn}}{\omega})\gamma_{po}\right]^2}, \quad (38)$$

$$J_{AE} = WK^2 \frac{\omega}{v_s} \frac{\mu_p\gamma_{po}(\gamma_{no}^2 + \omega^2/\omega_{Dn}^2)\omega_{cp}/\omega - \mu_n\gamma_{no}(\gamma_{po}^2 + \omega^2/\omega_{Dp}^2)\omega_{cn}/\omega}{\left[\gamma_{no}\gamma_{po} - \dfrac{\omega^2}{\omega_{Dn}\omega_{Dp}} - \dfrac{\omega_{cn}}{\omega_{Dp}} - \dfrac{\omega_{cp}}{\omega_{Dn}}\right]^2 + \left[(\dfrac{\omega}{\omega_{Dp}} + \dfrac{\omega_{cp}}{\omega})\gamma_{no} + (\dfrac{\omega}{\omega_{Dn}} + \dfrac{\omega_{cn}}{\omega})\gamma_{po}\right]^2} \cdot \quad (39)$$

The Weinreich relation (33) reduces to:

$$J_{AE} = \alpha W \frac{\mu_p\gamma_{po}(\gamma_{no}^2 + \omega^2/\omega_{Dn}^2)\omega_{cp}/\omega - \mu_n\gamma_{no}(\gamma_{po}^2 + \omega^2/\omega_{Dp}^2)\omega_{cn}/\omega}{\gamma_{po}(\gamma_{no}^2 + \omega^2/\omega_{Dn}^2)\omega_{cp}/\omega + \gamma_{no}(\gamma_{po}^2 + \omega^2/\omega_{Dp}^2)\omega_{cn}/\omega}. \quad (40)$$

Expressions (38) – (40) coincide with the results of Fink and Quentin[9] obtained for this special case.

3. Bipolar conduction without trapping and diffusion. In this special case the expressions (38) – (40) can be further simplified by setting $\omega/\omega_{Dn} = 0$ and $\omega/\omega_{Dp} = 0$. For the absorption coefficient we obtain

$$\alpha = K^2\frac{2\omega}{v_s} \frac{\gamma\omega\tau_c}{1 + \gamma^2\omega^2\tau_c^2}, \quad (41)$$

where

$$\gamma = \gamma_{no}\gamma_{po}/\gamma_a, \quad \gamma_a = 1 - v_a/v_s = 1 - \mu_a E/v_s,$$

$$v_a = \mu_a E = \frac{\mu_n\mu_p(n-p)}{\mu_n n + \mu_p p} E,$$

$$\tau_c = (\omega_{cn} + \omega_{cp})^{-1} = \left[e(\mu_n n + \mu_p p)/\varepsilon\varepsilon_0\right]^{-1}.$$

Here, τ_c is the dielectric relaxation time and v_a the ambipolar drift velocity. The analytical properties of α in this special case were analyzed by Greebe and van Dalen[11], Fink and Quentin[9], and Tsekvava[12]. The absorption coefficient α (eq.41) equals zero for three values of the applied electric field: E_n, E_p and E_a:

$$E_n = - v_s/\mu_n \qquad (v_n = v_s, \quad \gamma_{no} = 0), \quad (42)$$
$$E_p = v_s/\mu_p \qquad (v_p = v_s, \quad \gamma_{po} = 0), \quad (43)$$
$$E_a = v_s/\mu_a \qquad (v_a = v_s, \quad \gamma_a = 0). \quad (44)$$

In comparison with the case of one type of carriers we obtain for bipolar conduction an additional region where the acoustic wave is amplified, i.e. $\alpha < 0$ (minority carrier amplification). If electrons are minorities this region is $E_a < E < E_n$. For the absolute value of the ambipolar drift mobility we always have $|\mu_a| \leqslant \mu_n$, $|\mu_a| \leqslant \mu_p$, therefore the inversion point E_a lies outside the interval (E_n, E_p).

The acoustoelectric current density reduces to

$$J_{AE} = \frac{WK^2 e}{v_s \varepsilon \varepsilon_0} \frac{(\mu_p^2 p \gamma_{no} - \mu_n^2 n \gamma_{po}) \gamma_{no} \gamma_{no}}{\gamma_{po}^2 \gamma_{no}^2 + (\gamma_{no} \omega_{cp}/\omega + \gamma_{po} \omega_{cn}/\omega)^2} . \tag{45}$$

J_{AE} equals zero also for three values of the applied electric field as mentioned in ref. 3:

$$E_n = - v_s/\mu_n \qquad (v_n = v_s), \tag{42}$$

$$E_p = v_s/\mu_p \qquad (v_p = v_s), \tag{43}$$

$$E_{pn} = v_s \frac{\mu_n^2 n - \mu_p^2 p}{(p\mu_p + n\mu_n)\mu_p \mu_n} , \qquad (\mu_p^2 p \gamma_{no} = \mu_n^2 n \gamma_{po}). \tag{46}$$

Eqs. (42), (43) und (46) lead to the inequality $E_n \leqslant E_{pn} \leqslant E_p$. The authors of ref.3 proposed the use of these critical values E_n, E_{pn}, E_p, in addition to the measurement of conductivity, to determine the four carrier parameters n, p, μ_n, μ_p. It must be emphasized that this is only possible if traps have no influence on the acoustoelectric effect. In the presence of traps these critical values could be altered and some inversion points may even disappear. This will be demonstrated here by means of numerical calculations according to eq. (31).

4. Bipolar conduction in the presence of trapping, without diffusion and without an applied electric field. In this case we have $v_n = v_p = 0$, $\gamma_n = \gamma_p = 1$, moreover $\omega/\omega_{Dn} = \omega/\omega_{Dp} = 0$. The functions N_i and P_i in eqs. (29,(32) reduce to $N_1 = P_1 = 1$, $N_2 = P_2 = 0$, $N_3 = f_{nr}$, $P_3 = f_{pr}$. For the absorption coefficient (eq.(28)) we obtain:

$$\alpha = \frac{K^2}{v_s} \frac{\omega_{cp} + \omega_{cn}}{1 + (\omega_{cp} + \omega_{cn})^2/\omega^2} = \frac{K^2}{v_s \varepsilon \varepsilon_0} \frac{\sigma}{1 + \sigma^2/(\varepsilon \varepsilon_0 \omega)^2} , \tag{47}$$

where

$$\sigma = \sigma_p + \sigma_n = e\mu_p p + e\mu_n n. \tag{48}$$

The absorption coefficient does not depend on the trapping factors in this special case.

The acoustoelectric current density is given by the following simple expression

$$J_{AE} = \frac{WK^2 e}{v_s \varepsilon \varepsilon_0} \frac{f_{pr}\mu_p^2 p - f_{nr}\mu_n^2 n}{1 + \sigma^2/(\varepsilon \varepsilon_0 \omega)^2} \tag{49}$$

and for the modified Weinreich relation we obtain

$$J_{AE} = \alpha W (f_{pr}\mu_p \sigma_p - f_{nr}\mu_n \sigma_n)/\sigma . \tag{50}$$

Eqs.(49),(50) extend the result of ref.1 to the case of trapping. The acoustoelectric current is zero, when the condition

$$f_{pr}\mu_p^2 p = f_{nr}\mu_n^2 n \tag{51}$$

is satisfied, instead of $\mu_p^2 p = \mu_n^2 n$, $\tag{52}$

describing the case without traps. It is seen from eqs.(49),(50) that in this special case the acoustoelectric current density may be considered as a sum of contributions from holes and electrons in the direction of sound propagation:

$$J_{AE,p} = f_{pr}\mu_p(\sigma_p/\sigma)\alpha W, \tag{53}$$

$$J_{AE,n} = -f_{nr}\mu_n(\sigma_n/\sigma)\alpha W. \tag{54}$$

The contributions of both types of carriers are weighted only with the real part of the trapping factors. These trapping factors represent the mobile part of the charge bunched by the propagating acoustic wave: $f_{pr} \leqslant 1$, $f_{nr} \leqslant 1$. Therefore, the partial acoustoelectric currents in the case of trapping are always smaller than in the case without trapping.

5. The acoustoelectric Hall effect for bipolar conduction in the presence of trapping without diffusion and without an applied electric field.

In a bipolar semiconductor sample with the Ohmic conductivity $\sigma = \sigma_p + \sigma_n = e\mu_p p + e\mu_n n$ a magnetic field $\vec{B}$ perpendicular to the Ohmic current density $\vec{J} = \sigma\vec{E}$ gives rise to a transverse electric Hall field $\vec{E}_H$:

$$\sigma\vec{E}_H = -(A_p\mu_p\vec{J}_p - A_n\mu_n\vec{J}_n) \times \vec{B}, \tag{55}$$

where $\quad \vec{J}_p = \sigma_p\vec{E} = (\sigma_p/\sigma)\vec{J}, \quad \vec{J}_n = \sigma_n\vec{E} = (\sigma_n/\sigma)\vec{J} \tag{56}$

are the Ohmic current densities of holes and electrons, respectively, under the influence of the applied electric field $\vec{E}$ and A_p, A_n are the Hall factors depending on the scattering mechanism. A weak magnetic field is assumed: $\mu_n B \ll 1$, $\mu_p B \ll 1$. Substitution of eq.(56) in eq.(55) leads to the familiar expression for the Ohmic Hall effect

$$\vec{E}_H = -(\mu_H/\sigma)(\vec{J}\times\vec{B}) = -\mu_H(\vec{E}\times\vec{B}) \tag{57}$$

with the Hall mobility

$$\mu_H = A(\mu_p\sigma_p - \mu_n\sigma_n)/\sigma . \tag{58}$$

Here we have assumed for simplicity both Hall factors to be equal: $A_p = A_n = A$.

The Hall effect of the acoustoelectric current can be described in analogy to the Hall effect of the Ohmic current. Under the influence of the sound wave both types of charge carriers flow in the same direction. Contrary to the Ohmic Hall effect they are deflected by the Lorentz force to opposite sides of the sample. The transverse electric Hall field in this case is given by

$$\sigma\vec{E}_{HEAC} = -A(\mu_p\vec{J}_{AE,p} - \mu_n\vec{J}_{AE,n})\times\vec{B} . \tag{59}$$

Substituting the partial acoustoelectric current densities from eqs.(53),(54) valid in this special case and using eq.(50) we obtain:

$$\vec{E}_{HEAC} = -A\frac{f_{pr}\mu_p^2\sigma_p + f_{nr}\mu_n^2\sigma_n}{(f_{pr}\mu_p\sigma_p - f_{nr}\mu_n\sigma_n)\sigma}(\vec{J}_{AE}\times\vec{B}) = -\frac{\mu_{HEAC}}{\sigma}(\vec{J}_{AE}\times\vec{B}). \tag{60}$$

Eq.(60) is also an extension of the result of ref.1 for the case of traps. The measurements of $\vec{E}_{HEAC}$, $\vec{J}_{AE}$, and $\vec{B}$ yield a relation for n,p, μ_n and μ_p also in the case of traps. As seen from eq.(60), measurements of the acoustic energy density are not necessary to determine the value of μ_{HEAC}. Other relations may be obtained by measuring the conductivity σ (eq.(48)) and the ordinary Ohmic Hall effect (eqs. (57), (58)).

The sign inversion of the Ohmic Hall effect occurs for $\mu_p^2 p = \mu_n^2 n$. On the other hand, inversion of the acoustoelectric current appears for $f_{pr}\mu_p^2 p = f_{nr}\mu_n^2 n$ (eq.(51)) in the case with traps and for $\mu_p^2 p = \mu_n^2 n$ (eq.(52)) without trapping. Therefore, the difference between the inversion points of J_{AE} and μ_H indicates the influence of traps on the acoustoelectric current, provided that $f_{pr} \neq f_{nr}$.

EXPERIMENTAL RESULTS AND SOME NUMERICAL EXAMPLES

Photosensitive CdTe single crystals were grown from the melt with excess Te using the Bridgman method and were compensated by $CdCl_2$, as described by Höschl et al.[13]. The crystals were p-type in the dark with a resistivity of 10^9 Ωcm at 300 K. The conductivity could be changed by many orders of magnitude and converted to n-type by increasing photoexcitation with monochromatic band gap light of λ=854 nm. The propagation direction of the longitudinal acoustic wave was along the piezoelectric active $[111]$ direction. The ultrasound frequency was chosen between 10 and 30 MHz. Acoustoelectric measurements were performed as described in refs 1,10. From our previous investigation on bipolar photoconduction in CdTe[14,15], the assumptions $\tau_{rec,n} \gg 2\pi/\omega$ and $\tau_{rec,p} \gg 2\pi/\omega$ used in deriving eqs.(28),(31), (33) are satisfied at all intensities of photoexcitation. It is easily to prove that in CdTe with $v_s = 3.43\times10^5$ cm/s (ref.2) and $\mu_n = 600$ cm^{2}V^{-1}s^{-1}, $\mu_p = 60$ cm^{2}V^{-1}s^{-1} for 300 K the inequalities $\omega \ll \omega_{Dn}$, $\omega \ll \omega_{Dp}$ are also satisfied. The diffusion of charge carriers, therefore, can be neglected in the discussion of our acoustoelectric results in the case of trapping without an applied electric field (special cases 4 and 5 mentioned above). Fig.1a presents the measured Ohmic Hall voltage U_H and the open circuit acoustoelectric voltage $U_{AE} = E_{AE}L = J_{AE}L/\sigma$ as functions of the photogeneration rate G due to photoexcitation. E_{AE} is the acoustoelectric field in the direction of sound wave propagation and L the sample length. The corresponding coefficients μ_H and μ_{HEAC}, defined by eqs. (57),(58) and (60), respectively, are shown in fig. 1b. The acoustoelectric quantities U_{AE} and μ_{HEAC} change their sign at a lower generation rate than the Ohmic Hall effect. According to eq. (49) this means, that the real parts of the trapping factors f_{pr} and f_{nr} are different: $f_{pr} < f_{nr}$, i.e. the hole trapping dominates. The ratio f_{pr}/f_{nr} must be determined from the condition that the calculated dependences $\mu_n(G)$, $p(G)$ and $n(G)$ show no singularities[4,5,6]. These functions are presented in fig.2 with the ratio $f_{pr}/f_{nr} = 0.69$. The Hall factor was assumed to be A = 1.3 as in ref.2. For p-type CdTe the hole mobility μ_p could be measured in the dark and was assumed to be independent of the generation rate. Fig.2 shows that the electron mobility depends only weakly on photoexcitation. The electron density increases proportionally with the generation rate, the generated hole density $(p-p_o)$ increases with $G^{0.5}$. This behaviour can be explained by a model containing three kinds of defects, as previously shown in refs. 14,15.

Measurements with sound frequencies of 10 and 30 MHz yield $f_{pr}/f_{nr} =$ 0.67 and $f_{pr}/f_{nr} = 0.91$, respectively. Theoretical considerations[8] show that the real parts f_{pr} and f_{nr} tend to 1 with increasing frequency. In our case this means, that the variation of f_{pr} with frequency is greater than that of f_{nr}.

The acoustoelectric current as a function of the applied electric field can be calculated using the determined values of n,p, μ_n and μ_p at different generation rates (fig.2). For a qualitative demonstration of this dependence, the calculation with the general eq.(31) was performed for a set of trapping factors (table 1). The results are shown in fig.3. For a better graphical presentation of acoustoelectric current curves at photoconductivities, which differ by three orders of magnitude, values of the acoustoelectric field $E_{AE} = J_{AE}/\sigma$ were calculated and are shown in this diagram. Increasing numbers on the curves mean increasing generation rates. The trapping factors used for this calculation result from a model of two traps for electrons and holes. The densities of these traps are 10^{14} cm^{-3}, the energy levels are 0.4 eV from the band edges and the capture cross sections for holes are 1.1×10^{-13}cm^2 and for electrons 3×10^{-14}cm^2 (ref.5).

The following features are predicted by this calculation. Only one inversion point E_p exists at very low generation rates, three inversion points E_p, E_{pn} and E_n are present at intermediate photoexcitation and only one inversion point E_n exists at high generation rates. This is a consequence of the influence of traps on the acoustoelectric effect.

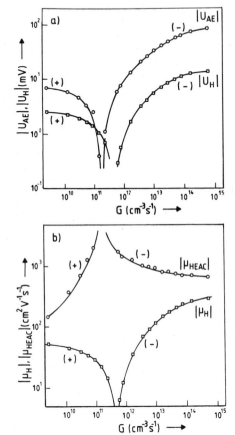

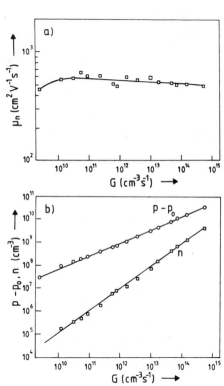

Fig.1. (a)Experimental Hall voltage U_H and acoustoelectric voltage U_{AE} as functions of the generation rate G due to photoexcitation. (b) Dependence of the Hall coefficients μ_H and μ_{HEAC} on the generation rate G. CdTe single crystal dimensions $3.9\times1.5\times0.8$ mm^3, sound frequency $\omega/2\pi$= 23 MHz, temperature T=295 K, (ref. 5,6,16).

Fig.2. (a) Dependence of the electron mobility μ_n on the generation rate. (b) Generated hole density $(p-p_0)$ and electron density n as functions of the generation rate G. The hole mobility is μ_p = 44 cm^2 V^{-1}s^{-1}. Calculations were performed using eqs. (48),(58) and (60) with measured values of σ,μ_H and μ_{HEAC}(fig.1b) and f_{pr}/f_{nr} = 0.69, (ref.5,6,16).

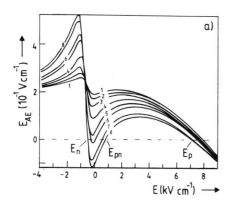

a)

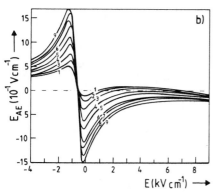

b)

Fig.3. Calculated dependence of the acoustoelectric field E_{AE} on the applied electric field E for different values of the photoexcitation. Calculations were performed with eq.(31) for the parameters listed in table 1,(ref.5,6).

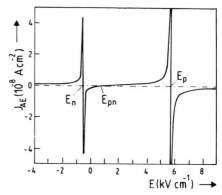

Fig.4. Calculated field dependence of the acoustoelectric current density without trapping . This curve is calculated using eq.(31) with $f_{nr}=f_{pr}=1$, $f_{ni}=f_{pi}=0$, $\mu_n = 600$ $cm^2V^{-1}s^{-1}$, $\mu_p = 60$ $cm^2V^{-1}s^{-1}$, $n = 3\times10^6cm^{-3}$, $p = 10^8cm^{-3}$. The values of K^2, ε, v_s, W, ω and T are the same as used for the results of calculations presented in fig.3,(ref.5).

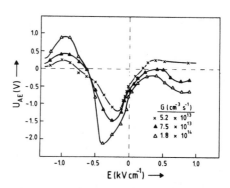

Fig.5. Experimental dependence of the acoustoelectric voltage U_{AE} on the applied electric field E at different generation rates. Sample dimensions, sound frequency and temperature as in fig.1, (ref.6,17).

331

Table 1. Parameters used for the calculation of the acoustoelectric field $E_{AE} = J_{AE}/\sigma$ as a function of the applied electric field with eq. (31), presented in fig.3: $\varepsilon = 10$, $K^2 = 3 \times 10^{-4}$, $\omega/2\pi = 23 \times 10^6 s^{-1}$, $v_s = 3.43 \times 10^5 cm/s$, $W = 10^{-5}$ W s/cm^3, $T = 295$ K, $\mu_p = 44$ cm^{2}V^{-1}s^{-1}, $f_{nr} = 0.65$, $f_{ni} = 0.48$, $f_{pr} = 0.52$, $f_{pi} = 0.49$; the values of n,p and μ_n were taken from fig.2 and are given below [5,6]

Curve No.	μ_n (cm^2/Vs)	n (cm^{-3})	p (cm^{-3})	Curve No.	μ_n (cm^2/Vs)	n (cm^{-3})	p (cm^{-3})
Fig.3a				Fig.3b			
1	454	6.3×10^4	8.1×10^7	1	483	7.6×10^6	7.4×10^8
2	563	1.7×10^5	1.4×10^8	2	574	1.1×10^7	1.1×10^9
3	568	3.4×10^5	1.9×10^8	3	551	2.8×10^7	1.8×10^9
4	597	7.2×10^5	2.8×10^8	4	570	6.7×10^7	3.2×10^9
5	597	1.7×10^6	4.2×10^8	5	525	1.4×10^8	4.4×10^9
6	550	2.8×10^6	5.0×10^8	6	515	4.0×10^8	8.3×10^9
7	504	5.5×10^6	6.4×10^8	7	493	6.7×10^8	1.1×10^{10}
8	483	7.6×10^6	7.4×10^8	8	500	1.2×10^9	1.7×10^{10}
				9	477	2.3×10^9	2.4×10^{10}

Without traps three inversion points must always appear, as seen in fig.4. The experimental results shown in fig.5, however, differ strongly from this theoretical predictions for the ideal case without traps. For high photoexcitation rates two of the expected inversion points E_{pn} and E_p disappear. The experimental curve has only one inversion point E_n at high excitation. Additional evidence of trapping gives the shape of the measured U_{AE} (E) curves in fig.5. The theoretical curve shows pronounced extrema close to the inversion points E_n and E_p in the absence of traps (fig.4), whereas the presence of traps leads to smooth curves with extrema far off these inversion points (fig.3), as seen also for the experimental curves in fig.5. Considering the inversion point E_n it is seen that the extrema of the calculated curves in the case of trapping (fig.3) are shifted away from this inversion point. The same behaviour was also observed in the experiments (fig.5).

The theory which includes trapping of charge carriers explains the measured dependences of the acoustoelectric voltage on the applied electric field in a better way.

CONCLUSIONS

We have demonstrated the possibility of application of acousto-electric measurements for the characterization of bipolar photoconductors on the example of the wide gap II-VI semiconductor CdTe. The influence of carrier trapping on acoustoelectric effects was outlined. The use of acousto-electric measurements enables us to obtain information not only on densities and mobilities of carriers in bipolar semiconductors, but also on the trap parameters.

ACKNOWLEDGEMENTS

It is a pleasure for me to thank Dipl.-Phys. M.Adolf and Dipl.-Ing. H.Dignus for acoustoelectric measurements and to Prof.Dr.U.Birkholz, Dipl.-Phys.N.Link and Dipl.-Phys. M.Rick for many stimulating discussions.

REFERENCES

1. B.-P.Kietis, R.Reksnys, and A.Sakalas, Ohmic and Acoustoelectric Currents in the Hall Effect for the Investigation of Bipolar Conductivity, <u>Phys. Status Solidi (a)</u> 39: 601 (1977).

2. B.-P.Kietis, R.Reksnys, A.Sakalas, and P.Höschl, Investigation of the Bipolar Photoconductivity in CdTe, <u>Phys.Status Solidi (a)</u> 43: 705 (1977).

3. R.Baubinas, B.-P.Kietis, R.Reksnys, and A.Sakalas, Bipolar Photoconductivity and Acoustoelectric Current in CdTe, CdS and CdSe, <u>Phys.Status Solidi (a)</u> 50: K63 (1978).

4. M.Adolf, J.Rosenzweig, and U.Birkholz, Bipolar Photoconductivity and Hall Effect of the Acoustoelectric Current in CdTe, <u>Phys.Status Solidi (a)</u> 85: K81 (1984).

5. J.Rosenzweig, Acoustoelectric Phenomena in the Case of Bipolar Photoconduction in CdTe, <u>Habilitation Thesis</u>, University of Karlsruhe (1987).

6. J.Rosenzweig, Characterization of Photoconducting CdTe Using Acoustoelectric Methods, <u>J.Crystal Growth</u> 86: 689 (1988).

7. A.R.Hutson and D.L.White, Elastic Wave Propagation in Piezoelectric Semiconductors, <u>J.Appl.Phys.</u> 33: 40 (1962).

8. C.A.A.J.Greebe, The Influence of Trapping on the Acousto-Electric Effect in CdS, <u>Philips. Res.Rept.</u> 21: 1 (1966).

9. M.Fink and G.Quentin, Acoustoelectric Phenomena in Piezoelectric Semiconductors. Influence of Minority Carriers, <u>Phys.Status Solidi (a)</u> 4: 397 (1971).

10. P.D.Southgate and H.N.Spector, Effect of Carrier Trapping on the Weinreich Relation in Acoustic Amplification, <u>J.Appl.Phys.</u> 36: 3728 (1965).

11. C.A.A.J.Greebe and P.A.van Dalen, Ultrasonic Amplification by Minority Charge Carriers in Piezoelectric Semiconductors, <u>Philips Res.Rept.</u> 24: 168 (1969).

12. B.E.Tsekvava, Acoustic Effects in Semiconductors in the Presence of Two Types of Carrier , <u>Sov.Phys.Semicond.</u> 7: 795 (1973).

13. P.Höschl, P.Polivka, V.Prosser, and A.Sakalas, Preparation of Cadmium Telluride Single Crystals for Nuclear Detectors, <u>Czech.J.Phys.</u> B25: 585 (1975).

14. J.Rosenzweig, Photothermoelectric Power and Photo-Hall Effect in CdTe in the Case of Bipolar Photoconduction, <u>Doctoral Thesis</u>, University of Karlsruhe (1981).

15. J.Rosenzweig and U.Birkholz, Bipolar Photo-Hall Effect in CdTe, <u>J.Crystal Growth</u> 59: 263 (1982).

16. M.Adolf, private communication, 1987.

17. H.Dignus, private communication, 1987.

SOME PHYSICAL PROPERTIES OF $AgInS_2$

FILMS OBTAINED BY SPRAY-PYROLYSIS TECHNIQUE

Mushin Zor

University of Anatolia
Physics Department
Eskisehir-Turkey

INTRODUCTION

Spray-pyrolysis (SP) is one of the methods to obtain semiconducting films. It is the spraying of a solution containing a mixture of the components in the appropriate ratios on to a hot surface. Then pyrolysis takes place and a film starts developing on the substrate. The rate of growth depends mainly on the substrate temperature, the relative concentration of the components and the flow rate of the solution. Atomization is of prime concern in this method in order to obtain a flat surface.

A SnO_2 transparent and conducting film was first prepared with this technique in the 1940's[1]. Semiconducting CdS and CdSe films were later produced in the 1960's[2,3] and then ternary, quaternary and quinary compounds followed in the 1970's and 1980's[2-18].

Here we present X-ray diffraction, absorption and conductivity measurements of $AgInS_2$ films produced using spray-pyrolysis at different substrate temperatures.

EXPERIMENTAL WORK

Aqueous dilute solutions of $AgNO_3$, $InCl_3$ and $(NH_2)_2CS$ were separately prepared and then mixed with the appropriate ratios such that the resultant solution to be sprayed would contain Ag:In:S = 1:1:2 ratios, respectively. The system used for spray-pyrolysis is shown in Fig. 1. Ordinary glass of 1mm thick was used as substrate material and the total area was about 10 cm x 10 cm. The substrate temperature was varied between 125°C and 335°C, but a preset temperature was kept constant throughout the growth of each film with deviations of about ±1°C. The flow rate and the pressure for spraying were set at 3.5ml/min and 4 p.s.i. respectively. N_2 was used as the carrier gas.

RESULTS AND DISCUSSION

The $AgInS_2$ films prepared at substrate temperatures below 240°C and

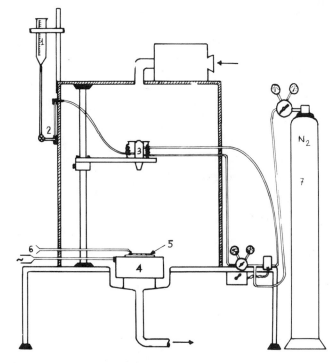

Fig. 1. The set-up for spray-pyrolysis. (1) Solution
for spraying, (2) flow-metre, (3) spray-head,
(4) heater, (5) substrate, (6) thermo-couple,
(7) N_2 tube.

above $275^{\circ}C$ showed amorphous structure whereas those grown at temperatures
in between had a polycrystalline structure belonging to the orthorhombic
phase of $AgInS_2$ crystals. Fig 2 shows X-ray powder diffraction spectra of
the films grown at various substrate temperatures. The orthorhombic phase
can easily be identified in Fig. 2b. The peaks corresponding to the (120)
and (121) planes are seen to disappear in the films grown at $270^{\circ}C$.
However the peak for (002) showed a considerable increase.

Electron micrographs taken with the scanning electron microscope show
that the surface of these films is rather rough as seen in Fig. 3. The
one obtained at substrate temperature of $125^{\circ}C$ had shown some cracks and
pieces of the film could be easily peeled off. The films grown at higher
temperatures were seen to stick very well on to the glass surface.

The absorption spectra at room temperature of these films are shown
in Fig. 4. The noisy part of the spectrum above about 750nm was caused by
the IR lamp in that region. The fundamental absorption edge shown in
Fig. 4b is quite abrupt for polycrystalline films. On the other hand for
the films produced at lower and higher substrate temperatures the
absorption increased considerably at longer wavelengths. The insets in
the figures show the variation of the square of the absorption coefficients with the energy of the incoming radiation. This fit shows
that $AgInS_2$ films have a direct band gap with a value of 1.97 eV as
derived from the insert graph. The films with amorphous structure did
not show any well defined band gap, although the ir optical band gap
was smaller than that of the polycrystalline films.

336

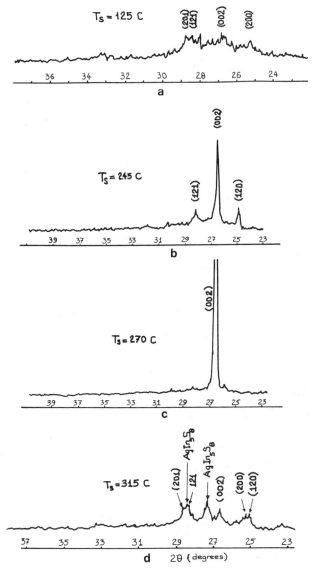

Fig. 2   X-ray powder diffraction of AgInS$_2$ films grown at different
substrate temperatures; (a) 125°C, (b) 245°C, (c) 270°C
and (d) 315°C.

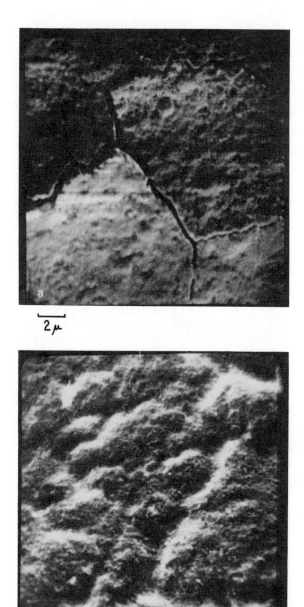

Fig. 3. The electron micrographs of the $AgInS_2$ films grown at substrate temperatures (a) 125°C, (b) 270°C

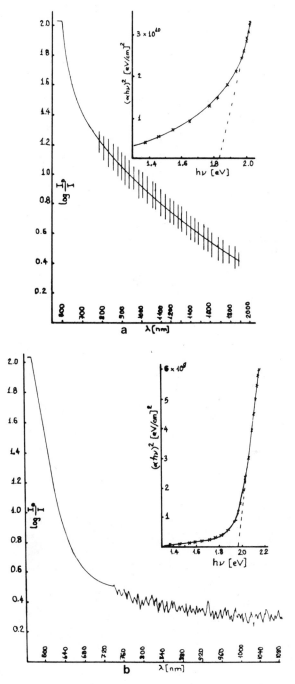

Fig. 4(a) and (b). Optical absorption for the AgInS$_2$ films
grown at different substrate temperatures
(a) 125°C, and (b) 270°C.

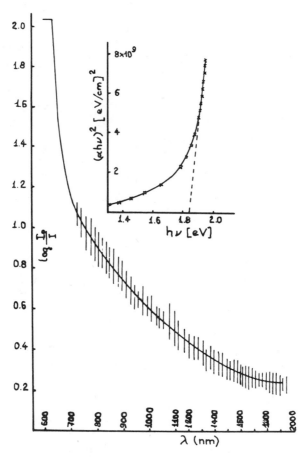

Fig. 4 (c). Optical Absorption of AgInS$_2$
grown at 335°C.

Most of the electrical measurements were taken with the samples in
the dark. All of the films measured in this way exhibited n-type
conduction. Indium metal was used to form electrodes and from the voltage-
current measurements it was observed that it made an ohmic contact with
the AgInS$_2$ films. The resistivity at room temperature varied between
1 x 10^3 ohm-cm to 2 x 10^4 ohm-cm for the films produced at substrate
temperatures between 170°C and 295°C. For substrate temperatures greater
than 295°C, the resistivity showed a notable decrease, down to 8.9 ohm-cm
for the films grown at 335°C.

The variation of the conductivity with respect to the temperature for
a 10 Volts potential difference applied across the metal electrodes is
shown in Fig. 5 for the film grown at 270°C. The activation energy
involved in conduction was found from the Arrhenius plot to be equal to
0.36 eV. The activation energies for other samples varied between 0.30 eV
to 0.36 eV and these energies are probably the ionization energies of
donor-like centres present in the SP grown AgInS$_2$ films.

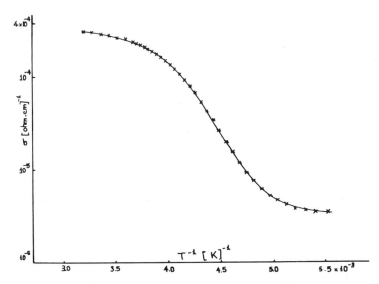

Fig. 5.   The Arrhenius plot of the conductivity of $AgInS_2$
film grown at 270°C.

$AgInS_2$ belonging to the group of I-III-VI$_2$ compounds is an analogue
of CdS. $AgInS_2$ crystals can be grown by other methods in two phases which
are tetragonal and orthorhombic.   The films produced by SP method had the
orthorhombic structure and it was not possible to obtain a tetragonal
phase[6,16].

The energy band gap of the orthorhombic $AgInS_2$ crystal is 1.96 eV
which is almost equal to the band gap of the SP grown polycrystalline films.

The observed decrease in the resistivity of the SP grown films at
high temperatures was attributed to the likely formation of the In_2O_3
compound. It was not possible to predict the nature of the donor-like
centres but in this kind of compound non-stoichiometry, vacancies,
interstitials would certainly play a role in the n-type conduction.

The optimum substrate temperature for growing $AgInS_2$ films by SP
method was found to be between 240-275°C.

ACKNOWLEDGEMENT

The author is indebted to Prof. Dr. Ö Öktü of Middle East Technical
University (Ankara, TR) for his generous support and very valuable
advice during this work and to the colleagues in Physics Department in
Hacettepe University (Ankara, TR.).

REFERENCES

1.   Br. Patent 632256 (1942) to H.A. McMasters, Lebley-Owers Ford Glass Co.

2.   J.E. Hill and R.R. Chamberlin, U.S.Pat., 3 148084 (1964).

3.   R.R. Chamberlin and J.S. Skarman, J. Electrochem. Soc., 13:86 (1966).

4.   B.R. Bamplin and R.S. Feigelson, Thin Solid Films, 60: 141 (1979)

5.   B.R. Bamplin, Prog. Crystal Growth Charact., 1: 395 (1979)

6. M. Gorska, R. Bealieu, J.J. Loferski and B. Roessler, Thin Solid Films, 60: 141 (1979)

7. F.M. Micheletti and P. Mark, Appl. Phys. Lett., 10: 136 (1967).

8. C. Wu and R.H. Bube, J. Appl. Phys., 45: 648 (1974).

9. Y.Y. Ma and R.H. Bube, J. Electrochemical Soc., Solid-State Science and Tech., 124: 1430 (1977).

10. R.S. Feigelson, N. N'Diaye, S.Y. Yin and R.H. Bube, J. Appl. Phys., 48: 3162 (1977).

11. J.C. Manifacier, M. De Murcia, J.P. Fillard and E. Vicario, Thin Solid Films, 41: 127 (1977).

12. T. Feng, A.K. Ghosh and C. Fishman, Appl. Phys. Lett., 35: 266 (1979).

13. O.P. Agnihotri and B.K. Gupta, Jap Journal of Appl. Phys, 18: 317 (1979).

14. H.L. Kwok and W.C. Siu, Thin Solid Films, 61: 249 (1979)

15. H.L. Kwok and Y.C. Chan, Thin Solid Films, 66: 303 (1980).

16. M. Zor, The Habilitation Thesis, Hacettepe Univ., Ankara, Turkey (1982).

17. H. Onnogawa and K. Miyashita, Jap. Journal of Appl. Phys., 23: 965 (1984).

18. S. Kolhe, S.K. Kulkarni, M.G. Takwale and V.G. Bhide, Solar Energy Materials, 13: 203 (1986).

AUTHOR INDEX

SUBJECT INDEX